The Reach of Science

THE REACH OF
SCIENCE

Henryk Mehlberg

UNIVERSITY OF

TORONTO PRESS · 1958

To my Wife

Preface

THIS MONOGRAPH is a study in the philosophy of science. It does not
preach a scientistic creed nor does it purport to show that science is a
"sacred cow." It centres around a single question and aims at accumulating
a reasonable body of evidence in support of a definite answer. The question
is concerned with the potentialities and the limitations of science and
can be formulated as follows: What theoretical and practical problems
could possibly be solved by applying scientific method? To what problems is this method inapplicable? The answer asserts the inherent
universality of science: if a problem is solvable at all, it can be solved,
in principle, by applying scientific method.

The book is divided into three parts of approximately equal length,
intended to provide together the requisite body of evidence in support
of the universality of science. Part I is largely preparatory and gives
the indispensable preliminary comments on the crucial concepts involved
in the investigation: the concepts of science, of scientific method, of
scientific knowledge, and of the nature of scientifically unsolvable problems. Science is viewed as a growing body of socially relevant and
reliable information, that is, information the truth of which is secured
to a reasonable degree by appropriate methods. Scientific methods are
treated as integral parts of science in any phase of its development,
rather than as techniques for promoting science. A scientific solution to
a problem is taken to be a pertinent answer in conjunction with a method
for establishing the truth of the answer. As a result of these preliminary
analyses, the concept of truth emerges as essential both in understanding
science and in determining its scope. In this Part, a good deal of attention
is given to isolating the philosophical (epistemological) problem of the
scope of scientific knowledge from the cluster of related, yet funda-
mentally different, sociological queries about the social impact and the
social value of such knowledge.

Part II examines the three groups of fundamental scientific procedures: fact-finding methods, law-finding methods, and methods of theory-formation. The object of this survey (and the reason for its frequent departures from conventional accounts of scientific methodology) is to define the scope of all these methods in terms of a concept that has come to play an increasingly important part in contemporary philosophy of science: the concept of the empirical verifiability of a proposition (that is, the possibility of using observational data in order to find out the truth-value of the proposition). The conclusion reached as a result of these methodological considerations is that the fundamental procedures of contemporary science can be applied to a given question only if it is answerable by a verifiable proposition or by a system of such propositions. Hence, scientifically solvable problems are problems with verifiable solutions.

Part III aims at establishing the principle that the sum-total of problems with verifiable solutions not only accounts for the actual scope of scientific method but also delimits the widest conceivable extension of this scope: if a problem has no verifiable solution then it has no solution whatsoever. In conjunction with the aforementioned conclusion of the methodological survey, this "principle of verifiability" (asserting that solvable problems are coextensive with those admitting of verifiable solutions) leads directly to the thesis of the universality of science. The principle is obviously a reformulation of the so-called verifiability theory of meaning which, in one form or another, underlies most contemporary views of science. In this new formulation, however, the emphasis is on the concept of truth rather than on the concept of meaning. Another departure from the numerous traditional versions of the verifiability theory of meaning lies in the fact that the principle of verifiability is neither taken for granted, nor considered self-evident, nor assigned the role of an expedient convention, but is rather deduced from definitions of truth and of verifiability. The definitions of these two concepts as related to science and the corresponding justification of the principle of verifiability take up most of Part III. Its final section outlines the implications of the inherent universality of science; they include the rejection of phenomenalistic and similar restrictions on the scope of scientific knowledge and the assertion of the legitimacy of metaphysical views.

In spite of its concentration on a single issue, the investigation covers much of the ground debated in the contemporary philosophy of science; for example, the main problems in scientific methodology are dealt with in Part II. The book may therefore be of interest to students of the philosophy of science even if they are not especially concerned with the possibility of circumscribing the potentialities and limitations of science in a systematic way. Much of the book, and, certainly, the core of the argument, should be accessible to the general reader interested in the philosophical implications and presuppositions of science. With this reader in mind, I have tried to avoid expandable technicalities in the illustrative material and in terminology wherever this was possible without weakening the argument or making the presentation too lengthy.

A few words, finally, about the origin of this study. The main external motivation—I mean motivation external to the interest in the problem of determining the scope of science—came from a previous study of mine in which an attempt was made to account for the part that the principle of verifiability has come to play in contemporary science and its philosophy (*Science et positivisme*, 1948). The question arose as to how an appropriate version of this principle could be derived from well-established premises instead of being admitted as a mere postulate. In trying to answer this question, I have gradually come to realize that a substantiation of the principle of verifiability becomes possible only if the emphasis in the wording is shifted from the concept of meaning to that of truth, and, moreover, that this reformulation and derivation of the principle would also provide a basis for establishing the inherent universality of science. In the present study I have tried, to the best of my ability, to use the new formulation to justify the principle of verifiability and to determine the scope of scientific knowledge accordingly. Needless to say, while considering this dual programme promising, I am well aware of the shortcomings in its implementation, most of which were certainly avoidable, even in a first attempt.

I should like to take this opportunity to express my gratitude for assistance towards publication from the Humanities Research Council of Canada and the Publications Fund of the University of Toronto Press. I wish to thank the officers of the Press for their consistently co-operative attitude and for the valuable help given in their Editorial Department. To Messrs. J. W. Creighton and D. Gautier, formerly my students at the University of Toronto, I am grateful for assistance in preparing the text for publication. The Editors of the *Journal of Philosophy* have kindly permitted me to use materials from my paper "The Range and Limits of Scientific Method" published in volume 51 (1954).

H. M.

The University of Chicago
July 1957

Contents

PART ONE

The Concept of Scientific Knowledge

CHAPTER ONE

Cognitive and Social Aspects of Science

---

1 THE PROBLEM

A DETERMINATION of the reach of science amounts to distinguishing what science can do from what it is incapable of doing. The solution of this problem would indicate all the theoretical and practical questions that can be answered satisfactorily by recourse to scientific method, by separating them from questions inaccessible to this method. Questions of the first kind may be said to constitute the range of science; those of the second illustrate inevitable restrictions on the applicability of scientific method and should therefore be considered to define the limits of science. This separation of theoretical and practical problems susceptible to a scientific solution from those that are not can thus be viewed as tantamount to determining the range and the limits of science. Our investigation into the reach of science can therefore be subdivided naturally into two themes: the range and the limits of science.

It is worth realizing at the outset that the problem just formulated is concerned neither with the social value of scientific knowledge nor with its social impact, but with the scope of the knowledge that science can provide; the epistemology, not the sociology of science is involved. The epistemological problem refers to the ability of science to acquire new knowledge by successfully applying scientific method to theoretical and practical problems; the scientific knowledge available at a given moment can simply be identified with all the solutions of theoretical and practical problems man has reached up to this moment through his mastery of this method. The sociological problem, on the other hand, is concerned with the social repercussions of the spread of scientific knowledge and the value ascribable to science on such grounds. Yet, although distinct, the epistemological and the sociological aspects of science are closely interrelated. The meaning of science to society depends obviously upon the kind, amount, and dependability of the information science can provide, that is, upon the scope of scientific knowledge. Any epistemological investigation into the scope of scientific knowledge is bound to affect in some way our views of the present and future impact of science on society. Since, at this juncture, the social impact of science is both controversial and increasingly pervasive, a discussion of the epistemological status of science is likely to carry with it disputable social and ideological implications, owing to this connection with the corresponding sociological problem. It goes without saying that such implications are very prejudicial to an unbiased, objective discussion of the relevant facts and theories.

There is little doubt that man is at present in a precarious and critical situation because of the social repercussions of science. Views differ widely as to how the crisis can be met.[1] According to some schools of thought, our only chance of overcoming our present difficulties depends upon a clear realization of the incompetence of science to solve the problems raised by these difficulties and, consequently, upon a search for non-scientific ways of dealing with them. Other schools of thought, on the contrary, maintain that the only way of bringing the crisis under control is to promote and accelerate the progress of science, particularly in the field of the humanities. New scientific knowledge of man, if acquired in time, would, supposedly, help to solve problems that now confront us. The epistemological problem concerning the cognitive potentialities of science is thus obviously relevant to, and entangled with, this sociological controversy.

Apart from the bias involved in the issue concerning the social effects and value of science, other difficulties seem to confront our attempt to determine its range and limits. Trying to circumscribe in advance all the theoretical and practical problems that could be solved by scientific methods looks like indulging in a prophecy of the most gratuitous kind: it would hardly occur to anybody to delimit the prospects and potentialities of musical composition or architectural construction. The fact is that the decisive phases in the history of any art are due to unpredictable and irreplaceable interventions of creative artists and no amount of pondering over the theoretical possibilities of music or architecture would enable anyone to anticipate, or to set limits to, what future men of genius

---

[1]Cf. B. Russell, *The Impact of Science on Society* (1951), J. B. Conant, *Modern Science and Modern Man* (1952), F. A. Hayek, *The Counter-Revolution of Science* (1952), D. R. Owen, *Scientism, Religion and Man* (1952), J. S. Fulton, *Science and Man's Hope* (1955).

may accomplish in this field. The determination of the range and limits of science is likely to raise similar difficulties. It seems impossible, indeed, to ascertain in advance whether or not a particular problem can be solved by applying scientific methods without anticipating, to some extent, its future solution. This would imply an absurdly conceited assumption; for example, to make out in advance whether the problem concerning the atomistic structure of energy, matter, and electricity could be solved by field-theoretical methods, one would have to anticipate the content and ultimate fate of Einstein's ideas on this topic.

Moreover, our problem concerning the range and limits of science is formulated in terms vague and ambiguous ("scientific method," "theoretical" and "practical" problems, the "possibility" of solving a problem by scientific method, etc.); can any sensible attempt at reaching a satisfactory solution be made unless and until the ambiguity and vagueness are removed and the problem itself is restated in clearly defined terms? Numerous attempts at clarifying crucial concepts occurring in several philosophical problems have been made in contemporary philosophy for the purpose of reinterpreting these problems in a precise way and thus facilitating their solution. Such attempts, however, have too often resulted in interpreting away the philosophical problems themselves. The same danger is obviously present here: how should we reinterpret and reformulate the problem concerning the range and limits of science so as to remove the ambiguity and vagueness of the initial wording, without making the problem itself either clearly unsolvable or trivial?

As a matter of fact, we shall find in the course of our detailed discussions that the ambiguities of the problem can be removed without suppressing it, that the problem can be extricated from its entanglement with contemporary ideological issues and the relevant facts and theories surveyed and examined without bias, and, finally, that the alleged similarity between an attempt at determining the potentialities and prospects of art on the one hand and the task of setting limits to science on the other hand is very superficial and does not affect our problem. The difficulties described above will thus turn out not to be insuperable. In this introductory section I should like to circumscribe these difficulties and to show that they are less formidable than they look at first sight. This will enable us to remove some troublesome ambiguities in the current wording of the epistemological problem and to locate it more accurately by bringing out its differences from the controversy over the social aspects of science.

Which are the most troublesome ambiguities? In the first place, the very concept of "the" method of science requires an explanation. Science uses many general methods, some of them deductive, others inductive, others statistical, experimental, mensural, or observational, but none of these procedures are identifiable with "the" scientific method. There are, also, more special scientific methods, for example for solving algebraic equations, for determining the atomic weight of chemical elements, for ascertaining the authenticity of an historical document or a work of art, for dating monuments of the past by utilizing radioactive properties of carbon isotopes, the life-time of uranium or geological stratification in conjunction with palaeontological data, etc. But again we have to ask, what is meant by "the" method of science? This is a crucial concept in an investigation of the range and limits of science and some indispensable comments on this score will be made in chapter II of this Part. Here we need only point out that a definition of "the" scientific method will have to depend in turn upon the concept of science. This central concept, however, is far from being unambiguous at present. Thus, there is a strong tendency to restrict science to the *natural sciences* and to exclude therefrom both the logico-mathematical disciplines and social and humanistic studies. Yet problems transcending the procedures of natural sciences may, perhaps, be successfully attacked by resorting to the methodological devices of the humanities. The range of problems solvable by the method of science depends obviously upon what "science" denotes; the meaning to be attached to "science" is therefore vital for the outcome of our whole discussion. We shall consider this question in the next chapter.

Another ambiguity implicit in the current formulation of our problem is less obvious, but hardly less misleading. I am referring to the *range* of science, that is, the class of problems that *could* be solved by scientific methods. The concept of possibility involved in this formulation is both important and vague, if used without special precautions. This may be illustrated by the following example. In the social sciences, several procedures have been devised and successfully applied to determine the trends of public opinion in a given country with respect to some particular issue; how people feel about the likelihood of war breaking out, the result of a forthcoming election, the entertainment value of a moving-picture, can be reliably determined by polling, interviewing, or submitting a suitable questionnaire to a fair sample of the population. Such problems are therefore solvable by statistical scientific methods and should be located within the range of science. Yet one may wonder whether all problems of this kind can be attacked in this way, whether, for example, the attitude of a nation towards its "leader" could be ascertained by the procedures just referred to if the nation happened to live under a totalitarian régime. The hesitation one feels in answering this question is due, of course, to the ambiguity of the word "possible." In one sense, it is *not* possible to apply the standard procedures for measuring public attitudes in the second case; the required population-sample would not be available, since no individual is likely to take the chances involved in disclosing his attitude to such a "touchy" subject. In another sense, one may well feel that it *is* (theoretically) possible to determine the public attitude with respect to a touchy issue since it is at least possible to devise procedures that would make polling practically safe for those involved and thus provide data for a statistical generalization.

There are also problems that cannot be solved either by presently available scientific procedures or by a combination of these and of merely discoverable methods. Thus, the statistical distribution of colour-blindness in a given human group can be ascertained by ordinary procedures and is accurately known, at present, in so far as contemporary society is concerned. But the question as to whether two persons endowed with normal colour vision have exactly similar sensations when perceiving the same coloured object under the same normal circumstances, turns out, on closer inspection, to be unanswerable by any imaginable scientific methods. Indeed, no method, whether scientific or not, whether presently available or not, could enable anyone to compare exactly the colours observed by two different persons, since the comparison by a particular investigator amounts to his ascertaining whether on seeing, remembering, or imagining these colours against a common background, he simultaneously has the feeling of their similarity or dissimilarity; the comparison therefore presupposes the identity of the observer. The intersubjective similarity of sensations of colour is therefore beyond the range of science, whether this range is defined in terms of presently available scientific methods, or in terms of any combination of conceivable scientific methods.

Thus, when discussing the range of science, we may have three different questions in our minds: (1) What problems have already been solved by some combination of scientific methods? (2) What problems can be solved by presently available methods, on the understanding that all that is required to reach a solution is for some competent investigator, or group of investigators, to set out to apply a suitable combination of presently available scientific methods? (3) What problems are solvable, at least theoretically, by scientific methods, the implication being that in order to reach a solution, it may be necessary to invent some new methods in addition to those now available? In other words, there are questions which science has already successfully answered, others which science can answer by resorting to its contemporary methodological arsenal and, finally, questions which science could possibly answer by suitably enlarging its supply of methods of inquiry. How is the concept of the range of science to be interpreted in view of these three groups of scientifically solvable problems? Needless to say, the answer to this question is in itself (i.e., isolated from our investigation), completely arbitrary. But we shall see later on that, within the context of this investigation, there are good reasons for choosing what may be called the *inherent* or the *essential* range of science, in contrast to its *extrinsic* or *accidental range*. In other words, we shall concentrate our discussions on the range of problems that could possibly be solved by any conceivable scientific methods, regardless of whether these methods happen to be available at the present moment; such questions may be said to be accessible to the method of science because of the very *nature* of this method, irrespective of the historical vicissitudes of science. Accordingly, we shall consider as setting limits to science only such problems as are inherently inaccessible to scientific methods, or unsolvable by any combination of such methods, either actually existent or merely discoverable. We shall, of course, have to refer time and again to the *accidental* range of science, that is the sum-total of problems which have been solved or can be solved by applying methods actually available at a given moment of scientific evolution. But this does not detract from the fact that the general problem of the reach of science, as discussed in this study and as involved, I believe, in the majority of present discussions, refers to the inherent, not the accidental range of science.

The explanation just outlined of the two concepts of the range of science implicit in our problem is but preliminary. A detailed analysis is of crucial importance and will be attempted at a later stage (§ 36). We shall see, then, that the ambiguity of "a possible scientific solution to a theoretical or practical problem" covers more than two meanings and that more qualifications are required in order to avoid frustrating the discussion because of the vagueness and obscurity of the relevant terms. However, the above distinction between the inherent and the accidental range of science should suffice at this stage. It disposes of the alleged analogy between an admittedly hopeless attempt at determining the potentialities and prospects of art and the problem of establishing the range and limits of science. Once the problem about science is referred to its *inherent* range and limitations, the analogy breaks down completely, since a determination of the inherent range of science has obviously nothing in common with prophesying its future development. To carry out this determination one has to ascertain which problems are inherently solvable, not which problems *will be* scientifically solved at some future moment.

Yet even if a person were prepared to share the conclusion we shall draw from this investigation (viz., that the method of science is inherently universal since problems which cannot possibly be solved by the scientific method cannot be solved by any other method either, because they have no solution) he would still not be committing himself to believe in a continual progress of scientific knowledge. New scientific knowledge has always been counterbalanced to some extent by knowledge simultaneously lost. Thus, there is no doubt whatsoever that quite a few facts relating to a given historical period were better known by those who lived during the period than by their most learned successors; such loss of knowledge due to the gradual disappearance of witnesses could obviously not be fully made up by the most comprehensive recording, since a contemporary comprehensive recording would constitute a new feature to be recorded in turn, and so on *ad infinitum*. There is nothing to indicate that this continual loss of knowledge would cease to accompany incessant acquisition, even on the now optimistic assumption that no political catastrophe would put an early end to scientific progress and stop it dead at some point of its development.

One may also refer to the continual shifts in scientific interest which make certain theories seem futile and stop any advance in the corresponding field almost instantaneously. This has often happened in the past and there is no reason why it should not happen in the future. The seventeenth-century mathematicians were enthusiastic about the theory of combinations and harboured excessive hopes as to its prospects; yet, apart from the fundamental facts then established, the theory is now practically dead. For some time metallurgy was but a branch of engineering; then it became a respectable part of physics and was of central interest for a while. Now it is gliding back into a relatively neglected state. The question as to which problems about metals are inherently solvable by scientific methods has, obviously, little to do with such shifts in scientific fashion.

These examples show that the inherent range of science is identifiable neither with the past nor with the future development of science; it pertains to the logical analysis of science, not to a prophetic anticipation of its future. The fact that a scientific theory, say elementary group theory, is logically complete because its axioms are sufficient to settle any and every relevant problem (i.e., either to prove or to disprove any state-

ment that belongs to it by virtue of its subject-matter) does not warrant
the prediction that any particular group-theoretical statement will eventually come to be either demonstrated or refuted. Similarly, the inherent
universality of science, its ability to cope in principle with any and every
solvable problem, would not imply that any particular problem will
actually be solved by scientific methods in some more or less remote
future. Hence, there is no connection between trying to determine the
inherent range of science and attempting to predict its development.

There is another way of approaching the supposed similarity between
circumscribing the inherent range of science and determining the potentialities of art. The question whether a solution of a particular problem
obtained within a particular science is correct can be answered by
applying objective methods, capable of eliciting common assent in all
competent investigators, whereas for the solution of a musical or architectural problem—if one chooses to use the term "solution" for these
problems also—there are no such publicly applicable criteria at all. For
example, the solution proposed for a mathematical problem by a mathematician can be checked by any competent investigator. Such a check
can even be performed mechanically at present. This does not mean
that the invention of the solution is mechanizable as well. Indeed, in
most interesting cases, scientific invention requires at present and will
always demonstrably require (§ 5)—regardless of any conceivable
advances in logical and servo-mechanical methods for verifying the
correctness of scientific solutions to theoretical and practical problems—
human qualities like perseverance, skill, knowledge, a minimum of good
luck, and sometimes genius. Invention is no more mechanizable in
science than in art. But the assessment of a scientific invention is both
mechanizable in principle and publicly verifiable. This implies that
scientifically admissible solutions have to meet specifiable requirements
and there is thus provision for the possibility of investigating the existence of solutions defined by these requirements, without determining
the content of such solutions. In other words, owing to the connection
between scientific problems and solutions established by the public verifiability of these solutions, it is possible to determine the solvability of
problems by scientific methods without having to indicate the relevant
solutions.

The situation in the arts is diametrically different. Of course, the
finding of an aesthetic solution, say the composition of a symphony or
the painting of a picture, is as little predictable and mechanizable as the
solving of a scientific problem. But the point is that there is nothing in
the case of art which corresponds to the possibility of checking the

correctness of a solution to a scientific problem. As a matter of fact, the only problem to be "solved" by the creative artist is how to bring into existence a thing of beauty. The correctness of any particular solution to such a problem cannot be checked mechanically or dedeuced logically from a set of assumptions. There is thus no specifiable connection between artistic problems and their solutions which might offer a clue as to the range of problems which could possibly be solved. That is why any attempt at determining the potentialities of art becomes absurd and has nothing in common with the possibility of solving the corresponding problem about science.

2   THE SOCIAL FUNCTION OF SCIENCE

In order to extricate the epistemological question about the cognitive reach of science from the controversy over the social repercussions and value of science, let us first comment briefly on this sociological issue. It is concerned with what science will, or possibly could, mean to society, over and above what it now does mean. The present impact of science on human society is considerable and seems to gain momentum steadily. If one is justified in making an extrapolation for the future from the present trend, one may anticipate that our lives and those of our successors will be affected in an ever increasing degree by science. This means that, in an ever increasing degree, we shall hold beliefs based upon scientific information, and that our food, our clothing, our housing, our transportation, our entertainment, our ways to fight disease and to wage war will tend to rely more and more exclusively upon facilities which are the result of the application of scientific theories. It is important to realize that such facilities are often automatically controlled at present and that they are likely to become entirely self-controlled in the future. In other words, a push-button life on an unprecedented scale appears likely. It seems as if, in order to get along in this coming scientific age, one will have merely to push the right button at the right time. In order to get at the right button, one may need perhaps some "pull," but no more than this pseudo-mechanics of push and pull would be involved in living and getting along in this scientific world.

The scientist himself, main architect of this world, would be able to rely increasingly on automatically controlled devices. Automatic observing instruments, automatic computers, electronic brains, would make his observations, his computations for him and even do the bulk of his thinking. He would thus become, in the long run, virtually superfluous; the only job left to him, as to his fellow men, would be supervision of

gadgets. Living in a scientific age would come to this. Science would
solve all the problems facing man by endowing him with appropriate
mechanical contrivances; the very task of advancing science would be
delegated to these contrivances.

Such appears to be the present trend and its projection into the future.
But this is, perhaps, mere appearance. When one tries to get at the essential realities of human life behind these push-button contrivances, one
cannot help wondering whether, in spite of all the revolutionary changes
which science has brought about, man himself has changed appreciably,
either in his mind or in his body. One cannot help wondering whether
the essentials of life are altered by science at all or ever will be altered:
science, by its very nature, may be unable to alter them. There are things
which it can do, and things which it possibly cannot do. What it has
done already is uniquely great, and what it will or may do is unpredictable and perhaps unimaginable at present. Yet all these scientific
achievements do not preclude the existence of insurmountable limits to
the social potentialities of science. I am not referring here to gratuitous
and fancy questions as to whether, owing to science, interplanetary travel
will materialize, man's individual life-span will be stretched up to im-
mortality, or man's nature will change so that superman emerges. Establishment of possible social limitations of science depends rather upon
discovering whether there are genuine social problems facing man which
are scientifically unsolvable because they transcend scientific methods
and push-button techniques; for example, can the scientific method
solve the problem of reconciling individual freedom with the requirements of an efficient assembly-line economy, or of promoting the social
cohesion of human groups by supplying them with publically verifiable
information, without recourse to authoritarian mass-indoctrination?

Before discussing what science might possibly do for man, let us first
characterize briefly what it is actually doing, that is, the functions it
actually discharges in human society. Perhaps the simplest way of
describing these functions is to say that science provides society with
reliable information of lasting value—that it is a news-agency comparable in principle to other news-agencies operating at present. The
ordinary news-man tells us what is the temperature in Toronto today,
and the scientist tells us what is the temperature of the sun. The ordinary
newsman reports on the size of the reservoir in the projected St. Lawr-
ence Seaway, and the scientist tells of the size of the atom or of the
universe. Like other news-agencies, science has developed elaborate
techniques for collecting relevant news, for checking it and evaluating
it, for commenting on it, and even for processing and doctoring it, in

order to make it fit smoothly into patterns of editorial policy called scientific laws and theories. As in ordinary news-agencies, the news analyst and commentator (called theoretician in science) is more highly thought of than the fact-finding reporter, the "observer." Einstein's comments are more important than the facts found by Michelson and Morley; the comments of Kepler and Newton were more important than the facts discovered by Tycho de Brahe.

There are, of course, quite a few differences between science and other news-agencies. For instance, in contrast to the policy pursued by other agencies, the scientific reporter is permitted and even encouraged to interfere with the events he covers; if successful, he is promoted to the coveted rank of experimenter. More important, the information provided by science compares very favourably, as far as its reliability is concerned, with the news supplied by other agencies. Science is, of course, a human undertaking and by no means infallible. But its techniques of verifying its information are so elaborate and efficient that there are virtually no disagreements over scientific information—indeed, in ordinary parlance, "scientific" has come to mean "reliable." Professional standards of ethics are high among scientists: information that proves unreliable is always withdrawn from circulation—a practice not too common in other news-agencies. Correspondingly, if accumulating evidence tilts the scales in favour of a news item that has previously been withdrawn, it is put back into circulation—and the swing of the pendulum often goes on. Thus, in the seventeenth century, the view that light consists of waves was put forward by Huyghens. A few decades later, Newton assembled an impressive body of evidence in support of the theory that light consists of particles. In the nineteenth century, new facts discovered by Yong, Fizeau, Hertz, and other observers, in conjunction with theories elaborated by Fresnel and Maxwell, strengthened the wave view of light, which seemed then almost definitively established. Yet, in our century, other new facts related to the interaction of light and matter and their theoretical elaboration by Einstein have resuscitated the particle view, without disposing, however, of the contrary evidence accumulated in the preceding centuries. Apparently, because of the impressive bodies of evidence supporting both views, the scientists feel now that they have to give equal consideration to two mutually incompatible theories. No attempt has been made to conceal or to minimize the uneasy situation.

There is also a remarkable difference between the technique of spreading scientific information and that of ordinary news-agencies. The latter reach the public at large directly and aim at this public. Scientific information, however, is conveyed at several levels of diminishing accuracy and increasing digestability; information eventually reaching the public at large is oversimplified and distorted. At the highest level of accuracy, it is conveyed in original scientific papers and aims at the expert: an estimated 35,000 periodicals discharge this function. At the next level, scientific information, partly digested and watered down, appears in treatises and monographs destined for the serious student; the zest and sparkle of the original contribution have almost vanished in the transition. The departure from the original increases at the next level of college compendia. Finally, at the stage of high-school textbooks, semi-popular expositions, and popular magazines the genuine scientific information is usually distorted beyond recognition. It is at this stage, nevertheless, that scientific information plays a decisive role in shaping mankind's outlook. This is one of the appalling shortcomings of our uneasy time: science, admittedly of vital importance to everybody, is actually intelligible to only a few.

However, the main difference between science and other news-agencies resides neither in the higher reliability of scientific information, nor in the particular techniques of spreading such information. It is rather connected with the scope of scientific information and the uses to which it is capable of being put. The ordinary news-agencies are concerned exclusively with single, particular, contemporary, and local facts, with the here and now. The scientist reaches out for the remote regions of space and time and is most anxious to find out laws governing all the facts irrespective of where and when they happen. Indeed, it is his knowledge of universal laws which enables him to get hold of remote facts. He is able to find out that the temperature of the sun is about $6,000°C$ at its surface and $20,000,000°C$ at the centre because he is familiar with the distribution of temperature which must prevail in any gaseous globe under given conditions of chemical composition, thermo-nuclear transformation and radiation-pressure. He does not tell us only of the size of this atom, or of that atom, but of any atom, wherever and whenever it be. The news-man finds out the size and the capacity of the projected reservoir: geometry establishes laws connecting size and capacity of any solid, irrespective of time and place. This greater scope of scientific information accounts also for the more lasting interest attached to it. Few people care today to read the newspaper of yesterday. But Euclidean geometry, containing universal laws which govern the connection of size and capacity of solids and similar matters, has been read and utilized for over twenty centuries. That is why the social function of science can be characterized as that of supplying society with reliable information of lasting interest.

The contrast between the perennial information provided by science and the ephemeral news supplied by other agencies does not weaken their basic similarity: socially speaking, science *is* a news-agency. This is apparent from the whole age-long record of services science has rendered to society by providing it with valuable information, and of decisive progress achieved in science by successfully applying scientific methods to socially important problems. The history of science can only be understood in terms of its social function.

It hardly needs to be pointed out that no hypothesization of science is entailed in such a view of its social function. At bottom, there is no real entity, to be located in space and time, observable by suitably equipped investigators, discharging a definite function in human groups and going by the name of science. The term "science" is but a convenient abbreviation, a device for describing the goals, activities, tools, achievements, and potentialities of certain human agents, namely the scientists; whatever can be justifiably said about science, is translatable into equivalent statements concerning scientists in their professional capacity. The translation may often be awkward and unhelpful, but it is always feasible. In particular, establishing the range and limits of science amounts to determining the range of problems that scientists can or could solve. Similarly, the social function of science is the function discharged by scientists in their professional capacity. In order to discharge it, they have to collect, check, record, and make available to society information of lasting value.

3 THE INTRINSIC VALUE OF SCIENCE

We have characterized, in the preceding section, the social function of science as that of an agency supplying society with reliable information of lasting value. What does the value of scientific information consist in? Why is it perennial? The value or disvalue of a given object in a given respect and to a given individual or group consists in its ability to satisfy, or to frustrate, a given need of this individual or group. The same object may satisfy several needs of the same person and frustrate several of his other needs; it will then be positively valuable in several respects to this person and negatively valuable in several others. The over-all value of the object to the person is determined by the balance of all its values and disvalues to him. Thus, food, housing, clothing, entertainment, have value to those who are hungry, homeless, who need protection from the weather, or are anxious to get rid of boredom and worries. The value of an object does not coincide with its *actually* satisfying the corresponding need, but rather with its *ability* to meet this need. Potential, not actual, satisfaction of needs is coextensive with value: food is valuable to one who is not actually hungry at a given moment, because it is capable of relieving his hunger. Accordingly, the values of science to society do not reside entirely in the uses to which scientific information is actually being put; they comprise, in addition, the potential or possible uses of such information with a view to satisfying some definite social needs. The social disvalues of science come from scientific information either preventing, or being capable of preventing, the satisfaction of definite social needs. The over-all social value of science, if determinable at all, would consist of the final balance of all such values and disvalues.

Two classes of scientific values can be distinguished: (1) *intrinsic* values, those which science possesses intrinsically, because of its ability to satisfy directly specifiable needs, and (2) *instrumental* values, ascribable to science in view of its ability to produce or to contribute to the production of things which can serve, in turn, to obtain intrinsically valuable objects. Since, socially speaking, science is a body of information, its intrinsic values are values ascribable directly to information, regardless of any other uses to which it may be put. The need for obtaining, or having, scientific information regardless of any instrumental use comes simply from our desire to know, that is, from the human form of the powerful exploratory drive proper to all higher animals. The needs which scientific information may satisfy or frustrate indirectly, independently of this exploratory drive, give the instrumental values and disvalues of science.

What, then, are the main intrinsic values and disvalues of scientific information? What are the credit and the debit sides in the over-all balance of science viewed as a body of information, independently of its instrumental role in the satisfaction of human needs? Contemporary science is often credited with having secured reliable information about the size, age, and eventual destiny of the universe around us, about the randomness of cosmic history, the origin, evolution, and cosmic distribution of life, and quite a few other theoretical problems of comparable importance to man. Man has been craving from time immemorial a reliable solution to the most striking puzzles presented by his environment. Before science had made enough headway to attack these puzzling problems, he had recourse only to makeshift answers. Of course, a lasting satisfaction of his exploratory drive and a consistent attitude of confident understanding towards his environment could have been arrived at only if a correct solution of the puzzles had been obtained by applying methods of investigation of a degree of reliability comparable to that of science. As a matter of fact, science has succeeded in making man's pre-scientific environment less puzzling, not only by making his knowledge thereof less fragmentary, but, primarily, by creating a new and more comprehensive human environment. The bounds to knowledge of the pre-scientific world were set by the natural limitations of man's perceptual apparatus, by his rudimentary ability to process logically his perceptual data, and by the comparatively small amount of knowledge bequeathed by past generations. In other words, this primitive environment consisted entirely of what was directly visible, audible, tangible, rememberable, or easily transmissible from generation to generation; it did not include the very remote in either space or time, the very large or the very small, etc. More generally, he was limited in regard to the multiplicity of observable qualities and relations, such as colours, sounds, shapes, distances, sizes, weights, temperatures, speeds and so on, which are the threads of the variegated fabric of the universe around us and range over wide intervals; only an infinitesimal segment of each interval was known to him prior to the emergence of science. Man was thus "suspended between the infinitely small and the infinitely large" in more respects than Pascal had realized.

Science has overcome the puzzles of a primitive environment by extending all these intervals and patiently discovering an immensely wider environment, an environment mainly beyond the reach of the old and crude exploratory devices, but nevertheless as reliably knowable as its pre-scientific core. Moreover, the objects and events within the wider environment have turned out to be governed by simple laws which provide for an explanation of observed events and a prediction of future events. It is these laws that have made the new environment, inclusive of its pre-scientific core, intelligible: illusions, hallucinations, dreams, diseases, catastrophes, and other unusual terrestrial or celestial phenomena have ceased to puzzle man, since the explanation and prediction of them have been rendered possible by the integration of the whole primitive environment within a new, coherent framework, governed by precise laws. Life itself, with its diverse manifestations, of which only an infinitesimal temporal slice was included in man's primitive environment, has gained much in intelligibility since it has been provided with a billion years of palaeontological background and its controlling laws have been integrated to a considerable extent with the universal laws of physics and chemistry, valid both within and without the realm of life.

While man's environment has been thus widened and thereby brought nearer to intelligibility, the scientific picture or representation of it has gradually taken on the form of an organized system of facts, laws, and theories, more suitable for collective use and transmission by teaching to future generations, and conducive to a unifying and internally consistent over-all synthesis. This advance of science towards unification has made so much headway up to the present that virtually the totality of laws governing inanimate nature in its multifarious mechanical, thermal, electromagnetic, optical, chemical, and related manifestations has become deducible (theoretically at least) from a relatively small set of basic assumptions filling a few printed pages.

The human needs whose satisfaction gives scientific information its intrinsic value fall into two distinct, though related, categories: science answers questions that have always bothered man; and, in addition, it constructs a coherent, comprehensive, and self-contained world-picture which, though it does not answer any single question directly, seems to be the source of an intellectual satisfaction at least comparable in degree to that obtained from scientific solutions of particular problems. It would certainly be a grave mistake to underestimate the intrinsic values of scientific achievements of the first kind. After having successfully attacked, by methods of its own, the perplexing ontological problems mentioned at the beginning of this section, science has been able to remove to a considerable extent the intellectual discomfort and puzzlement of man. It is now in a position to tell man where he stands; it enables him to know himself and the place he occupies in the universe around him. In this wide field, of course, science is not at present alone. It shares the field with religion and with philosophy. Moreover, what science has achieved so far in this field is far from satisfactory, as will be seen presently. Yet there is little doubt that in regard to the "ultimate" questions mentioned above, scientific information is felt at present to be an essential clue which neither religion nor philosophy can afford to disregard, and which they actually do take into consideration and will probably utilize in an ever increasing degree.

However, the construction of a comprehensive, coherent, and self-contained scientific world-picture seems to exceed in value such recent scientific forays into the ontological jungle. We not only desire to get answers to certain particular questions which happen to be of interest to us; we are anxious, as well, to build up an over-all view of the universe which will give us some feeling of security about the meaning of our existence. It seems essential to be able either to ascribe to the human mind an essential role in the universe and locate man within a context of significance and purpose, or to view man as just a swarm of molecules originated by the accidents of biological competition, on a planet that originated by an accident of stellar evolution out of a sun that owes its existence to an accident of cosmic dust formation, within a galaxy originated by accidental condensation, in a universe governed by laws of chance and accident. This alternative hinges upon the nature of the scientific world-picture, not upon the scientific or extra-scientific solutions of particular problems. Moreover, the creation of a self-contained scientific world-picture provides for the explanation and prediction of all observable phenomena and includes, as a corollary, the possibility of a scientific solution to age-old puzzles. By being more comprehensive, this second achievement of science should bring more weight to the credit-side of the balance.

There is, nevertheless, an impressive debit-side in the balance we are constructing for science. Scientific knowledge of a crucial sector of reality, namely man himself, of the facts and laws that govern human individual and social behaviour, is admittedly lagging behind the supply of information obtained by scientific method in other fields. Thus there is little doubt that, in contrast to the relatively mature state of physics, chemistry, and astronomy, and the rapid advance of biology, the scientific method has yielded so far comparatively poor results in the social sciences and the humanities. The lag in the sciences of man is apparent from the largely controversial state of expert opinion in respect to the bulk of the relevant problems and also from the disappointingly small predictive and controlling power of the available theories. However, the present relative backwardness of the sciences of man is due to an accidental, not an inherent limitation of the scientific method. This follows from the very fact that this method has by no means been entirely unsuccessful in the sciences of man. An impressive number of relevant facts have been mustered by carefully planned and controlled observation, for example in sociology. In quite a few cases, especially in psychology, general laws adequately supported by observational evidence and providing for fairly accurate prediction of future phenomena have been established. Practical successes in applying theories even to complicated cases are also undeniable: for example, subtle concepts of the Keynesian theory have been successfully applied in societies with formidable economic structures. These achievements, even if by no means comparable to those of the natural sciences, show, nevertheless, that the scientific method is not intrinsically inapplicable to social and humanistic problems and that the relative backwardness of the sciences of man is due to an accidental failure of the scientific method to yield, so far, results as satisfactory as those obtained in other fields.

I do not say, of course, that the present state of any field of human knowledge is *historically accidental,* nor do I deny that the backwardness of the social sciences could be accounted for in historical terms. It is *logically accidental*, since the extant achievements of the scientific method show that there is nothing in its nature which makes it inapplicable to man. In this investigation, however, we shall attempt to determine the logically essential, not the logically accidental limitations of science (§1). That is why we shall be able to disregard the gap in the scientific world-picture corresponding to the lag in the humanistic and social sciences, which points to merely accidental limitations of science.

Yet, one may wonder whether, in regard to a central concept of humanistic and social thought, the present incompetence of science is accidental and temporary: few people doubt the *value of science*, but is there a *science of values*? A mathematician may prove that a theorem is correct, but he could not possibly prove that the theorem is important or interesting. Importance and interest are value-predicates, beyond the scope of mathematics, even when applied to mathematical propositions. The importance of a mathematical proposition is not a suitable topic for a new mathematical proposition and hence beyond mathematics. The same applies to ethical value-statements. There is, for instance, the ethical truth that suffering must not be inflicted unnecessarily. This condemnation of cruelty is certainly valid. But the sciences officially dealing with suffering, say physiology and psychology, could not possibly produce any evidence concerning the ethical value of cruelty. Physiology and psychology may teach us how suffering can be inflicted, how it can be increased or decreased under suitable organismic and environmental conditions, but they could not possibly infer from their numerous findings concerning suffering that it ought not to be inflicted unnecessarily. This fact shows that physiology and psychology are not competent to solve all the problems concerning the phenomena of pain and suffering, for example problems concerning ethical aspects of suffering.

Strangely enough, this alleged inability of science to settle questions of value affects to a considerable degree science itself. I have said that the main task of science is to provide society with reliable information of lasting value. What information is of lasting value? The scientific method serves to make sure that the information the scientist offers is reliable. But how should we go about finding out whether it is valuable or scientifically significant? There is an infinity of questions that could be raised in any department of scientific knowledge. But reliable answers to such questions, even if available, are not necessarily items of scientific knowledge. Consider geography, for instance; it describes, with a certain accuracy, the surface of the earth. Yet too detailed a description of the earth would be felt to be scientifically insignificant—a mountain range deserves the geographer's attention, whereas an anthill does not. At what point does the geographical significance of an area start? The geographer answers such questions by relying on his sense of values, and so does the historian, the biologist, the physicist, the mathematician —indeed every scientist has to select, out of the infinity of questions which might be asked concerning his particular subject-matter, those which are significant, important, relevant, or susceptible to any other related value-predicate.

Several arguments can be adduced in support of the view that the inability of science to solve problems of value is not merely accidental. Science aims at objective knowledge, valid for every competent investigator—it seeks after truth and only after truth. Hence, it cannot, so the argument goes, ascribe values to things, because, in so doing, it would have to take sides and to forsake its objectivity. By its very nature it must cling to facts and disregard values. In particular, the scientist may make value-judgments about science—but they never occur within science. Science provides information, not evaluation. As H. Poincaré has put it, scientific theories are couched in the indicative, not in the imperative mood;[2] science tells us of what there is, without making recommendations as to what ought to be done about it. One may paraphrase his remark by pointing out that, as a matter of fact, value-predicates do not occur in the vocabulary of the most mature sciences. Consequently, value-problems could not even be stated, let alone solved, within the scientific framework.

Indeed, one may continue, scientific information is neutral as to the ethical or aesthetic or any other values involved in any of its problems. Man lives and dies according to the laws of physiology and of psychology, thrives and suffers in accordance with them; these scientific laws can be used in order either to kill him or to cure him, either to torture him or to make him happy. The ethical value of the various uses to which physiological and psychological information can be put is beyond science. Gas chambers, concentration camp facilities, mass destruction devices are based on scientific premises—as are sanatoria, social medicine, mass enlightenment. A scientific age will turn out to be a victory and not a disaster for man if, and only if, those extra-scientific values which are decisive for the uses it is put to, are provided from without science.

Francis Bacon taught that scientific knowledge is power at a time when modern science was at its very beginning. He was called "the demagogue of science." Yet his view proved prophetic beyond any

[2]Cf. H. Poincaré, *Dernières Pensées* (1913), p. 225.

expectation. Science is, at present, a formidable power. But, it is the power to do either what is good or what is evil. The choice between the different ways of using scientific power is beyond science. Hence science is not to be praised when this choice is inspired by ethical principles and not to be blamed when motives are immoral. After all, only human agents are praiseworthy or blameworthy according to what they do with the tools available to them. Science is not an agent, but a tool and is therefore responsible neither for the triumphs nor for the defeats of the scientific age. Such responsibility transcends science.

There are two issues involved in the above indictment of science and the restrictions on its competence. Whether scientific information is but a tool which can be used by human agents for either ethically valuable or ethically non-valuable purposes is one question; whether science has reliable and relevant information to offer concerning values is another question. It is obvious that even if science is competent to offer reliable information about values, this does not guarantee that such information will be utilized for valuable purposes. *Video meliora, proboque et deteriora sequor,* is a truth applicable to every human being: it is not sufficient to know what is good in order to be prepared to take action that is likely to bring it about. But this psychological feature of man is no shortcoming of science. As far as human society is concerned, the aim of science is to provide information, and not to implement it; one therefore cannot blame science for failing to fulfil what is clearly not its task. Of course, one cannot praise it either; neither the value nor the disvalue of science is involved in its neutrality in regard to the adoption of value-policies.

On the other hand, the question as to whether reliable information concerning values can be obtained by scientific methods is certainly part of the general problem concerning the reach of science. If value is, as suggested in the above discussion, coextensive with potential satisfaction of human needs, there is no apparent reason why problems about values should transcend scientific method. In view of the interrelatedness of values and of human needs, it is not surprising that value-predicates should be absent from the naturalist's vocabulary: their proper place is obviously in the conceptual apparatus of the social sciences and the humanities.[3] As a matter of fact, psychologists and economists have often succeeded in investigating valuational phenomena and in determining the value to be ascribed to definite objects in respect to definite human

[3]Cf. M. Weber, *The Methodology of the Social Sciences* (1949), and F. Kaufmann, *The Methodology of the Social Sciences* (1944).

needs. Granted that scientific findings concerning the ability of objects to satisfy human needs are not very satisfactory at this moment. But this is owing rather to accidental circumstances than to an inherent short-coming of science. In the first place, the over-all value to be determined by weighing all the particular values and disvalues involved may not be effectively determinable because the relative weight of the relevant needs is not defined accurately enough; the vagueness of the concept of over-all value is responsible for the difficulty, which is inherent in the nature of the problem, not of the scientific method. The same problem could not be solved by applying non-scientific methods. The only way of obtaining a solution would be to define more precisely the concepts involved and replace the problem with a more precise one. In the second place, the evaluation of an object may be within the range of common-sense procedures, and technical scientific devices may not be necessary. Science is but a refinement of common sense; in very many cases, it is not impossible, but simply unnecessary to resort to science in order to solve a problem concerning values. Since no recourse to science is being taken, these problems are viewed, unjustifiably, as transcending science.

Actually, the theoretical possibility of resorting to scientific methods is apparent in several problems of this type. We have seen, for example, that the question of the value ascribable to a particular mathematical result is outside mathematics. But this does not mean that it is outside science as a whole. Another scientific theory may well prove to be competent to deal with this question, just as the history of mathematics deals with mathematics, without being a part thereof. Actually, when somone, not its author, claims that a mathematical proposition is im-portant or interesting, he simply tries to convey that it contains a solu-tion to a problem that has bothered competent mathematicians and that they too were prepared to make efforts in order to reach such a solution: there was a *need* for a solution in the scientific community. Thus the statement, although referring to a mathematical proposition, is obviously concerned with the needs of mathematicians, and I do not see any reason why mathematical needs should prove intrinsically inaccessible to a scientific investigation.

I do not claim, of course, that such an investigation would necessarily make the evaluation of mathematical achievements more understandable, nor, least of all, that it would modify and rationalize the mathematician's scale of values. The point is that the investigation *could* be carried out within the framework of science. It may prove unrewarding, since the mathematician's common sense is perfectly sufficient for him to evaluate mathematical results. Granted that, in such cases, he has to exercise his *esprit de finesse*, rather than his *esprit géométrique*; nevertheless, these two factors, distinguished by Pascal, go very well together.

Let us add that the above analysis which tends to locate statements about values within the reach of science does not apply automatically to all value-statements. Statements about values are concerned with whether a given object is valuable to a given individual or group in a given respect. On the other hand, value-statements are those which involve some value-predicate, like "good," "evil," "beautiful," "ugly," etc. "X is valuable to Y" is a statement about values, "X is beautiful" is a value-statement. "X is valuable to Y because X is beautiful" is a compound statement involving both a value-judgment and a judgment about values. It seems obvious that the very nature of a value-predicate consists in the ability of the corresponding value-statement to enter such combinations with statements about values.

In other words, a predicate P can be defined as a value-predicate if, by its very meaning, the ascription of P to an object X entails that X is valuable. This definition tells us whether any given predicate is a value-predicate but it neither specifies the meaning of the particular predicate, nor does it explain why a group of value-predicates is labelled "ethical," or "esthetic," etc. The definition of the class of ethical predicates and the explanation of the meanings of individual ethical predicates are a central task of ethics.

In this context, it will suffice to point out that since, by definition, value-judgments stand in an entailment relation to judgments about values, the scientific status of the latter may ensure under suitable conditions a similar status of the former. This question will be touched upon (§33) in connection with our discussion of empirical verifiability. We shall then be in a better position to evaluate the gap in science corresponding to its present failure to provide satisfactory information about value-problems.

There is another item for the debit-side of our balance, in addition to this gap in presently available scientific knowledge which the social sciences have not filled: the disturbing fluctuation and lack of stability even in the parts of the scientific world-picture supplied by the most advanced theories of physics, astronomy, cosmology, and biology. Crucial features of the scientific outlook predominating at a given moment are at the mercy of what may seem a slight shift in the ever changing observational evidence and exploratory techniques. Thus the recent construction of a new telescope, with double the diameter of the largest previous instrument, was sufficient to upset estimates of the size and age of the universe. A quarter of a century ago, the discovery that light-beams passing close to the sun were deflected from their original course by a slightly larger amount than that predicted on Newtonian principles caused Einstein's revolutionary idea of a curved, finite, though unbounded non-Euclidean space, organically merged with time into a single cosmic medium, to be substituted in the generally accepted world-picture for the Newtonian universe with its infinite Euclidean space, absolutely distinct from time.

This sensitivity of the scientific world-picture to changes in the momentarily available observational evidence is by no means accidental. It seems to inhere in the nature of scientific method, which includes a fundamental principle (sometimes referred to as the Principle of Total Evidence) to the effect that every item of relevant observational evidence available at a given moment ought to be taken into consideration in assessing the degree of acceptability (or of empirical confirmation) of any scientific theory. Since the total observational evidence available at any moment keeps changing, both because of continuous improvements in exploratory techniques and because of the unceasing search for new significant facts, the scientific theories acceptable at a given moment must be continually readjusted to these changes, or replaced with completely new theories when the changes grow too large to be accommodated by mere readjustment. Thus, the Principle of Total Evidence makes necessary, owing to the incessant changes in this evidence, a never ending readjustment of the world-picture constituted by the basic scientific theories acceptable at a given moment. This lack of stability, due to the vicissitudes and upheavals of the scientific world-picture, seems hardly compatible with any appreciable reliability of the momentary scientific world-view, and is definitely incompatible with the ascription to it of any degree of certainty. The perennial quest for certainty, one of the manifestations of man's essential need for security in the face of the universe around him, cannot attain its goal by scientific methods, since the world-picture provided by these methods is necessarily fluctuating and uncertain.

Yet the fact that the scientific world-picture is fragmentary and transitory is not its main drawback. Its most disturbing feature would remain, even if the conspicuous gaps (e.g., those relating to social sciences) were filled in and even if, moreover, contrary to anything that has happened in the past, the accumulating observation data turned out to fit consistently into the available scientific theories, so that there would be no need to readjust these theories continually or replace them with radically new ones because of the Principle of Total Evidence. The

disturbing feature of scientific theories I am referring to is their ostensibly ineradicable ambiguity; it seems possible to give to the basic theories
any pre-assigned interpretation out of an infinite number of equally
admissible interpretations, and no clue is provided either by the internal
structure of the relevant theory or by the observational evidence used
to support it as to which interpretation is warranted.

Thus, the view that time has neither beginning nor end seemed to be
firmly embedded in basic scientific theories; the fact that the time-
variable t was always supposed to vary over the whole range of real
numbers, from minus to plus infinity, was the algebraic counterpart of
the conception of the unbounded nature of time in both past and future.
Yet, by simply putting $t = \log \cdot \text{nat} \cdot T$, a new time-variable T is
introduced, with $T = 0$ corresponding to $t = \text{minus infinity}$. In other
words, the new time-variable T corresponds to a time with a beginning,
but without an end. The point is that there is no telling which time-
variable is the "right" one: that is, which one refers to the time about
which science has reliable information to offer.[4] It seems rather disturbing that an ostensibly basic difference in human outlook should depend
entirely upon a simple algebraic manipulation.

A similar situation prevails in the scientific picture of cosmic space.
Instead of a finite, unbounded, curved, and expanding space, now preferred by the cosmologists as providing the smoothest approximation to,
and the most convenient framework for, all the relevant observational
data, the old-fashioned Euclidean, infinite space, with no curvature at
all, could be used as well: no inconsistencies would arise in regard to
these observational data, provided that certain other assumptions, not
concerned with either space or time, but rather with purely physical
matters, were suitably modified. Once more, there is no deciding between the two competing views of cosmic space, as far as the Principle
of Total Evidence is concerned. Reasons of convenience, the desire to
make the computations easier and more manageable, may and actually
do play a part in the preferences manifested by the scientists; but this
does not change the basic fact that the crucial, spatial aspect of the
scientific world-picture, though predominantly interpreted at present in
a definite way, could actually be interpreted in an infinity of other ways.

Thus, independently of gaps and incessant upheavals, the scientific
world-picture suffers from a basic and permanent ambiguity: it can be

---

[4]Several advantages have been pointed out favouring t or T, especially by the
late Professor Milne, who has advocated a time with a beginning. Similar ideas
were put forward by L. Chwistek. Cf. E. A. Milne, *Kinematic Relativity* (1949)
and L. Chwistek, *The Limits of Science* (1936).

given an infinite number of interpretations differing from each other in regard to the bearing they have on the general human outlook, though all of them derive the same amount of support from the sum-total of the scientific data available and from the scientific principles serving to interpret such data.

In this investigation, we are attempting to determine the competence of science to solve theoretical and practical problems, not the intrinsic values of scientific solutions. We shall not, therefore, have to strike the over-all balance for science. Yet, since the epistemological problem concerning the scope of scientific knowledge and the sociological question of the value ascribable to such knowledge are obviously interdependent and intertwined, it is important to ascertain in what respects the social merits and defects of scientific information affect the epistemological issue with which we are concerned. Thus, the credit side for society in the balance of science shows its ability to construct a coherent, comprehensive, and self-contained world-picture, over and above its competence to solve single problems of particular interest. This is, of course, an important aspect of the sociological problem which must be taken over into the epistemological problem: the fact that the over-all synthesis of science is still within the field of science, and does not require any extra-scientific agency, be it metaphysical or mystical, is certainly relevant in a discussion of the reach of science.

Similarly, the debit side for society in the balance affects in several ways the epistemology of science. We have discussed three items that have an adverse effect on the value of science:

(1) There are the gaps in the sciences of man and the doubts as to the competence of science to deal with human values. This is a fundamental aspect of our central issue and we have already pointed out that such gaps in the scientific knowledge of human nature and of human values are due rather to the extraneous, accidental circumstances under which the method of science has come to be applied than to the nature of this method. We shall therefore be able to disregard such gaps in this study, the main topic of which is concerned with the inherent, rather than with the accidental potentialities of science.

(2) On the other hand, the fact that empirical scientific theories always fall short of attaining a degree of reliability approaching absolute certainty can hardly be denied. We shall come to the conclusion later on (§38) that even though there are essential, insurmountable limits of reliability proper to any empirical method, the failure of science to obtain absolute certainty in the solutions it offers of empirical problems is inherent, not in the nature of scientific method, but rather in the nature

of empirical problems, which never admit of absolutely certain solutions by any method. The scope of scientific method will prove to be independent of whether and to what extent reliable scientific knowledge is infallible knowledge. Hence, two items in the indictment of science are unlikely to have a decisive bearing on the epistemological problem concerning the competence of science to deal with any pre-assigned theoretical or practical problem.

(3) The third item is the inevitable ambiguity of the scientific world-picture and the troubling possibility of interpreting the basic theories of science in various ways implying different ontological outlooks. This charge suggests a much more serious limitation of the reach of science. Of course, we were prepared in advance not to identify the inherent range of science with the problems that are solved at present by its methods, because new problems may come to be solved in the future; moreover, the sum-total of problems that are at least theoretically solvable by such methods exceeds even the aggregate of those already solved together with those ever to be solved. Yet one might be inclined to take it for granted that once a scientific theory is well established, all its statements are validated by the very fact that the theory itself has been validated. The third item in the indictment shows graphically that this wholesale dogmatic acceptance of well-established scientific theories is impossible. It just does not make sense to put on a par all the statements contained in a theory, even if the theory enjoys the most impressive support from the observational data available. The fact that the whole interpretation of a fundamental theory may be changed by a single algebraic manipulation, without the observational standing of the theory being affected, suggests that a scientist expounding the statements included in it does not necessarily intend all these statements to be taken literally. Some of these statements may be literally interpretable and adequately justifiable by the available evidence; for example, any statement of fact, reliably verified by actual observations, say about the present number of living human beings. Some other statements, unjustifiable if literally interpreted, may be adequately supported by the evidence available provided a definite "figurative" interpretation be put on them. This applies, for example, to astronomical estimates of interstellar distances. According to P. W. Bridgman,[5] such estimates can be well established by relevant data, provided "distance" in astronomical contexts be interpreted "operationally" and not in accordance with our usual understanding of this term, as applied to terrestrial objects. Finally, a third class of statements included in a basic scientific theory may turn

[5]Cf. P. W. Bridgman, "On the Nature and Limitations of Cosmical Enquiries," *Scientific Monthly*, vol. 37 (1933).

out not to be justifiable at all by the available evidence, regardless of the interpretation put upon them; that time had a beginning, that space is, or is not, Euclidean, belong in this last category, if it be true that such statements can be removed from the theory under consideration by a simple algebraic device, without impairing any of its vital functions.

It is, then, evident that an insight into the reach of science presupposes such a tripartite classification of the statements that make up the fundamental scientific theories, for the simple reason that problems solvable by statements of the first kind must be located within the range of science, whereas questions answerable by statements of the third kind are not solvable by scientific methods, and the status of problems corresponding to the second category of statements depends upon the interpretation. Such a tripartite classification of statements included in basic scientific theories, though certainly of decisive importance for a judicious adjustment of man's outlook to the changing scientific world-picture, is not automatically provided by science itself; it is a basic problem in the philosophy of science. We shall carry out this classification in connection with a central concept in the contemporary epistemology and philosophy of science, viz. the concept of empirical verifiability (§5): empirically verifiable statements will turn out to account for the first category, and empirically unverifiable statements for the third category; statements of the second category will have to be interpreted so as to ensure their empirical verifiability.

If correctly carried out, the tripartite classification of scientific statements would separate the cognitive part of science from its auxiliary, symbolic, and merely technical devices. The classification may therefore be of interest to those who are anxious to evaluate science. Yet we shall not use this break-up of scientific statements to assess the value of scientific knowledge, or, least of all, to take sides with either worshippers or detractors of science. Our only reason for summarily discussing the value of science while trying to determine the cognitive reach of science, is the desire to separate the two issues and to isolate some aspects of our epistemological problem which are implicit in its sociological counterpart. The tripartite division of scientific statements involved in the determination of the value of science is a case in point. It will turn out to affect decisively the solution of our epistemological problem.

4 THE INSTRUMENTAL VALUE OF SCIENCE

We shall now comment on the value ascribable to science independently of its ability to meet directly the human need for knowledge by supplying man with relevant and reliable information. Man not only

craves to know the answers to puzzling problems concerning the world
he lives in and to promote science because of its ability to provide such
answers: he is also faced with questions about what course of action he
ought to take in order to attain his objectives; he needs information for
practical purposes, as a guide to action. Such information, on which a
man acts successfully because it both prompts and enables him to take
a course of action that helps him to bring about the objective he pursues,
is also supplied in several cases by science. The value that accrues to
science on this behalf is to be distinguished from its intrinsic value, as
previously discussed, since different needs are met by scientific information in these two cases. In other words, man may need information
either to satisfy his exploratory drive or to guide him in action so that
he may attain his objectives. Accordingly, scientific information may be
valuable to him, either intrinsically, in the first case, or extrinsically, in
the second.

The ability of science to be a practical guide to human action is not
the only way in which it becomes extrinsically valuable to man. By
supplying a man with reliable information, science extends and modifies
the sum-total of his beliefs; every item of reliable information he obtains
from science as an answer to a problem which bothers him becomes a
new belief. And the aggregate of all such beliefs that a man entertains,
as a result of the information he obtains from science, is valuable not
only because it satisfies his needs for knowledge and for practical
guidance. The very fact of his entertaining new beliefs makes him a
new man; a society whose members acquire new beliefs through the
agency of science becomes thereby a new society. These changes in
human beliefs brought about by science may be desirable or undesirable
depending upon whether they satisfy or frustrate human needs. Accordingly, they contribute to the over-all value of science, independently of
the theoretical and the practical value it already possesses. This type of
scientific value, conditional on the changes in human beliefs may be
termed ideological;[6] though still extrinsic, it is no longer practical.

In this section, we shall comment briefly on the over-all extrinsic
value of science, be it practical or ideological. In other words, we shall
discuss the value of science both as a guide to action and as the source
of far-reaching social changes traceable to changes in human beliefs
brought about by science. Once more, this sociological question concerning the extrinsic value of science is but auxiliary in the context of
our investigation; it serves to separate our main epistemological issue
from its sociological entanglements and to isolate some epistemological
points of view implicit in the sociology of science.

[6]Cf. A. Grünbaum, "Science and Ideology," *Scientific Monthly*, vol. 79 (1954).

The credit side in the balance of extrinsic values ascribable to science is too obvious to warrant discussion. What is involved here is mainly the practical role of scientific information as a guide to action. Science is simply the decisive factor in man's practical prospects brought about by the accelerated growth of his technological power and by the corresponding advance of his control over his expanded environment. Significantly, this accelerated advance started about three centuries ago, that is, shortly after science itself began to snowball. Both advances have been continually interrelated: science kept making headway by solving scientific problems arising out of the needs of technology and, by the same token, kept contributing to technological progress.

The debit side is more difficult to ascertain. It seems to affect mainly the ideological value of science. The surge of science and its ever increasing impact upon man's outlook and his way of life are often held responsible for the grave dangers implicit in man's present situation. The spread of scientific information is supposed to have considerably weakened the efficacy of a comprehensive body of moral, religious, political, and philosophical beliefs which has been, so far, largely responsible for the social cohesion of human groups and the spiritual security of human individuals in the face of the surrounding universe. Allegedly, science, while depriving man to a large extent of this vital body of beliefs, has not succeeded so far in creating an alternative, comprehensive system of its own, capable of helping man to achieve a comparable degree of social cohesion and individual security. Indeed, the failure of science to replace this body of beliefs and the corresponding "system of values"[7] is claimed to be neither accidental nor temporary, but rather inherent in its very nature, since science is allegedly value-free; though capable of weakening a social system of values by undermining its theoretical presuppositions, it is not in a position to create a new system of values. It is contended that the ideological and spiritual vacuum created by the partial disappearance of a vital system of beliefs and of the associated values, with no compensation for this social loss, is a decisive feature of the present critical situation, and it is blamed entirely on science.

In particular, the present means of mass destruction are, of course, the product of scientific activity, and their presence is allegedly rendered more dangerous by the partial collapse of moral standards of conduct resulting from the spread of the theory of the historical and cultural relativity of all norms of ethical behaviour,[8] which is also imputed to

[7]The correspondence between a body of beliefs and a system of values is explained later on in this section; see p. 34.

[8]Cf. E. Westermarck, *Ethical Reality* (1932) and A. Edel, *Ethical Judgment: The Use of Science in Ethics* (1955).

science. In a similar vein, the fact that small groups of individuals are now able to attain dictatorial control over substantial fractions of mankind is also blamed on science, because the facilities for handling large human populations are based ultimately on scientific information; moreover, it is claimed, groups that are in a position to take advantage of such facilities in order to establish or to maintain dictatorships, are unlikely to be deterred from so doing by ethical considerations now rendered scientifically obsolete (i.e., relativistically unjustifiable). It is contended that those who are not aware of the intrinsically limited potentialities of science are prevented from looking beyond it for means and methods to cope with this critical situation which the social impact of science has created.

I think that most of the foregoing remarks referring to the dangers created by the social impact of science are warranted. One has to recognize that science has actually failed so far to create a socially effective system of beliefs and values of its own. One must also acknowledge that, in spite of all the human well-being science has created, and all the misery it has prevented, it would hardly make sense to strike a final balance by weighing these benefits against its detrimental consequences. It may well be true that hundreds of millions of people have been saved from crippling disease or starvation or death owing to the activities of science, whereas the number of those who have suffered because of scientifically devised facilities is much less: thus, the number of people tortured or killed during the last war is generally estimated as being of the order of some tens of millions, against the tens of tens of millions who have benefited from the discoveries of science. The arithmetical balance would be in favour of science. But who would dare to exalt or even to justify this balance in the face of the tens of millions to whom science's impact did spell doom? There is little human sense in this accountancy of martyrdom.

In the light of such facts one wonders whether it is worth while to look for some flaws in the above indictment of the social repercussions of science. There are flaws, of course. Thus, one may well doubt the alleged connection between the spread of scientific information and the ensuing feeling of insecurity of human beings with regard to the surrounding universe. Since scientific knowledge of the universe is admittedly superior to knowledge obtained or obtainable from other sources as far as its predictive and explanatory power is concerned, an increase of our scientific knowledge can but increase, in the long run, our feeling of security and trust in regard to the world we live in; we certainly feel more confident when we understand and can predict the events which occur in our environment than when our surroundings are completely unintelligible and unpredictable. The history of pre-scientific ideas about disease, agricultural crops, astronomical phenomena, and the whole record of prejudice and superstition supply ample corroborating evidence.

One may also wonder whether the alleged weakening of a socially vital body of beliefs can be accounted for by the spread of scientific information, since the actual spread of such information is too slow to have brought about a universal change in human outlook. As a matter of fact, the dispersion of scientific information through society is as little uniform as the advance of national systems of production; national economic systems of the "atomic" type coexist at this very moment with Palaeolithic economies in certain backward countries, with all the intermediary stages being represented somewhere. There is a similar lag in the hold of scientific information on various human groups. I have pointed out already that the basic theories of contemporary science are presently accessible to an almost infinitesimal minority of mankind and there is little likelihood that this situation will change in the foreseeable future. Indeed, the fact that almost all of mankind is unable to understand the single most dynamic factor in the social world, namely science, is another discouraging feature of the present situation, certainly out of keeping with the democratic spirit of the time. The same fact, however, makes one question whether science has had the alleged impact upon a body of beliefs that play a decisive part in society at large. How can society be deprived of its beliefs if it is unable to understand the information which is offered to it?

These few critical remarks concerning the allegedly detrimental effects of science may remove some misgivings, but they cannot possibly detract from our basic acknowledgment that the spiritual vacuum created by the advance of science is a decisive factor in man's present plight. It is true, as has just been said, that the spread of scientific information throughout society and the ensuing modifications in a socially vital body of beliefs have been so slow that, apart from relatively small human groups whose ways of thinking are dominated by scientific information, the vast majority of mankind has been almost completely unaffected by them. Nevertheless, there is little doubt that countries where economic advance has reached the "atomic" level play a decisive role in shaping man's destiny, whereas the backward economies prevailing in other countries prevent them from appreciably influencing the course of human affairs. In view of this strategic importance of economically advanced nations, and of the parallelism of economic and scientific progress, it is very probable that human groups whose controlling beliefs are largely dominated by science will affect much more efficiently the affairs of mankind than other groups less susceptible to the impact of science. The spiritual vacuum brought about by the virtual monopoly enjoyed by scientific information, and the allegedly ensuing partial collapse of moral standards of conduct, would therefore be little less threatening if they proved to be confined to those groups which, though numerically small, nevertheless shape the course of human affairs.

Yet, whatever the extent of the beneficial and detrimental effects of science, we have to realize that, in the context of this investigation, our objective is not to strike a final balance and to determine the over-all extrinsic value attributable to science, but rather to isolate the bearing of this sociological issue upon the epistemological problem concerning the cognitive potentialities of science. There are obviously epistemological aspects to this sociological controversy. Science is accused of having brought about a critical situation in human society by undermining a vitally important body of beliefs associated with an equally important system of values, without being able to replace this body of beliefs or to provide an alternative basis for this system of values. To what extent does this accusation affect the range and limits of science? What would the inability of science to create, by methods of its own, a system of social values similar to the system undermined by the spread of scientific information amount to? The existence of a system of values supported by a given ideology in a given society means that members of this society hold the beliefs that make up this ideology and are induced thereby to experience needs that correspond to the values of the system. Thus, a definite distribution of values in a class of objects conditional upon their ability to satitsfy the relevant needs of a social group is induced by the prevalence of a particular ideology in this group. Now an ideology may be weakened by science either because its tenets are incompatible with scientific information or because the evidence supporting these tenets is invalidated by such information; the tenets of the ideology are either refuted by scientific information or thereby shown to be gratuitous. Given such a situation, that is, a clash between scientific information and the ideology underlying a system of values, the inability of science to maintain the system of values in spite of the disappearance of the supporting ideology can be traced to two sources:

(1) The ideology may simply include the beliefs concerning the values under consideration; these beliefs enable the followers of the ideology to pass the relevant value-judgments. But science, being allegedly value-free, that is, incapable of providing information con-

cerning values, would not be in a position to enable the group to pass such judgments. This possibility is but a duplication of the view denying science competence in value-problems which we have discussed in the previous section. There is no need to resume this discussion here.

(2) Suppose the ideology does not include value-judgments but rather creates in its followers needs which correspond to the values of the system. In this case the values would simply disappear with the ideology. The question then arises as to whether science is able to re-create the same system of values without recourse to the ideology. Determining a course of action to be taken by a human individual or group in order to bring about the desired objective is a typical problem of human policies. The objective, in our particular case, is the coming into existence of a system of needs inducing a given system of values, on the understanding that the ideology of the social group which has been responsible so far for this system of values has been undermined by the impact of science. Now it is certainly the case that the spread of scientific information in a society does often bring about new needs and, accordingly, new values, hence, science is not intrinsically incapable of creating values. Yet the question whether the particular system of values corresponding to a given ideology can be created by scientific methods independently of this ideology is not settled thereby. More generally, the question as to whether science is competent to solve all *policy problems* is obviously an important aspect of our epistemological problem. A negative answer to this question would entail cognitive limitations of science distinct from those set by its allegedly value-free nature. The scientist would then not only be prevented, by the very nature of scientific method, from solving value-problems concerning the beauty of a work of art, the nobility of an action, or, for that matter, the importance of a scientific idea; he would also be unable to find out whether this or that should be done to achieve a particular objective. We shall have to discuss in detail the restrictions on the universality of scientific method which would result if science were found incompetent to solve problems of human policies and human values. And this is the main angle from which the social impact of science will have to be considered in the present investigation.

5   THE PRINCIPLE OF VERIFIABILITY IN SCIENCE

We have tried so far to clarify the epistemological problem concerning the inherent range and limits of science by extricating it from related problems in the sociology of science and by removing some troublesome ambiguities in its wording. Let us conclude these preliminary considerations by outlining a tentative solution to our problem. It is obvious that a determination of the inherent range of scientific method can only be attained by submitting this method to a thorough analysis; this will be the object of Part II of this study. Similarly, a delimitation of problems that are intrinsically inaccessible to scientific method requires, in addition to an inventory of conditions to be met by problems solvable by this method, a systematic investigation of problems that do not meet such conditions; this is the topic of Part III of this inquiry. At this preliminary stage, it may be helpful to outline, however sketchily, the result of these subsequent discussions, in order to locate the problem more accurately and to indicate the general direction of the road which might bring us closer to a solution.

We shall have to distinguish two basic types of scientific procedure. The scientist who tries to solve a particular problem aims both at discovering its solution and at establishing the dependability of this solution: there are scientific procedures of discovery and other scientific procedures of verification and proof. These two types of procedure differ conspicuously from each other in their scope.

The history of science shows abundantly that, in spite of all the advances of scientific technique, there never has been a systematic method of discovery, that trial and error remain, after all, the basic procedure used by all investigators. Is this shortcoming accidental or inherent? Well, mathematical logicians of the last two decades (Church, Turing, Tarski) have shown that no infallible method of discovery could ever be found, because such a method does not exist. It has been demonstrably established, for instance, that apart from the most elementary mathematical theories, and even within elementary arithmetic, there could not possibly exist a method whose correct application to any meaningful problem would lead infallibly, in a finite number of predetermined steps, to its correct solution. Cases where such a method does exist are exceptional and, in a sense, trivial. This inherent shortcoming in methods of discovery within science has of course to be granted. However, the point is that a lack of systematic methods of discovery does not carry with it any limitation upon methods of verification and proof. There is no doubt, either, that the latter form the gist of the scientific procedure. Regardless of how an individual investigator did discover a solution to a scientific problem, his solution will not become part and parcel of science unless and until its correctness has been verified by other competent investigators and can be rechecked by any competent investigator. The basic law of hydrostatics and the law of gravitation may have occurred to their respective inventors while one was taking a bath and the other was watching a falling apple; the laws themselves became integrated into science only because they were found to be adequately supported by the observational evidence and because their conformity with the observational evidence can be verified by every investigator.

The gist of scientific method is therefore verification and proof, not discovery. This method consists essentially of a set of procedures enabling one to ascertain, by appropriately combining observation and logical process, whether some particular solution to some particular problem is correct, regardless of the way the proponent of the solution got hold of it. Applying scientific method to this problem consists in finding out, by logical inference, what bearing the accumulated observational evidence has on the correctness of some particular solution to this problem. Hence, a theoretical problem would be intrinsically inaccessible to a scientific approach only if no conceivable amount of observational evidence and logical processing could have any bearing whatsoever upon the correctness of any of its solutions. There are many problems of this kind. A trivial example is afforded by the question as to whether every material object doubles its size every second. It goes without saying that no answer to this question could ever be justified by valid inference from observational data, because the universal swelling would affect the yardsticks along with any other object and therefore remain undetectable. There are other problems, by no means trivial, and yet intrinsically inaccessible to a scientific approach because the correctness of their solutions cannot be tested by any conceivable observational data subjected to any conceivable logical process. The realization that some particular problem—for example, that of absolute simultaneity between distant events—is of this type has often marked a decisive advance in science, for instance, in Relativity and Quantum Theory. Such problems have therefore attracted the attention of investigators into the foundations of science and led to a convergent result in the main trends of contemporary philosophy of science, regardless of whether they go by the name of operationalism or neopositivism, pragmatism or instrumentalism, conventionalism or empiricism: theoretical problems on whose solution no conceivable amount of observational data processed by any conceivable logical procedures has any bearing whatsoever, are cognitively meaningless. Propositions answering such problems are cognitively meaningless propositions. This so-called Principle of Verifiability, asserting that propositions which are intrinsically incapable of being either confirmed or refuted on observational grounds are devoid of cognitive meaning and must not be admitted in any scientific theory, amounts actually to a claim for the inherent universality of science and the non-existence of any limits inherent in its method: since, according to this principle, problems intrinsically unsolvable by the scientific method are cognitively meaningless, all cognitively meaningful problems must be intrinsically accessible to the method of science. Thus no inherent limits could be set to the method of science. The reason for this scientific monopoly would not be the potential omniscience of the scientist, but rather the unintelligibility of the questions to which his method is inherently inapplicable.

The Principle of Verifiability has thus a definite bearing upon the reach of science and offers a convenient framework for our discussion of this issue. As it stands, however, this principle raises more questions than it claims to answer. It bans unverifiable statements from all empirical sciences, from psychology, biology, chemistry, and especially from physics, which seems increasingly to absorb the sum-total of our knowledge of nature. The tendency to apply this principle underlies all recent developments in fundamental scientific theories, and, most of all, in Relativity and Quantum Theory. Yet a glimpse of the historical vicissitudes of the Principle of Verifiability discloses the difficulties raised in a precise formulation of it, and by its implications and its proof. All epistemologies inspired by this principle, for example the positivistic epistemology of science, are anxious to eliminate from science any and every statement which seems to be unverifiable and to smack of "metaphysics." Yet by definition, such epistemologies tend to go too far and to condemn theories and disciplines which come to flourish in the subsequent evolution of science and which thus in turn provide a striking refutation of the narrow-minded epistemologies that opposed their development. Thus, August Comte,[9] responsible both for the very idea of a "positive" science free from any "metaphysical" impurities and for the criterion of verifiability destined to differentiate positive laws from metaphysical statements, was induced on behalf of his positivistic epistemology, to condemn (along with such idle assumptions as the luminiferous ether and the electrical fluids) a vast array of fertile and strictly verifiable studies, including molecular theory and the calculus of probability.

At the turn of the nineteenth and twentieth centuries the objections to the allegedly unverifiable assumptions of atomic physics led once more to a clash of followers and opponents of the positivist epistemology. And, once more, the evolution of science has conspicuously refuted the

[9]Cf. A. Comte, *Cours de philosophie positive* (1839).

narrow-minded positivistic opposition. Molecular and atomistic theories have triumphantly invaded the whole of empirical science. The frame of mind of contemporary physicists may best be illustrated by the position taken in regard to the reality of molecules and atoms (i.e., at bottom, in regard to the verifiability of molecular and atomic hypotheses) by an investigator whom nobody could suspect of condoning "metaphysics" or of sponsoring unverifiable hypotheses: in referring to the atom, Professor Bridgman writes "we are now as convinced of its physical reality as of our hands and feet."[10]

Similar tendencies can be noticed in the evolution of contemporary neopositivism. At the beginning, L. Wittgenstein,[11] who seems to have started the ball rolling, confined admissibility and meaning to directly verifiable statements expressing immediate observational data; virtually the sum-total of scientific knowledge was thus declared unverifiable and interpretable only in devious symbolic ways. A more liberal position was then taken up by K. Popper;[12] apart from directly verifiable statements, scientific hypotheses that are *refutable* by finite sets of such statements were also declared admissible. This "liberalism" was carried further by R. Carnap;[13] his account of the positivistic epistemology tended to grant admissibility to virtually all the types of statements occurring in conventional scientific theories. C. G. Hempel[14] has submitted these successive phases of the neopositivist philosophy of science to a searching analysis and has explored several alternative conceptions of empirical verifiability.

The conclusion to be derived from this survey of the history of the Principle of Verifiability is obvious: it must not be given too restrictive a formulation, since this would not merely free science from "metaphysical" impurities but would curtail science beyond repair. Yet the problem concerning the reach of science would not be solved by merely rewording the principle. Granted that a suitably re-formulated Principle of Verifiability would ban from science all unverifiable statements without curtailing science by the same token. But one may still wonder on what grounds such recalcitrant statements are to be declared inadmissible? What is the justification of the Principle of Verifiability?

[10]P. W. Bridgman, *The Logic of Modern Physics* (1927), p. 49.
[11]L. Wittgenstein, *Tractatus Logico-Philosophicus* (1921).
[12]K. Popper, *Logik der Forschung* (1935).
[13]R. Carnap, "Testability and Meaning," *Philosophy of Science*, vol. 3 (1936) and vol. 4 (1937); "The Methodological Character of Theoretical Concepts," *Minnesota Studies in the Philosophy of Science*, vol. 1 (1956).
[14]C. G. Hempel, "Problems and Changes in the Empiricist Criterion of Meaning," *Revue internationale de philosophie*, vol. 11 (1950); "The Concept of Cognitive Significance," *Proc. Amer. Acad. of Arts and Sciences*, vol. 80 (1951).

The extensive discussions carried out so far in connection with this principle have been concerned with its precise formulation and the extent of its implications, rather than with its justification. These discussions have aimed at defining the concept of verifiability which is involved in the principle and at determining the range of philosophical and scientific statements supposedly banned by it, rather than at justifying the principle itself. No proof has been attempted of the claim that unverifiable statements are "inadmissible" and devoid of "cognitive value." In this context, however, we shall have to discuss this rather neglected question as to why and to what extent unverifiable statements are inadmissible, since only such an investigation of this principle will enable us to determine its bearing on the reach of science.

It goes without saying that a satisfactory definition of the crucial concept of verifiability is essential for any substantiation of the Principle of Verifiability; the indispensable comments on this score will be made later on. As to the concepts of "admissibility" of statements and of the "cognitive value" to be ascribed to them, their use is hardly satisfactory in an epistemological discussion. Epistemology is concerned neither with ascribing values to statements, nor with making recommendations as to their admissibility. We propose therefore to reinterpret the Principle of Verifiability to make it assert that *unverifiable statements are neither true nor false* and to see whether this contention can be deduced from appropriate definitions of verifiability and truth.

Such a contention, if correct, would constitute a justification of the gist of the Principle of Verifiability in its usual formulation, because it amounts to saying that when a person makes an unverifiable statement, he is neither right nor wrong, he neither adds to his knowledge nor subtracts therefrom. If knowledge may be conceived as the totality of *true* statements which the knower is able to make on adequate evidence available to him, then unverifiable statements would not constitute knowledge according to the modified wording of the principle, because they are not true. No theory that consists entirely of such statements would represent a body of knowledge. This interpretation does not imply that scientific theories should refrain altogether from utilizing unverifiable statements (as they actually do not), nor does it prevent such statements from playing an important, or even an indispensable, role in theories which contain knowledge. It does entail, however, that theories which include both verifiable and unverifiable statements contain knowledge only so far as their verifiable statements go.

The justification of the Principle of Verifiability which we shall outline has nothing in common with the view that unverifiable statements are meaningless. If correct, this theory of meaning would provide, of course, for an alternative justification of the principle and for a rejection of any use of unverifiable statements. However, I shall take no advantage of this approach since the verifiability theory of meaning seems to me utterly untenable.[15] It asserts that any unverifiable statement is meaningless and unintelligible, even if it is correctly constructed out of meaningful terms. This is hardly compatible with the fact that unverifiable sentences actually used in scientific discourse and in ordinary language do possess all the three dimensions of meaning proper to expressions occurring in cognitive contexts. They are believed or disbelieved by people who speak the language in question and are therefore *psychologically meaningful*. They stand to each other and to verifiable statements as well in logical relations of compatibility, entailment, etc. and hence are *logically meaningful*. They are often about definite extra-linguistic entities, they refer to definite objects, and therefore are *referentially meaningful*.

Unverifiable statements, moreover, sometimes play an essential part in scientific theories. Far from being meaningless strings of words, they represent important constituents of science. In order to illustrate this point, it may be noticed that the currently cited examples of unverifiable statements can be classified under three categories, according to the kinds of concepts they respectively utilize. These statements involve either specifically philosophical concepts of no interest to the scientist, or scientifically relevant concepts in scientifically irrelevant contexts, or scientifically relevant concepts in scientifically relevant contexts. Unverifiable assumptions concerning the nature and existence of things in themselves, of the absolute, of entelechies, of disembodied spirits, fall under the first category. Hypotheses to the effect that every material object

[15] I take it that the verifiability theory of meaning, by asserting that a statement is meaningless unless it is verifiable, may serve as a reason for the "inadmissibility" of unverifiable statements, which constitutes the Principle of Verifiability. Should this principle be construed, as sometimes happens, as asserting that the meaningfulness of a statement is equivalent to its being verifiable, then it would simply coincide with the verifiability theory of meaning. The latter's untenability, instead of making it inadequate for supporting the principle, would refute the principle itself. To substantiate the Principle of Verifiability, we can neither identify it with, nor rest it on, this theory of meaning.

Our interpretation of the Principle of Verifiability could be brought closer to its usual formulations if we construed the principle as asserting that only verifiable statements are "cognitively meaningful." The cognitive meaningfulness of a statement would then be identified with its possessing a truth-value. This terminology, though suggested by some authors, is hardly satisfactory, since unverifiable statements play an essential part in scientific theories and must therefore be granted some "cognitive meaning."

doubles its size every second, or that all material processes are accelerating their tempo at some given rate, are examples of scientifically irrelevant assumptions couched in scientifically relevant terms. The third class of unverifiable statements consists of those actually used by the scientist. The presence of such assumptions in science was implicitly asserted in Duhem's denial of the existence of an *experimentum crucis in physics*[16] and in Poincaré's classical discussion of what he calls the conventional nature of geometry and of mechanics.[17] Their presence could be shown even more convincingly with the resources of contemporary logical theory.

At this stage, it may suffice to point out the admittedly unverifiable nature of *single* geometrical propositions. Euclidean geometry (or, for that matter, any other system of geometry used by the physicist and the cosmologist), interpreted as describing the observable behaviour of material objects and of light-beams, is an important constituent of empirical science. Nevertheless, *single* geometrical propositions are intrinsically incapable of being either confirmed or refuted by observational data. For instance, statements asserting that there is a point between any two points, or that a segment of a straight line joining two points is the shortest distance between them, could be shown to be neither provable nor disprovable by any conceivable observational data, simply because they neither refer to observable objects nor can be derived from statements concerning such objects.

Now unverifiable geometrical propositions have of course the property of leading, in conjunction with other geometrical and physical propositions, to verifiable consequences. This circumstance accounts for their main function in empirical science. By utilizing geometrical theorems along with physical laws and particular observational data, one often succeeds in drawing verifiable conclusions, for example in predicting whether and when two observable bodies will coincide, or in determining the weight of a body whose geometrical properties are approximately known. However, this property of leading in conjunction with other non-geometrical statements to verifiable consequences establishes only a very indirect link between geometry and the body of all verifiable statements; it does not by any means ensure the verifiability of single geometrical propositions. As a matter of fact, any statement, however speculative or remote from observational data, has this property of leading, in conjunction with *some* premisses, to verifiable consequences. From the unverifiable assumption that there are demons and the auxiliary premiss

---

[16]P. Duhem, *La Théorie physique, son object et sa structure* (1906).
[17]H. Poincaré, *Science et hypothèse* (1913).

that the existence of demons would entail the existence of suffering, the verifiable (and, unfortunately, verified) proposition that suffering exists certainly follows. This consequence does not make the existence of demons verifiable. Similarly, the ability of geometrical propositions to lead, in conjunction with other physical assumptions, to verifiable conclusions concerning, for example, the observable spatial behaviour of large-scale material objects, does not confer verifiability upon the geometrical statements themselves.[18]

The unverifiability of single geometrical propositions does not prevent them from being psychologically meaningful, since they have been understood and even firmly believed by sixty generations of scientists. They are also logically meaningful, because they stand in logical relations to other geometrical propositions and to verifiable statements as well. Some geometrical propositions, such as those containing the concepts of finite solids, volumes, and surfaces, are referentially meaningful, for in their usual interpretation they refer to definite material objects or to kinds of material objects. They certainly play an important part in science. To declare them meaningless is therefore hardly satisfactory. We shall accordingly reject, in what follows, the verifiability theory of meaning and interpret the Principle of Verifiability as asserting that unverifiable statements are neither true nor false. In other words, statements whose truth-value could not possibly be ascertained, even with probability, by utilizing any conceivable observational evidence, have actually no definite truth-value.

Our major reason for reformulating the Principle of Verifiability is neither the need to account for the presence of unverifiable statements in essential scientific theories, nor the desire to remove value-problems from epistemological discussions, but rather the possibility which the new wording provides of justifying the principle by deducing it from definitions of truth and verifiability. Two other advantages of reformulating the principle may be mentioned. It has been argued recently[19] that the very task of providing statements with a general criterion of significance which the principle is supposed to fulfil, is faced with serious

[18]This does not put geometrical propositions on a par with assumptions concerning the diabolical origin of suffering. The role of unverifiable assumptions in empirical theories can be characterized in a simple way, without resorting either to the verifiability theory of meaning or to the view that only the whole system of geometrical and physical proportions is empirically verifiable, whereas single propositions of the system are not. Moreover this latter, "wholistic" view (Duhem, Reichenbach, Quine) is not compatible with the verifiability theory of meaning; if single geometrical propositions were meaningless because they are unverifiable, then systems of such meaningless propositions would be meaningless too.

[19]J. L. Evans, On Meaning and Verification, *Mind*, January, 1953.

difficulties. Similarly, C. G. Hempel has been led by his aforementioned investigations to doubt whether any version of the Principle of Verifiability can provide a sound criterion of meaning. Such difficulties do not arise if the Principle of Verifiability is construed as denying to unverifiable statements the possession of a truth-value, but not of meaning. The principle was also held to be self-defeating, since it seems to entail its own meaninglessness. This fundamental doubt concerning the epistemological status of the principle is removed by the new wording (cf. § 38).

On the other hand, we have to acknowledge, of course, that any attempt to justify the Principle of Verifiability would hardly be of interest if it required the abandonment of a basic logical law. Yet the new wording of the principle is at variance with the Law of Excluded Middle. The principle claims that unverifiable statements are neither true nor false, while the logical law stipulates that every statement is either true or false. To meet this difficulty one might appeal to the possibility of constructing an artificial and consistent language whose statements are not subject to the Law of Excluded Middle. In this investigation, however, we are anxious to justify the applicability of the Principle of Verifiability to actual scientific statements, couched in actual, ordinary language, by showing that anyone who correctly understands the adjectives "true" and "false" cannot help realizing that the unverifiable statements of this language are neither true nor false. Since there is little doubt that the Law of Excluded Middle is valid for ordinary language, we shall first have to dispose of the ostensible incompatibility of the Principle of Verifiability with this law (§ 28) and then proceed to the main argument by showing that the unverifiability of any statement is basically due to the complete vagueness of some of its constituent terms and that such vagueness, in turn, prevents the statement from being either true or false. This will require an analysis of the concepts of truth (§ 30), of vagueness (§ 37), and of verifiability (§§ 32–5), before an attempt can be made to establish the connection between the verifiability of statements and the definiteness of their truth-value, a connection which entails the inherent universality of science.

CHAPTER TWO

The Logical Structure of Science

---

6 A CRITERION OF SCIENTIFIC STATUS

LET US NOW continue this preliminary elucidation of the import of our problem by clarifying its key concepts. In any attempt at determining the reach of science, the very concept of "science" is of course crucial. Yet there is by no means a unique way of construing it in accordance with its use by either scientists or philosophers of science, let alone in accordance with popular and semi-popular interpretations. Roughly speaking, by "science" (as of some particular moment) one understands the sum-total of all the particular or special sciences (as of that moment). And a particular science, so considered, consists of all the problems which belong to it by virtue of their subject-matter and have been solved within its framework by applying methods proper to this science. Hence, in order to circumscribe the problems that can be solved by science, we shall have to find out under what conditions a particular problem whose subject-matter locates it within some particular science is considered as having been solved therein. In what sense did Descartes solve the problem of optical refraction, Newton the problem of the motions of celestial bodies, Darwin the problem of the origin of biological species, Maxwell the problem of the nature of light?

The answer seems rather obvious in the simplest problems. Thus the question as to how the path of a light-beam penetrating from the air into water is changed at the boundary surface of both media, is considered as having been solved by a *law* which Descartes and Snellius put forward independently of each other. What are the significant features of this solution? Why does it exemplify a scientifically satisfactory solution to the problem of the dependence of the angle of refraction of the light-beam upon its angle of incidence?

Before both investigators tried to find a solution, a body of reliable data was available specifying the value of the angle of refraction for a sizable sample of angles of incidence. These data were felt to constitute a body of relevant evidence which would have a decisive bearing upon the correctness of any solution that might be put forward. They explain also why the particular solution of Descartes and Snellius, requiring a constant ratio of the sines of both angles for one and the same light-beam, was considered correct:

(1) First of all, the law of refraction met the obvious requirement of being a logically admissible (i.e., pertinent and unambiguous) answer to the question under consideration. Since the range of logically admissible answers to any question is determined by the very meaning of the question, finding an answer meeting this first requirement does not call for any special investigation. In the problem of Descartes and Snellius, for example, any real function of a real variable determining the angle of refraction for any pre-assigned angle of incidence would constitute a logically admissible answer, provided both the independent and the dependent variables of the function have values compatible with the definition of an angular quantity. A logically admissible answer may be either true or false; if it happens to be coincidentally true, this fact is still not sufficient to grant it the status of a scientifically satisfactory solution; a solution that is satisfactory from this point of view must obviously be accompanied by a body of evidence that adequately supports its correctness. We thus obtain two more conditions to be fulfilled by a scientific solution.

(2) The logically admissible answer put forward by Descartes and by Snellius has been supported by the available data.

(3) The reliability, number and variety of these data were felt to constitute an adequate support for the correctness of the answer.

Apart from these three essential features, there are several accidental circumstances in our example. Thus, the evidence had already been available before the solution was looked for by the investigators. This is exceptional; as a rule, prior to the search for a solution, only a few data, rather suggesting than adequately supporting the solution, are available. Once the solution has been put forward, further data are collected in order to complete the body of evidence. It is also accidental that no invention of instrumental equipment, or of a mathematical formalism, or of an empirical theory was necessary in order to set forth the solution or to establish its justification by the body of evidence. In more complicated cases, the solution of a theoretical problem may require some or all of these three additional factors. Newton's solution of the problem concerning the motions of celestial bodies would have been impossible without the new mathematical formalism of the theory of fluxions (differential calculus) and the new empirical theory of mechanics, both invented by Newton himself; the instrumental equipment was already available. The modern theory of atomic phenomena would have been impossible without the mathematical theory of infinite matrices (or, alternatively, of functional spaces), without Bohr's transitional theory with its planetary atomic model and correspondence principle, and, finally, without all the instruments devised to observe individual events at the atomic and sub-atomic level (scintillation screens, Geiger counters, Wilson's cloud-chambers, etc.).

Yet, in spite of the extreme simplicity of the Descartes-Snellius solution (with all three additional factors vanishing simultaneously), it seems to me that it illustrates sufficiently the two basic features of the scientific solution of any problem. It shows that a scientific solution to a theoretical problem consists essentially in formulating a logically admissible answer in conjunction with a body of evidence which adequately supports the correctness of this answer. The answer itself, if formulated in a single statement, can also be viewed as a piece or a bit of pertinent information; the evidence accompanying the statement makes for its reliability. Thus a scientifically satisfactory solution to a problem is in substance a reliable bit of information which forms a logically admissible answer to this problem.

As a rule, however, scientific information is not offered in single bits or statements. A whole package of information is submitted at a time, consisting of several statements, in conjunction with a body of evidence in support of the package. Kepler's laws, for instance, offered a solution to the problem of planetary motions, with the observations of Tycho de Brahe as the supporting evidence. A more important example of a scientific solution consisting of a whole package of information is afforded where relatively few explicitly formulated statements are presented but there is a theoretically infinite set of propositions which can be derived from these statements by a specifiable logical procedure. Such packages of information involving theoretically infinite sets of propositions are called *scientific theories*. Their use is essential in the whole technique of scientific knowledge. A theory is not merely a convenient space-saving device for conveying virtually infinite information, while actually offering only an infinitesimal part thereof, together with a recipe for obtaining the remainder out of the explicitly formulated sample. The most fundamental problems of science can be solved only by theories, not by single bits of information. Thus the question of how the multiplicity of biological species is to be accounted for is pertinently answered by Darwin's theory; a single statement could hardly convey a logically admissible answer. The question concerning the nature of space can be answered, for example, by the sum-total of propositions provable within Euclidean geometry, that is, by an entire theory, not by a single statement. The question of the nature of light is answerable by Maxwell's theory, and not by a single statement of his, or of any other theory.

The dual requirement of containing both a pertinent answer to the relevant question and a body of evidence supporting the answer applies naturally both to single statements and to whole theories if the solution they contain is claimed to be scientifically satisfactory. It is obvious, however, that when a scientific solution to some particular problem consists of a theoretically infinite set of statements, it is out of the question to produce directly a body of evidence supporting each statement included in the theory. Instead of producing such evidence, the scientist describes the procedure he has used to obtain the required evidence, on the understanding that the same procedure would enable any competent investigator to obtain such evidence. As a matter of fact, this device is also used, as a rule, with problems solvable by single statements—apart from the most elementary ones. For example, the author of a scientific paper reporting the discovery of an unknown biological species may reproduce, by way of evidence, the picture of a specimen of this species. But such a procedure is never complete, and is feasible only for the simplest statements of fact advanced as solutions of scientific problems. The procedure fails when the problems become slightly more complicated. If a new chemical substance has been isolated and its main parameters are described by the discoverer, he will not directly produce the evidence in his communication, but simply indicate the procedures he has used in order to isolate the substance, to determine its chemical formula, its density, its melting point, its boiling point, etc. In other words, in regard to each claim he makes, there is an indication of the method he has used to substantiate the claim, the implication being that the same method can be resorted to by any competent investigator in order to confirm, or to check on, the correctness of the description of the main parameters of the new substance. In this sense, the application of such typical methods as the mass-measurement of large-scale bodies, atomic and molecular mass-measurement, wave-length measurement, aims always at establishing the correctness of some particular solution to some particular problem, or, to put it in another way, at *validating* this particular solution. We shall therefore call such procedures *methods of validation* or *validating methods*. In the course of this inquiry, there will be ample opportunity to ascertain that these methods of validation, which aim primarily at obtaining a body of evidence adequately supporting the correctness of some particular answer to some particular problem, form the gist of the whole method of science.

Of course, the same validating method may be applied to several particular solutions of several particular problems, as illustrated by the aforementioned examples. A validating method usually applied to problems of a certain kind may not be mentioned at all if the context makes it sufficiently clear; a remark to the effect that "the usual procedures have been resorted to" to determine, say, the boiling point of the new substance is a stylistic device which will suffice ordinarily and may often be superfluous. This, however, does not detract from the basic fact that a scientific solution of any problem always involves, at least implicitly, a logically admissible answer in conjunction with a method of validation.

When the same validating method happens to apply to a whole range of important and unsolved problems, the invention of the method may carry much more weight in scientific strategy and tactics than the successful application of this or another method to some otherwise important problem. Thus, all the recently devised procedures for timing prehistoric events (radio-carbon methods, etc.) are of greater importance for the further advance of science than the results achieved by applying them to specific problems of prehistory or cultural anthropology, despite the intrinsic interest of these problems. This point may also be illustrated by the decisive progress in our scientific knowledge of the molecular, atomic, and sub-atomic world, owing mainly to the discovery of new methods of validation which are applicable to problems concerning individual events at this level (for example, the Stern-Franck method for determining the velocities of individual molecules, which has rendered possible a direct verification of Maxwell's law of velocity-distribution in gases under conditions of thermodynamic equilibrium). According to our way of interpreting the solution of a scientific problem, the invention of a validating method does not by itself provide a single solution; but it may facilitate the solution of a vast range of hitherto unsolved problems and its pragmatic importance will therefore be indisputable. Conversely, the solution of a theoretical problem (say, the nature of alpha-radiation, by Rutherford) may lead to the discovery of new validating methods (the method of nuclear research by alpha-bombardment). There is thus a historical interplay between advances in validating methods and progress in solving scientific problems. This historical interdependence, however, does not change the basic fact that a scientific solution to any problem always involves a logically admissible answer in conjunction with a method of validation.

Methods for validating scientific theories are a generalization of those serving to validate single statements, as we shall see in detail later on (§§26–7). Let us say, for the time being, that, in substance, the validation of a whole theory amounts to the validation of a number of single statements selected from among those included in the theory ("the consequences of the theory," to use the ordinary terminology), by applying the usual methods for validating single statements. It is assumed, however, that the statements have been selected for validation in accordance with certain requirements, and that the results obtained in validating these selected statements are then submitted to an additional logical treatment which eventually yields a validation of the theory itself. The role played by the instrumental equipment, often so characteristic of methods of validation concerned with single statements of fact, does not reappear in the additional operations required for validating whole theories. These operations, though by no means trivial or negligible, are purely inferential.

Thus, a particular science is not just a set of statements concerning its subject-matter, nor yet a theory or a cluster of theories consisting of such statements, since in neither case can the evidence in support of the relevant statements and theories be omitted. And as this evidence is mostly conveyed by way of a procedure describing how the data for it can be obtained by every competent investigator, that is, by indicating a validating method, any particular science is essentially a system of statements referring to the subject-matter of this science, in conjunction with the particular validating methods of these statements. Science does not merely discover information; it offers reliable information, information supported by adequate evidence. This means also that neither a single statement nor a body of statements can acquire scientific standing unless they are accompanied by a method for obtaining adequate evidence in support of whatever they assert. The methods of validation are part and parcel of any special science, just as the information it contains is part and parcel thereof. In other words, a special science, as of today, is the sum-total of all the problems which belong to it by virtue of their subject-matter and which have already obtained therein a satisfactory solution. Science, as of today, is the sum-total of all the special sciences as of today, that is, the sum-total of presently available solutions of any problem belonging to any special science. The bits or packages of information with their validating methods are the ultimate constituents of science.

While stressing the joint and co-ordinated role of information and validation, I did not intend, of course, to put on a par all conceivable bits and packages of scientific information, or all scientific methods of validation. We shall see that, on the contrary, the importance of a piece of scientific information depends in general on whether it is a particular fact, or a general law, or a whole theory, and that, similarly, there is a hierarchy, though a more complex one, categorizing the validating methods. Nor did I try to define science unambiguously in terms of this double role of information and of validation in the structure of science. Both information and validation have acquired, within science, many specific features, but basically their combination is no monopoly of science, since such a combination occurs whenever anybody puts forward an adequately supported claim instead of simply uttering a gratuitous assertion. The additional features of the scientific technique, such as organizing information in terms of bits and packages or, most important, utilizing scientific methods of validation (as exemplified, say, by the Principle of Total Evidence) will have to be taken into account in due time. The preceding remarks have aimed rather at dispelling the most troublesome misunderstandings and ambiguities associated with the concept of science, illustrated, for instance, by the frequent doubts as to whether such disciplines as economics or pure mathematics are sciences in any genuine sense. Since we have stressed the basic role of this duality of information and validation in the structure of science, we are going to dispose of these ambiguities by taking as broad a view as possible of the scope of science: whenever there is a tendency to grant the status of science to a set of statements or a cluster of theories, we shall include them within science as understood in this study, even if tendencies to the contrary, denying scientific eligibility to the statements and theories under consideration, are also noticeable, provided that our dual criterion of scientific status be met. No information is scientific unless it is offered with a method of obtaining a body of evidence which adequately confirms the correctness of this information.

7   THE STATUS OF MATHEMATICS

We shall now consider the main discrepancies in the meanings attached to "science," in the light of what has just been said about the dual role of scientific information and evidence. Empirical disciplines such as physics, chemistry, biology, psychology, which resort ultimately to observation and experiment in validating their claims, are always granted scientific status, whereas mathematics, whose laws are established without any recourse to either observation or experiment, is often located beyond the realm of science proper. This terminology, however, by no means prevails: publications devoted to "mathematical sciences" are well known and all the classical classifications of the sciences do include mathematics. At first sight, our basic requirement for eligibility to scientific status, viz. the association of any piece of information offered with a validating method (called "proof" in mathematics) seems to be met within mathematics to the highest degree possible. We shall see in some detail later on that a validating method consists in most cases of a suitable interplay of observational and inferential operations, with some complications arising whenever the use of instrumental equipment with its underlying theory is involved: one validates a statement by inferring it from premises which describe the results of actual observations. The validating methods of mathematics, which dispense, in principle, with both observation and instrumental equipment, may therefore be viewed as a limiting case, characterized by vanishing observational and instrumental components, in the general pattern of scientific validation. Consequently, following our basic criterion, we have to include mathematics within the scope of science proper, along with mathematical logic, which, in the last two decades, has come to play an increasingly prominent part in the field of mathematics, and has almost supplanted the traditional formal logic originated by Aristotle.

It should be noted that the mere inclusion of mathematics and logic within science proper does not force us to take sides in the fertile and far-reaching controversy over the nature of mathematics, its subject-matter, and the reasons for its unchallenged applicability in the empirical sciences. Yet we cannot disregard the peculiar role allotted to direct observation in the logico-mathematical method of formalization (i.e., of constructing "formalized axiomatic systems"). This method became prominent in the last decades and is often interpreted in a way that seems definitely incompatible with our dual requirement for scientific status; on the basis of this view, often referred to as the "formalist philosophy of mathematics," it seems impossible to ascribe to mathematical theories any informational content whatsoever, let alone a body of evidence supporting such content.[1] The contention is contained in the following formalist claims:

(1) Every mathematical theory can be formalized, that is, transformed into a formalized axiomatic system, without any change in its mathematical status;

[1] Cf. M. Black, *The Nature of Mathematics* (1933) and H. B. Curry, *The Formalist Philosophy of Mathematics* (1952).

(2) In any formalized axiomatic system, all the "well formed formulae" (i.e., all the counterparts, within the system, of meaningful statements) play their properly mathematical role independently of any significance and reference to subject-matter that they may (but need not) possess.

Thus the formalist view of mathematics is clearly at variance with our main reason, just stated, for including mathematics within the realm of science proper, namely, the fact that every mathematical theory conveys meaningful information whose reliability comes from the availability of an adequate validating method.

In order, then, to justify our own position, we must here briefly consider the formalist claims. Let us point out, in the first place, that this view of mathematics seems to have been mainly inspired by one of the most impressive results of contemporary foundational research: the availability, in every formalized axiomatic system, of purely observational criteria of the correctness of proofs. Such criteria enable anyone to check the correctness of any proof in any such system by making certain specifiable observations of the formulae included in the system and of certain specifiable directly perceivable relations obtaining among these formulae, without bothering about either the significance or the subject-matter of the relevant formulae. This well-established result seems to substantiate, at first glance, the formalist view. Since it is possible, within any formalized axiomatic system, to check any proof without considering either meaning or subject-matter of the formulae involved in the proof, why should these be relevant at all, from the point of view of the mathematician, whose main job is admittedly the construction of correct proofs?

I do not suggest that the existence of observational criteria of proof correctness in formalized mathematical theories is solely responsible for the formalist philosophy of mathematics. Other influences have certainly contributed to the emergence of this philosophical view, for example, the predominantly empiricist outlook of our era which makes it imperative to account for the very existence of mathematical knowledge, since such knowledge is both indubitably obtainable without observation and infallibly applicable to whatever may be observed. No recourse to man's alleged powers of intuition, distinct from his capabilities of observation and of logical thought, would be in keeping with this empiricist outlook. Apart from formalism, there are other empiricist attempts at explaining the existence of mathematical knowledge, for example, the view that such knowledge is tautological. This view would make mathematics less enigmatic, since man's ability to secure, without observing anything, information which is safely and usefully applicable to anything he may come to observe, ceases to be a puzzle when this information turns out to be tautological and, hence, not very informative. On closer inspection, an empiricist philosophy of mathematics based on its allegedly tautological nature proves more charitable to the mathematician than formalism: tautology is preferable to non-sense. Yet, though slightly more charitable than formalism, this explanation of mathematical knowledge lacks the support of foundational research which formalism enjoys, and conspicuously fails to do justice to the mathematician's cognitive attainments, hardly describable as roundabout ways of saying that $A = A$.

To evaluate the force of this support let us consider a very elementary example, say the formalization of the arithmetical theory of addition, in the field of natural numbers. The theory to be formalized consists of all the statements of the type $2+3=5$, $3+5+4=12$, $2+3=3+2$ etc. It is couched in a rudimentary language whose *vocabulary* contains but four undefined words or symbols: the numerals "0," "1" and the signs of addition and of equality "$+$," "$=$." In addition to these four words, the vocabulary of the theory includes a potentially infinite set of defined numerals "2," "3," "4," etc., that is, expressions introduced by, or occurring in, definitional equations[2] of the form: $1+1=2$, $2+1=3$, etc. A numerical expression is a finite sequence of numerals in which any two adjacent numerals are separated by the sign of addition. A meaningful statement or a *well-formed formula* of this language is any expression consisting of two numerical expressions separated by the sign of equality. The statement "$1=1$" is an *axiom*. All the *theorems* can be derived from this axiom by applying to it a finite number of times the *rules of inference* of the theory:

Rule No. 1: If $A = B$ is a theorem, then $A + 1 = B + 1$ is also a theorem; in other words, the symmetrical extension of a theorem is always a theorem.

Rule No. 2: If a defined numeral is substituted in a theorem for its defining expression, the resulting expression is also a theorem.

Thus, $2 + 2 = 4$ can be derived from the axiom by applying to the latter three times the first rule of inference, then lumping together the right side of the resulting equation by consecutively applying Rule

---

[2]A definitional equation is, by definition, any formula of the type "$N + 1 = M$," N being an already defined numeral, $N + 1$ the defining expression, and M the defined numeral. It is possible to replace this "inductive" definition of a definitional equation, which seems circular because of the reference to "already defined numerals," with a longer non-inductive definition involving no such pseudo-circularity.

No. 2 to the definitions of 2, 3, and 4 and, finally, by transforming the left side into $2 + 2$ by a twofold application of Rule No. 2. All in all, three applications of the first rule and five applications of the second rule yield a proof of $2 + 2 = 4$ from $1 = 1$:

$$1 = 1 \tag{1}$$
$$1 + 1 = 1 + 1 \tag{2}$$
$$1 + 1 + 1 = 1 + 1 + 1 \tag{3}$$
$$1 + 1 + 1 + 1 = 1 + 1 + 1 + 1 \tag{4}$$
$$1 + 1 + 1 + 1 = 2 + 1 + 1 \tag{5}$$
$$1 + 1 + 1 + 1 = 3 + 1 \tag{6}$$
$$1 + 1 + 1 + 1 = 4 \tag{7}$$
$$2 + 1 + 1 = 4 \tag{8}$$
$$2 + 2 = 4 \tag{9}$$

The proof of theorem (9) consists of the sequence of formulae (1)–(9), in this order. To check on the correctness of the proof, one has to make sure that every non-axiomatic formula in this sequence of formulae is the result of applying either Rule No. 1 or Rule No. 2 to another formula which occurs earlier in the sequence. It can be easily shown that all the valid formulae of the arithmetical theory of addition of natural numbers are end-formulae of such proofs and, conversely, that the end-formula of any such proof is a valid formula of this fragment of arithmetic.

In general, every formalized axiomatic system is uniquely characterized by the following four conditions:

(1) A list of all words or symbols to occur in the system (its *vocabulary*) is supposed to be drawn up, with no additional comments as to the meanings, or the objects of reference, of the words.

(2) A list of effective rules is available enabling anyone to decide in a finite number of predetermined steps whether or not any pre-assigned expression obtainable from the vocabulary (that is, any finite sequence of symbols occurring in List No. 1) is to be considered a *well-formed formula* of the system. In other words, the well-formed formulae of the system are, by definition, all those and only those finite sequences of words listed in the vocabulary of the system which meet the requirements contained in List No. 2.

(3) A selection of well-formed formulae of the system, to be called its *axioms*, is made explicit in a new list. Alternatively, an effective rule enabling anyone to decide in a finite number of predetermined steps whether any pre-assigned well-formed formula is axiomatic may replace this list.

(4) A list of directly observable relations among the well-formed formulae of the system is made available and said to constitute the *Rules of Direct Inference* or *Rules of Procedure* or the *Transformative Rules* of the system. Accordingly, a well-formed formula A is said to be directly inferable from, or a direct consequence of, the well-formed formula B if A bears to B a dyadic relation occurring in List No. 4. Thus $1 + 1 = 1 + 1$ is directly inferable from the formula $1 = 1$ (or, for that matter, the formula $3 + 1 = 2 + 1$ from the formula $3 = 2$), because the former formula bears to the latter the relation referred to as "symmetrical extension" in Rule No. 1 of this system. Similarly, a well-formed formula C would be said to be a direct consequence of the well-formed formulae D and E, if C bears to D and E a triadic relation occurring in List No. 4. Coincidentally, no triadic relation was involved in the rules of inference of the arithmetical system dealt with above.

Given a formalized axiomatic system defined by a set of four lists of the kind just described, a correct *proof* of a well-formed formula A is, by definition, any finite sequence of well-formed formulae of the system meeting the two following conditions:

(1) The last formula in the sequence is A;

(2) Every formula in the sequence is either a replica of an axiomatic formula or bears to some formulae which precede it in the sequence one of the relations occurring in List No. 4.

It is easily seen that the above elementary arithmetical system is uniquely characterized by a particular set of four lists of the kind described, and, more important, that any formalized system defined by such a quartet of lists offers the possibility of deciding whether or not any pre-assigned finite sequence of its well-formed formulae constitutes a correct proof of its end-formula; in order to decide, one has only to observe the formulae of the sequence, without considering their meaning or subject-matter. To appreciate the importance of this result, one has to bear in mind that *every* mathematical theory can be formalized, without any loss of generality or of practical value. The possibility of formalizing the whole of mathematics is not apparent from what has just been said. It is linked up with another essential feature of modern foundational research, a feature so intrinsically important that its connection with the method of formalization is likely to be overlooked: the possibility of reducing the whole of mathematics to pure logic. At present all mathematical concepts are held to be definable in terms of purely logical concepts, and all mathematical statements are held to be deducible from purely logical premisses. Both possibilities, that of reducing mathematics to logic and that of formalizing mathematics, were discovered by G. Frege.[3] Actually, his main purpose was to lay down a logical foundation for mathematics. The possibility of formalizing mathematics was a by-product of his efforts to reduce mathematics to logic. The interconnection of his two results is due to the fact that the two basic *logical* theories[4] are easily formalized; their formalization makes possible, in turn, the formalization of every discipline which is fully reducible to them, and this includes the whole of mathematics, according to Frege's main result.

As a matter of fact, formalization and reduction to logic are independent of each other. In the above trivial example, an elementary arithmetical theory is formalized, without being reduced to logic; conversely, in A. N. Whitehead's and B. Russell's monumental *Principia Mathematica*, a reduction of mathematics to logic is attained on an unprecedented scale, although strict formalization is neither claimed nor achieved. The important point is, however, that the possibility of formalizing the whole of mathematics (not a single, artificially isolated, and very elementary mathematical theory) has been so far established only because of the reduction of mathematics to logic: all mathematical theories are at present known to be capable of formalization, because all are known to be reducible to two effectively formalized logical theories. The existence of observational criteria of proof correctness, which enables the mathematician to verify proofs constructed within a mathematical theory without bothering about their meanings and subject-matter, is obviously independent of whether the theory is reduced to logic; the existence of such observational criteria follows from the basic structure of a formalized system constituted by the ordered quartet of the aforementioned lists. Formalization, not reduction to logic, is therefore directly responsible for the main argument of the formalist philosophy of mathematics: the possibility of verifying proofs independently of their significance and subject-matter. The reduction to logic is relevant in evaluating formalist claims only because it shows the scope of the formalist argument, viz., its applicability to the whole of mathematics; verifying any and every mathematical proof without taking into account the meaning or the subject-matter of the statements involved in the proof is possible because mathematics can be reduced to logic.

Yet, while fully appreciating the importance of observational criteria

[3] G. Frege, *Die Grundlagen der Arithmetik* (1881).
[4] These two theories are the propositional calculus and the predicate calculus. Roughly speaking, the former contains the logical laws governing the propositional connectives (such as "and," "if," "or" and "not") whereas the laws of the predicate calculus refer to the use of "quantifiers" (such as "every" and "some").

of proof correctness (which assure, for example, the public verifiability of mathematical results and make possible the application of servo-mechanisms to important segments of what used to require mental work by the mathematician), we may well fail to see why the formalist rejection of mathematical significance and mathematical subject-matter should follow therefrom. In order to substantiate this doubt let us observe first, in fairness, that the formalist view of mathematics, with its emphasis on the importance of observational criteria of proof cor-rectness, does not obliterate the distinction between mathematics and the observational sciences. The point is that, in empirical sciences, the observations made with a view to validating a particular statement are always concerned with the subject-matter of this statement, with what the statement is about; one has to look at the microscope, not at the statement "this is a microscope" in order to validate the statement. In formalized mathematical and logical systems the situation is different; to check on the correctness of a mathematical proof made up of an ordered set of statements, one has to watch the statements themselves, their arrange-ment within the proof, and certain other of their directly observable properties, but not what these statements are about. Thus the formalist is perfectly able to account for the undeniable difference between mathematical and empirical theories without relinquishing his basic tenet as to the decisive role of direct observation in the methods of mathematics proper.

Although formalism, through the new role which it assigns to observa-tion in the mathematico-logical field, is able to effect a *rapprochement* between this field and the empirical sciences, without obliterating the difference between them, its ban on significance and subject-matter in mathematical knowledge seems to me hardly justifiable. If one asks why a particular formalized axiomatic system S is defined by a specified set of four lists which establish, respectively, the vocabulary of the system, its supply of well-formed formulae, its supply of axiomatic formulae, and finally its rules of direct inference (determined by relations of direct deducibility among its well-formed formulae), there is only one possible answer: because such a combination of a vocabulary, a class of well-formed formulae made up of elements of the vocabulary, a subclass of axiomatic formulae included in the class of well-formed formulae, and a set of relations of direct deducibility among the well-formed formulae happens to reproduce faithfully enough the vocabulary, the class of meaningful statements, the subclass of axiomatic statements, and the class of rules of inference ascribable to some non-formalized axiomatic system S', or to some non-axiomatic mathematical theory S''. Both S' and S'' would cease to be mathematical theories at all if their crucial logical and mathematical terms were not construed in accordance with their respective logical or mathematical meanings, since these meanings are essential in correctly understanding and handling S' and S''.[5] Why, then, can one afford to disregard both significance and subject-matter within a formalized axiomatic system S, obtained by the formalization of S' or of S'', in spite of the fact that S is just a replica of the non-formalized system S' or S'' in which both meaning and subject-matter are relevant? The simple reason is that, in a formalized axiomatic system, the significance of all the formulae and their subject-matter has been sufficiently taken care of in the framing of the four basic lists of the system, so that there is no more need to take them explicitly into consideration, while performing within the formalized system the operations which have been devised to this very end. Similarly, a pedestrian who is exclusively interested in safety while crossing a street where there is heavy traffic, can afford to concentrate entirely on the alternating green and red colours of the traffic signals without paying any attention at all to the actual traffic, because he is aware that the alternating colours have been set up in perfect correspondence with the alternating phases of the traffic. A formalist philosophy of street traffic would have to maintain that the only thing that matters is the succession of the traffic signals; the chance presence or absence of actual traffic is completely irrelevant to the pedestrian.

Let us conclude from our few comments on formalism that mathematics is a science in the sense of being a cluster of validated theories, that is, of packages of information offered in conjunction with their proofs or methods of validation. The informational and the evidential components are therefore present in mathematics as in any other science. The distinctive nature of mathematical methods of validation consists in their vanishing observational and instrumental components, only an inferential component being left. The presence of purely observational criteria of proof correctness in every formalized axiomatic system (and this means, potentially, in every mathematical theory) can be accounted for by considering the characteristic structure of formalized systems, as related to their non-formalized counterparts, without rejecting mathematical significance or subject-matter. By accepting, to this extent, the significance and subject-matter of mathematical theories, one is not committed to accepting "logicism"[6] or "intuitionism,"[7] or any other

---

[5]In the above example the interpretation and subject-matter of a non-formalized or non-axiomatized arithmetic is essential for the corresponding formalized system to be mathematical at all.

[6]Cf. B. Russell, *Introduction to Mathematical Philosophy* (1919).

[7]Cf. A. Heyting, *Intuitionism* (1956).

ready-made recipe for interpreting these aspects of mathematical thought.[8]

8   THE STATUS OF EMPIRICAL SCIENCES

Law-finding versus Fact-finding Sciences

Another variation in the use of the term "science" is connected with the difference between "fact-finding" and "law-finding" studies, the implication being that only the latter are eligible for scientific rank. Physics and chemistry, for example, are trying to discover and establish universal laws which are applicable to any and every space-time region and valid always and everywhere. On the other hand, the political and economic history of mankind, the history of nations and civilizations and geography, are chiefly concerned with local events; their subject-matter consists of single, particular facts, rather than of general propositions expressing pervasive regularities of the universe.[9] Such fact-finding studies are often denied scientific status.[10]

---

[8]The view that mathematical methods of validation are purely inferential, since their observational and instrumental components vanish, is in agreement with the logicist position. However, this agreement does not include other items of the logicist philosophy of mathematics (e.g., the definability of all mathematical concepts in logical terms and the derivability of all mathematical assumptions from logical premisses). As a matter of fact, the inferential nature of mathematical methods follows directly from the so-called Deduction Theorem of mathematical logic, discovered by Herbrand and Tarski and implying that whenever a mathematical theorem T can be deduced from a set of assumptions A, there exists a provable logical law of which the conditional "If A then T" is a substitution instance. In other words, if all logical laws of the form "If C then D" were known, all possible mathematical results could be obtained by simple substitution. In this methodological sense, mathematics is theoretically a part of logic.

Such a qualified reduction of mathematics to the purely logical theory of inference, though certainly at variance with the letter of mathematical intuitionism, need not offend its spirit. The formalization of intuitionist mathematics has been achieved to a considerable degree and it may well be construed as implying the possibility of reconstructing every intuitionistic proof within the field of pure logic —the only difference being that the logic involved would be the intuitionistic one. This difference is considerably reduced by the well-known results of Gödel, Kleene, Mostowski and other investigators, concerning the intertranslatability of "classical" and intuitionist logics. Cf. S. C. Kleene, *Introduction to Metamathematics* (1952), pp. 492 ff.

[9]Although laws are always formulated in *general* propositions, we shall have to distinguish (in Part II, chapter Two) *universal* laws applicable always and everywhere, from *regional* laws which assert regular interconnections between particular facts within a limited space-time domain. Laws of biological evolution, of human psychology, and of economics are limited in this sense since they hold only within a region confined to the surface of the earth during a finite time-interval. Physical and chemical laws held to obtain in every space-time region are not limited, but universal.

[10]R. B. Braithwaite, *Scientific Explanation* (1955).

In any event, our dual requirement of providing relevant information in conjunction with adequate supporting evidence is certainly met by both the law-finding and the fact-finding disciplines, as well as by the sciences of an intermediary type which are interested in discovering and establishing both particular facts and general laws referring to their subject-matter. We have therefore to include, within the realm of science, all disciplines of a law-finding, a fact-finding, and an intermediate nature. In so doing, we can claim once more that the restricted use of "science" referred to in this sub-section has no monopoly and that our more comprehensive terminology could well be supported by several linguistic usages. It is customary, for example, to refer to the "social sciences" without implying that they have to be predominantly or exclusively law-finding.

Natural versus Social Sciences

The tendency to restrict the term "science" to law-finding studies, though traceable to several historical sources (e.g., to the Aristotelian view of the universality of knowledge), is primarily connected with the attempt to confine scientific status to *natural sciences* only.[11] The reason for this interconnection is the fact that natural sciences are predominantly law-finding, in contradistinction to the social and humanistic studies which are predominantly fact-finding. However, the distinction between natural sciences and the humanities is based, ostensibly at least, on their subject-matter rather than on the emphasis they put on general laws and particular facts, respectively. Restricting scientific status to natural sciences would therefore imply that only "natural phenomena" constitute the field of science, whereas the study of "cultural phenomena" out to be allotted to scholarship rather than to science proper. Thus the history of nations, civilizations, cultures, religions, arts, and sciences would be located beyond the province of science in view of their subject-matter rather than of their quest for particular facts.

Yet the criterion for differentiating the two groups of disciplines, namely the natural and the humanistic studies (inclusive of the social sciences), has never been clearly stated, in spite of the voluminuous literature which has grown up around this topic. The examples of the predominantly fact-finding natural sciences—geography, geology, and palaeontology—and of the predominantly law-finding social theories—economics, linguistics, and psychology—show clearly that the two classifications overlap. Moreover, the distinction between the natural sciences and the humanities is not based exclusively on the subject-matter of the relevant theories, since quite a few significant aspects of

[11]F. A. Hayek, *The Counter-Revolution of Science* (1952).

man's individual and social behavour, which definitely pertain to human "culture," are allotted to natural sciences; this applies especially to human psychology, individual or social, developmental or general, normal or abnormal. Since psychology is mainly interested in discovering the laws that govern human behaviour and utilizes particular facts only as evidential data for validating such laws, one may wonder whether the two criteria in conjunction, that is cultural subject-matter together with the quest for particular facts, would not suffice to separate the humanistic and social studies from the natural sciences. Yet although this device would bring us closer to the objective by disposing of a few troublesome counter-instances, it would still fail to account for the law-finding nature of such social sciences as economics.[12]

In any event, neither consideration of subject-matter nor the distinction between law-finding and fact-finding investigations, nor their combination, is relevant to the application of our dual requirement for scientific status, which is designed to determine the range and limits of the unique procedure underlying all known scientific methods, including those of the humanities and social sciences. If disciplines treating man's cultural activities offer information about their subject-matter, and provide methods for obtaining evidence which will adequately support this information, then they meet our criterion of scientific status. There is little doubt that that is what actually happens in the methodologically advanced parts of the humanities and the social studies, as the use made of textual evidence, the scrutiny of past monuments, the archaeological diggings, the measurements of public opinion etc., amply testify. There is no doubt either that other departments of these disciplines, less favoured in this respect, are striving vigorously to eliminate gratuitous or unverifiable contentions from the body of relevant information they offer. Consequently, we are driven once more to take up the "liberal" view of the applicability of the term "science," and to extend its legitimate use to the humanities and social studies as well as to the unquestionably scientific natural studies.

Pure versus Applied Sciences

We shall again have to adopt a conciliatory view in permitting the *applied* sciences to rank with their purer brethren as genuinely scientific

[12]It would seem, indeed, that the attempt to restrict the field of science to "natural" phenomena is based, at least in part, on peculiar English usage. Other closely related members of the Indo-European linguistic group do not share such usage; the French, for example, have their "Académie des Sciences Morales et Politiques."

disciplines. The boundary between pure and applied studies is not easy to draw and seems to depend rather on extraneous circumstances than on such intrinsic features of the disciplines concerned as subject-matter and method. A science is sometimes labelled "applied" if it presupposes and utilizes theories belonging to some other science. If this were, however, the basis of the dichotomy "pure *versus* applied," then we would have to consider mathematics as applied logic, theoretical physics as applied mathematics, experimental physics as applied theoretical physics, and engineering as applied experimental physics. This hierarchy of "applications" is hardly in accordance with the meaning of "applied science"; only engineering, the last rung in the hierarchical ladder, is ordinarily labelled "applied." And even in this case, there are remarkable fluctuations. For example, the physics of solids, relegated for a long time to the applied field, came to occupy a rather respectable position within the pure science, physics; it has recently been neglected but it has not been relabelled. I think that "pure science" is rather greedy: once it finds a topic within its bounds, it refuses to part with it, regardless of the ups and downs of scientific interest in this topic.

As a matter of fact, the classification into pure and applied sciences seems to refer mainly to the predominant interest of the scientist himself. The scientist feels somehow that the laws and facts of physics, chemistry, biology, and astronomy are intrinsically interesting to him and to his fellow scientists, whereas the formulae and the methods of applied science are likely to elicit interest in the scientific world only because of the services they may render in helping to solve some non-scientific problems (technological, medical, etc.). The requirement of providing relevant information in conjunction with adequate supporting evidence does not specify the kinds of motives which are supposed to prompt those engaged in discovering the information and the evidence. From our general point of view, we shall, therefore, have to include within the realm of science all studies, pure or applied, which meet our requirement.

Object-Science versus Meta-Science

We are going to consider only one more of the meanings attached to "science." This usage is less widely known than the preceding ones, but turns out to be of great importance in connection with the inherent range of science. The vast majority of the sciences which have, so far, reached a respectable standing deal with some particular subject-matter, which may be the animal kingdom (zoology) or the chemical processes in living creatures (biochemistry), but which does not include another special science, or a group of such special sciences. There are, however, some particular sciences the subject-matter of which does consist of another special science or of a group of special sciences. Examples are the history of physics, a special science with physics as its subject-matter; the sociology of science; and the theory of the logical structure of mathematics, or metamathematics, a discipline particularly flourishing at present.

We can therefore distinguish first-level sciences, or object-sciences, each dealing with a subject-matter which is not a science; second-level sciences whose subject-matter is a particular first-level science or a group of first-level sciences; and sciences of successive, higher levels.

This stratification of particular sciences into successive levels, depending upon their respective subject-matter, is of little help in classifying sciences existing at present, since there are but a few advanced sciences of the second level, and none on any more elevated level. Yet, in spite of the present vacuum on the higher levels of scientific inquiry, the very possibility of climbing above the first two levels turns out to be of great importance in evaluating the inherent range and limits of science: for example, there are good reasons supporting the view that a substantial fraction of philosophical problems are susceptible, in principle, to the method of science, provided this method be applied at a sufficiently high level (§ 13).

Moreover, the second-level sciences so far developed, undoubtedly meet our basic requirement, since they apply methods essentially similar to those of the object-sciences: for example, the history of science is using the ordinary empirical procedures of the historian and metamathematics resorts (with some slight refinements) to the very mathematical method of its subject-matter. We cannot, therefore, deny the label "scientific" to the relevant and reliable information provided at this second level. And we shall have to adopt a similar attitude to disciplines of any higher level, provided the basic requirement of scientific status be met thereon.

9   THE NATURE OF SCIENTIFIC METHODS

The method of science has twice been a subject of our preliminary discussions. In the preceding section, we noticed that any particular science can be viewed as a cluster of theories, each theory being in turn a set of statements associated with proper methods for their validation. From this point of view, science can be conceived as the sum-total of all particular sciences, and hence as an organized system of theories and their respective validating methods. The methods are as indispensable a part of science as are its theories.

On the other hand, we saw in § 1 that the inherent range and limits of science are definable in terms respectively of the class of all problems accessible to any actual or merely discoverable scientific method and of the class of problems inherently inaccessible to such methods. Thus, the concept of a validating method is as essential in ascertaining the reach of science as it is in describing the basic structure of science. In other words, what science means hinges upon the way it makes use of its tools in order to perform its basic function of supplying reliable information; and similarly, what science could possibly do and what it is incapable of doing hinge still upon the tools at the scientist's disposal (including those which he could possibly manufacture) and his ways of handling them, i.e., on scientific method, rather than on scientific information. That is why our whole investigation into the reach of science will have to concentrate mainly on an analysis of scientific methods of validation.

It is worth noticing that, even quite apart from our central problem, it is customary to refer to "the" method of science, as a single uniquely determined procedure, rather than to particular scientific methods applied in various departments of science. Is there more to this way of speaking than just a convenient abbreviation? Is "the" scientific method but the sum-total of methods used within science and does "applying the method of science" boil down to resorting to some appropriate selection out of the scientist's well-stocked supply? This question is more than a terminological one and has a direct bearing on our central problem. For if it were held that there is no specifiable procedure which deserves the status of *the* method of science, the endeavour to determine the range of science as a whole would consist in determining separately the ranges of the various methods and in attempting to unite all these ranges in a single assembly of scientifically solvable problems. A simple consideration shows the difficulties of such an approach.

Any particular scientific method, for example, the method of parallax for measuring astronomical distances, has a range of its own. It has been successfully used until now to determine the distances of several known stars; it may also be applied in the future to astronomical objects presently unknown. The inherent range of this method includes, of course, both kinds of problems, those already solved by resorting to it and those which could be so treated in principle but have remained unsolved because of some logically accidental circumstances. Now, if we raise a similar question in regard to all the numerous validating methods applied in contemporary science, we shall not obtain a clear-cut answer, because these various methods partly overlap and their joint range would constitute a logically accidental aggregate, hardly illuminating in connection with the inherent range of science. This applies even to the joint range of methods for measuring distances at the sub-atomic, the atomic, the molecular, the directly perceivable, the terrestrial, the planetary, the stellar, the intra-galactic, and the intergalactic level.[13]

Moreover, as has been pointed out in § 1, the inherent range of science includes problems solvable by presently available methods along with problems which could possibly be solved by devising new scientific methods in addition to those which are known at present. Thus, the location of a single sub-atomic particle is considered at present a scientifically solvable problem, in spite of the fact that no method is now available for treating it. The reason for placing this problem within the range of science is the fact that the well-established laws of nature do not preclude the existence of an efficient method for locating sub-atomic particles.[14] Thus, if our central problem were formulated in terms of particular validating methods, we would have to take into account, in addition to the ranges of actually used methods, the ranges of all possible validating methods and this seems an even less promising procedure. How could one determine the range of problems that could possibly be solved by applying as yet unknown methods?

Thus, in view of the great number of partly overlapping methods presently applied in various departments of science, and of the elusive nature of presently unavailable though theoretically discoverable methods, it would be extremely difficult to evaluate the combined problem-solving power ascribable to the heterogeneous assembly of all conceivable

---

[13]P. W. Bridgman, *The Logic of Modern Physics* (1927).

[14]A method for locating such particles could be obtained theoretically by constructing a "gamma-ray microscope." Microscopes using such radiation would suffice for the location of sub-atomic particles, for example electrons, because the linear dimensions of an electron, being about 10^{-13} cm., are well within the range of a gamma-ray microscope, provided the gamma rays used be sufficiently "hard." In "mental experiments" concerning the visibility of electrons through the agency of such microscopes and the "experimental significance" of the concept of electron (often associated with the possibility of observing it by means of theoretically constructible devices) a decisive importance is usually ascribed to the fact that the construction of such a microscope, or of similar devices, does not violate any well-established law of nature. This is why the presently non-existent method of indirectly observing the position of sub-atomic particles is considered as being possible, or discoverable "in principle." Cf. W. Heisenberg, *The Physical Principles of Quantum Theory* (1930) and J. von Neumann, *Die mathematischen Grundlagen der Quantentheorie* (1932).

scientific methods, regardless of whether they are in actual use, or could possibly be discovered. It seems more promising to subsume all such conceivable scientific methods under one comprehensive procedure, which might be termed "the method of science," and to attempt then to evaluate the reach of science in terms of this unique procedure. Such a procedure underlying all special scientific methods will be tentatively outlined towards the end of this chapter. Before proceeding to this uni-fication, we shall try to obtain some clues as to the nature of this pro-cedure by surveying the essential features of scientific methods in actual use. A few remarks about the nature of any method, whether cognitive or not, scientific or not, may serve to introduce our survey.

A *method* is a statement or a set of statements describing a repeatable sequence of *operations*, such that each individual sequence of operations so described would enable a human individual or group to bring about, either infallibly or in a fair proportion of cases, a repeatable event called the *objective* of the method. A method which comprises a repeatable sequence of operations is said to be *applied* by a human individual if this person performs an individual sequence of such operations with a view to bringing about an instance of the event called the objective of the method. If the objective of the method is always an event occurring in some individual object, then the method is said to be *applied to this object*. Thus, in order to drive a nail into a piece of wood, one may hit the head of the nail with a hammer several times in succession. The method consists, then, in a repeatable sequence of hits executed with the hammer in a specified way; the objective of the method is driving a nail into a piece of wood; the object of the method is any system con-sisting of a nail and a piece of wood.

The class of all objects to which a method could possibly be applied is the *range* of the method. In our example, the range is made up of all the couples consisting of a nail and a piece of wood such that the nail could be driven into the piece of wood by using a hammer in a way specified in the description of the method.

In a related, though somewhat different sense, anyone who carries out a repeatable sequence of operations with the intention of securing thereby the occurrence of a certain objective may be said to apply the method whose statement *would* relate this sequence to this objective, even if he has not been prompted to do so by familiarity with a descrip-tion of the method, or trained to perform such sequences of actions by following instructions based on such a description. In this broader sense, every repeatable sequence of operations associated with a specified repeatable event is a method, with this event as its objective, provided the operations and the objective be describable by a statement which would constitute a method in the first sense.

It is unnecessary for us to embark upon a detailed discussion of the many features that are essential to the various types of method. However, we must outline briefly those characteristics of particular importance to validating methods, which alone concern us here.

(1) Thus, every method involves a repeatable *sequence of operations,* either always or very often, bringing about, when performed, a repeatable event called the *objective* of the method.

(2) Every method requires further a person or class of persons who are said to be familiar with the method, or *competent to apply it,* if they are prepared and able to perform the sequence of operations comprised in the method in order to bring about the event called its objective. The same person may apply the same method several times, and several persons may attain their objectives by applying the same method.

(3) The operations involved in a method may require the handling of a specifiable set of objects which do not include the object to which the method is being applied. We shall say that such a set of objects constitutes the *instrumental equipment* of the method. The equipment must meet the requirements included in the statement of the method, but may otherwise vary from application to application of the method. One need not use the same hammer each time one drives a nail into a piece of wood.

(4) The objective, supposed to follow an individual application of the method in a certain fraction of all its applications (which fraction can coincide in limiting cases with the sum-total of all the applications), may fail to materialize on some particular occasion. An application which is followed by the occurrence of the event forming the objective of the method is a successful application of this method. The probability that a single application of a given method chosen at random will be successful may be called the *reliability* of the method. The reliability is the ratio of all the successful applications of the method to all its applications, in a sufficiently long run of attempts to apply it.

In principle, the reliability of a method can be characterized by a single number. Yet the probability that a particular application of a given method will be successful may be affected by the special circumstances under which the method comes to be applied: this probability will then differ from the reliability of the method. More particularly, the probability of some pre-assigned application of the method being successful may depend upon the choice of the person who applies the method, upon the physical, psychological, and social conditions prevailing in his environment, upon the choice of instrumental equipment and of the way of handling it, provided all such choices be compatible with the statement of the method. Rules which specify the circumstances left undetermined by the statement of the method (such as the choice of the equipment or of the way of handling it) and which may thereby increase the probability of successfully applying the method, constitute its *technique*.

(5) In some cases there exists a special theory determining the degree of reliability of a given method and how this reliability may possibly be enhanced by use of a suitable technique. Such a theory might provide, among its consequences, a law of the following type: "Whenever a competent person carries out a sequence of operations of the kind described in the statement of the method, an event called the objective of the method will occur with the probability *p*." A theory of this nature may be called the *theory underlying the method*. Of course, familiarity with such a theory is by no means necessary for application of the method (in either the broader or the narrower sense).

Let us now examine the precise manner in which these characteristics appear in validating methods. We have seen that the solution of a scientific problem consists in putting forward a statement or a system of statements which constitute a logically admissible answer to this problem and in describing a method of obtaining adequate supporting evidence for these statements. A validating method is thus a procedure which is destined to, and capable of, providing adequate evidence in support of a statement of some particular kind, or of a system of such statements. A method of validation is therefore a method in the general sense just described, its main distinctive feature being its objective, the production of a body of evidence supporting a statement or a system of statements. This particular objective carries with it several special characteristics of validating methods, which will be discussed in detail in Part II.

The operations involved in a validating method may consist of observations; measurements; experiments; deductive, inductive, or statistical inferences, etc. The person who applies a validating method is the investigator, and, in special cases, the *observer*. The instrumental equipment is referred to as the "experimental set-up," or the "observational set-up," or the "mensural set-up," as the case may be. With regard to measurement, the reliability of the validating method is called its precision, or its accuracy. The theory underlying a particular validating method, for example measurement, may include the physical theory of the instruments used, the statistical theory of errors, and, as a rule, a fragment of pure mathematics.

It is worth while to point out that the above definition of a validating method in terms of its particular objective has serious drawbacks when applied to more complicated situations. The objective of a validating method was said to consist in the production of a body of evidence supporting some particular answer to some particular problem. Now this objective is actually attained in the most typical and most easily understood cases of validation. "Evidence," however, is a complex, unwieldy concept, whose connotation is a mixture of several heterogeneous ingredients:

(1) In cases of purely inferential validation, the "body of evidence" is simply a set of premisses which are taken for granted or known to be true and which serve to validate the statement under consideration by permitting the deduction or inference of the latter from them.

(2) In the case of statements validated by direct observation, the "body of evidence" may simply coincide with the objects to be observed for purposes of validation. Evidence produced in court in support of some statement of fact concerning a single material object may consist in this very object being "produced," or, shown to those who desire to validate the statement.

(3) In more general cases of validation, involving observational and inferential operations, the body of evidence will include both kinds of items listed under (1) and (2), namely, a set of material objects capable of eliciting relevant observations (i.e., observations which, when reported on, will form part of the premisses) and, in addition, a set of statements taken for granted and destined to complete the list of premisses. The statement that forms a logically admissible answer to the problem under consideration will then be validated by being inferred from the completed list of premisses.

The concept of evidence is thus associated with various levels of inquiry, since it refers sometimes to cognitive items (namely, statements known to be true and therefore used as premisses) and sometimes to extra-cognitive entities, physical objects or events, which may be instrumental in eliciting relevant observational experiences. Moreover, this enumeration of the three kinds of "bodies of evidence" is far from exhaustive; it does not apply, for example, to introspective validation. That is why it is desirable to dispense, if possible, with the concept of evidence in defining the fundamental notion of validating method.

On closer inspection, we realize that such an alternative approach is actually available and that the concept of evidence then turns out to be derived from the concept of validating method. Hence it is advantageous to define a validating method, along with the whole chain of epistemological concepts derivable from it, independently of "evidence." We can stipulate that a validating method for problems of the kind K is any method which, given a particular problem P of the kind K and some logically admissible answer A to P, will produce either the new statement "A is correct" or "A is not correct," provided that in a fair proportion of cases this new statement be itself correct, that is, provided the method be sufficiently reliable. Once a validating method and its reliability have been thus defined without resorting to the concept of evidence, the latter can be defined in terms of the former. We need merely say that "to produce a body of evidence which adequately supports a given statement" means to have validated this statement by applying a reliable method of validation. Accordingly, in our previous definition of "a scientific solution to a theoretical problem," we should replace the concept of evidence with that of a *reliable validating method*; a problem is scientifically solved if some logically admissible answer to it has been validated by a sufficiently reliable method.

This shows that the objective of a validating method is more advantageously defined as the production of a *reliably correct answer* to any problem of a given kind K, than in terms of producing a body of evidence supporting an answer to each problem of the kind K. It goes without saying that the concept of evidence remains nevertheless extremely helpful in this whole inquiry, because it applies to the most typical cases. Yet the definition just formulated, which does not resort to "evidence," has a wider scope; when a statement (say, "this is green") is validated by a person who directly observes the object referred to in this statement, then a case of validation is certainly present. Yet the only "evidence" in support of the correctness of this statement is simply the fact that this person has been induced to make this statement by applying a reliable validating method, viz., direct observation.

10 TYPES OF SCIENTIFIC METHODS

Classifications of Scientific Methods

Apart from features shared by all validating methods, some important characteristics apply only to subclasses of such methods and thereby give rise to several classifications of these methods. In this section, we shall outline five classifications of validating methods which will prove useful in subsequent discussions.

(1) We have referred already to *actually available* validating methods, in contrast to *merely possible* or discoverable methods. A method is actually available if it has been successfully applied. A theoretically

possible method may either be described by a precise statement (as in "indirect observation of elementary particles by means of a gamma-ray microscope"), or be at least describable by such a statement. It goes without saying that the distinction between accidental and inherent range applies only to methods which are actually available. Discoverable methods have only an inherent range. But it is this inherent range which is important in connection with our central problem.

(2) Validating methods can be subdivided according to the kind of operations they involve. The basic split is between *non-inferential* or *direct* methods on the one hand, and *inferential* or *indirect* methods on the other. In this classification "inferential" acts as a dominant characteristic, whereas "sensory," "introspective," and other features of non-inferential procedures are merely recessive; a method is indirect if each of its applications involves at least one inferential operation. Sense-perception, sense-data observation ("inspection"), introspection, and direct memory of perceived, sensed, or introspected events, are the basic direct methods of validation within science. Of course, if there are such things as "empathy" or "ESP" or "telepathy," and if they are fairly reliable, they would have to be classified as direct procedures as well.

The inferential procedures are essential components of such validating methods as measurement and indirect observation. Inferential methods may be subdivided in turn into deductive and inductive procedures, with statistical inference as an important subgroup of the latter. Deductive methods are virtually omnipresent in science; they have attained a particularly high degree of efficacy and refinement in the validation of theories, especially those which have been axiomatized.

(3) Another classification of validating methods is based on the kinds of problems to which they are applicable. We have already pointed out that some problems are answerable by single statements, whereas the solution of others requires whole sets of propositions, called scientific theories. Problems answerable by single statements can be subdivided in turn, depending upon whether their logically admissible answers are statements of particular facts or of general laws. We thus obtain a tripartite division of all scientific methods of validation according to their objectives: such methods serve to validate either statements of particular facts, or statements of general laws, or finally whole theories (i.e., systems of law-like, or fact-like,[15] or mixed statements). Thus, for example,

[15] A fact-like statement specifies a property attributable to a specified individual, or a relation obtaining among a few specified individuals. A law-like statement of the attributive type may specify a property attributable to all individuals within a limited space-time region, or within the entire spatio-temporal medium; law-like statements of a relational type are defined in a similar way. This preliminary explanation will be elaborated in Part II.

measurement and indirect observation are fact-finding methods; inductive generalization, curve-fitting, extra- and interpolation are methods for validating law-like statements; the axiomatic method with its various adjustments to the empirical field exemplifies methods of theory-formation.

(4) It is obvious that very *special* methods of validation, restricted to a single theory or discipline, are of little importance in connection with determining the reach of science. We shall have mainly to examine *basic* scientific methods, i.e., methods of validation applicable to problems encountered either in every science or in a substantial fraction of all the sciences.

Deductive methods, used in every science, and inductive procedures, present in every empirical science, are basic in this sense, along with such methods as counting, measurement, experiment, etc. In contrast, intelligence tests and procedures capable of establishing the density of a substance, or its molecular weight, are special methods. The boundary-line separating the two groups is obviously often hazy. Yet traditionally, and for good reasons, methodology and philosophy of science have always been concerned with relatively basic methods; the theory of more special methods, applicable within a small group of sciences, is allotted to these sciences themselves.

(5) Another classification of scientific methods, of decisive importance for our argument, is based on the extent to which a particular method departs from a simple trial and error procedure. We have seen that, in a sense, the objective of any validating method is the production of a body of evidence in support of a particular solution to some particular problem. There is, however, an important ambiguity in the interpretation of objectives of validating methods; it is concerned with the distinction, within science, of *methods of discovery* from *methods of verification* and proof. This distinction can best be illustrated by considering the simplest validating methods: direct fact-finding methods which do not involve either inference or instrumental equipment. Thus, by simply having a look at the object referred to in the question as to whether the pointer of a particular apparatus coincides with a particular mark on its scale, we can obtain a reliably correct answer. The operation called "direct observation" (with the naked eye) constitutes, in this case, the whole validating procedure. The peculiarity of the case consists in the fact that, given the problem, the method both enables us to find the correct answer and supplies the necessary evidence in support of the answer. The situation would have been different if we had been asked to ascertain whether the pointer coincides with the zero mark of the scale. In this case, the answer to the problem is presented in advance

and the whole task of the method is to provide evidence for this answer. We may say that, in the first case, observation has been applied as a *method of discovery*, or as a *decision-procedure*, whereas in the latter case, it has served rather as a *method of verification*. Thus we realize that a primitive fact-finding method can be used either for discovery or for verification.

In more complicated cases, a validating method may be unable to perform functions of both discovery and verification; only the latter function can then be discharged. It may also happen that a method only suggests with a certain plausibility the correct answer to any problem located within its range, without at the same time being able to provide satisfactory evidence for the answer suggested. We shall then say that such a method is *heuristic*. We thus obtain a tripartite classification by distinguishing verificatory and heuristic procedures from those of discovery.

A glance at the more complicated methods used in science shows the applicability of this classification. Thus measurement (say, of length) is a fact-finding method of the discovery type (as a matter of fact, we shall find later on that no scientific fact-finding method is only verificatory, let alone only heuristic: all of them are basically methods of discovery). Whenever inferential procedures enter into the picture, however, either in the deductive or in the empirical field, the methods are, in general, only verificatory. Thus methods for solving algebraic equations of a degree lower than 5 are decision-procedures. But for the solution of algebraic equations of any higher degree, no decision-procedures are available nor can they be found, for the simple reason that the existence of such procedures can be shown to be incompatible with the very assumptions implied by the relevant problem. Thus, the non-existence of decision-procedures for problems solvable by methods which involve inferential operations is by no means logically accidental. Needless to say, with no method of discovery available for a class of problems, merely heuristic procedures become valuable; for example, the suggestion "to try an exponential function" as a solution for a type of differential equation that cannot be integrated by applying general decision-procedures is often helpful, although it may fail to ensure success.

Generally speaking, since the solving of a non-elementary problem in the empirical field involves the production of a body of evidence consisting of results of observations obtained by utilizing special instrumental equipment, and submitted, subsequently, to some appropriate logical process, a fully effective decision-procedure for empirical problems would have to meet the following requirements:

(1) Given any problem P located within the range of the method (which we shall suppose, for the sake of simplicity, to be of maximum reliability), the method has to enable any competent investigator to pick out, among all the logically admissible answers to the problem P, a single definite answer A which is the correct one.

(2) The method has to enable any such investigator either to construct or to otherwise secure the necessary instrumental equipment in order to make the observations required for ascertaining that the answer A is correct.

(3) The method has to specify the final observations to be made, after the equipment has been properly used.

(4) The method has to enable any competent investigator to construct, out of the final observations just referred to and the consequences of a pre-assigned sequence of theories $T_1, T_2, \ldots T_n$, a proof that the answer A is correct.

In other words, a method for solving empirical problems would be an effective decision-procedure in all the relevant respects if, and only if, it enabled every competent investigator to formulate the correct answer A to any problem comprised by the range of the method, to construct or secure the necessary apparatus, to attach it properly to the objects concerned, to make the relevant observations concerning the readings of the apparatus, and, finally, to construct the proof of A out of the propositions describing these readings and of auxiliary premises supplied by a set of pre-assigned theories. The most cursory glance at the history of science shows that no such basic method of discovery has ever been available; no major advance in the whole field of science has ever resulted from the application of such a method. We shall have to discuss later on the logically non-accidental nature of this situation and its bearing on the universality of science, viz., the fact that verification, not discovery, is the core of scientific method and that only verificatory methods are inherently universal.

The Unity of Scientific Method

We can now take up the question whether or not there is a universal type of validating procedure underlying all scientific methods, which we have just classified in five different ways. What typical procedure is common to heuristic methods as well as to methods of verification and of discovery: to all observational and experimental ways of solving problems, to methods of counting and of measurement, to inferential operations, either deductive or inductive or mixed, all of which are discernible in the gamut of scientific activities and which aim either at

getting hold of single facts, or at establishing universal laws, or at substantiating ramified theories?

To begin with, I think we may consider the purely inferential procedures prevalent in the logico-mathematical field as a limiting case of
the scientific method, characterized by a vanishing observational component. Similarly, at the other extreme, non-inferential procedures of
empirical science, such as direct sense-perception, or introspection, may
be taken into account as a limiting case of the general method characterized by a vanishing inferential component. Fact-finding methods
based on measurement or on indirect observation (observation aided by
instruments like microscopes or telescopes which improve upon the
sensitivity of man's sense organs) are located between these two extreme limiting cases, but rather closer to the non-inferential limit,
because they involve a sizable amount of observation and a specifiable
minimum of inference. In all these intermediary cases, regardless of
whether the perception involved is planned or not, whether the method
is experimental (involving a suitable processing of the object to be observed) or not, or whether the method essentially requires instrumental
equipment or not, the type of procedure is always the same and may be
illustrated by a report on a single measurement: "My objective was to
find out the value of the quantity Q for the object X. I have noticed
that the instrument Y was properly connected with the instrument Z
and the whole properly attached to the observed object X. I also made
sure that a microscope was inserted between my eye and the scales of
certain pre-assigned instruments and that, subsequently, a process of
interaction between the instruments and the object X was started by
turning on an electrical switch. Finally, I noticed a number of pointer
readings on the scales of the pre-assigned instruments; I substituted
these observed values (readings) for pre-assigned variables occurring
in the equations of certain pre-assigned theories $T_1, T_2, \ldots T_n$, and on
this basis I computed the value of the quantity Q for the object X." In
other words, a particular fact-like statement concerning the value taken
on by a given quantity for a given object has been validated by first
making a series of observations about the object investigated, the instruments used, their connection, and the final relevant pointer readings,
and then inferring from statements thus validated by direct observation,
and from theories relevant to the problem, the correctness of the statement that specifies the value of the quantity for the object under consideration.

Of course, the validation of a single statement will not always be
describable in this way. What differs from case to case is: (1) the part

played by observations; (2) the kind of instrumental equipment (if any); (3) the theories utilized (if any); (4) the logical processes of deductive and inductive inference (if any) which utilize as premisses the readings on the scales of relevant instruments, in conjunction with any intervening theories. In limiting cases, factors 2, 3, 4 may vanish and validation will still take place. This occurs in introspection, when there is neither instrumental equipment, nor inference, nor intervening theory. In the general case, however, all the four factors will play an essential part; none can be dispensed with without making the validation of the relevant statement impossible.

Yet to obtain a general oversimplified formula of what is involved in the validation of a single fact-like statement, we can take advantage of the fact that the instrumental equipment is utilizable only because of observations made on it, and that the intervening theories affect the result of the validation only by affecting the inferential processes entering the procedure. The validation of a single fact-like statement thus hinges eventually upon observation and inference—no other operations are essentially involved.

The validation of law-like statements ordinarily involves, in addition to the validation of a number of fact-like instances of this statement, inferential procedures of inter- and extrapolation, inductive generalization, etc. Thus the two basic operations of observation and inference still retain their role, the only difference being a comparative increase in the part played by inference and, consequently, a shift of the whole procedure towards the inferential pole.

The validation of theories, as a rule, seems to mean the validation of a number of single statements, included in the theory, which are suitably selected for this purpose and called consequences of the theory. To what extent a theory can be considered validated through the validation of some of its consequences is a question which will prove vital in our investigation. Yet the answer to this question in any specific case does not detract from the fact that the only possible way to validate a theory is to validate some of its consequences, irrespective of the additional requirements which these consequences may have to meet. We may say, therefore, that the basic procedure involved in validating single statements or theories is always the drawing of inferences from observations. Hence the question concerning the range and limits of science, in this oversimplified account, is the question of the range and limits of observation and of inference.

In other words, the scientific method, as applied to a given problem, which may be solvable either by a particular statement of fact, or by a general law, or by a theory, consists always in submitting the relevant observational evidence to some suitable logical processing in order to ascertain what bearing this evidence has upon the correctness of some particular answer to the problem. This basic procedure may therefore be described as *the* scientific method. We have characterized it so far only very vaguely and have used just a few examples in order to illustrate how special scientific methods can be subsumed under this procedure. A more precise characterization of the basic procedure and a justification for considering all scientific methods as special cases of this procedure will require a detailed analysis of actual scientific methods and a precise definition of the relevant types of observational and inferential operations.

We shall then be in a position to ascertain to what extent the reach of science is determined by the scope of observation and logical inference. Our conclusion, in oversimplified form, will be that the range and limits of science depend essentially upon what is observable and what is inferable therefrom. Scientific knowledge consists of the results of such observation and inference.

CHAPTER THREE

Scientifically Unsolvable Problems

---

11 UNSOLVABLE PROBLEMS WITHIN SCIENCE

THE LIMITS of science are set by scientifically unsolvable problems. In numerous cases, the impossibility of ever solving a given problem by applying the method of science has been established beyond reasonable doubt by means of this very method. This has happened both in the field of purely demonstrative and in that of observational disciplines. The classic examples of deductively unsolvable problems are afforded by the impossibility of squaring a circle, of trisecting an angle, of determining by purely algebraic methods the ratio of the circumference and the diameter of a circle, of solving by a general procedure algebraic equations of a degree superior to 4, etc. On the other hand, the impossibility of ever constructing a mechanical or a thermodynamical *perpetuum mobile*, of dispatching a physical signal with a velocity exceeding the speed of light, and of ascertaining whether any pre-assigned gravitational field is brought about by actually present masses or rather conjured up by an appropriate choice of a frame of reference (Equivalence Principle of the General Theory of Relativity), provides examples of scientifically insoluble problems within the observational field. Developments of the last three decades, both in the observational and in the demonstrative field, have led to the discovery of a considerably longer list of scientifically insoluble problems, often of a more sweeping nature than the comparatively few analogous results of earlier years. These developments seem to have contributed to an understanding of the potentialities of science, and were distinctly felt by students of the foundations of science to point to some inherent limits of scientific method.

The fact that the impossibility of scientifically solving certain problems has been established by resorting to the very method of science might be interpreted as justifying the view that such problems should be located within the range of science; the latter would just have to be redefined so as to include all problems which can either be solved by the method of science or be shown to be unsolvable by applying this method. This artificial device, however, would fail to settle the whole issue, since those problems whose scientific insolubility can be scientifically established do not account for the totality of problems which are either actually or allegedly beyond the range of science. That is why we shall have to survey the whole domain of scientifically insoluble problems regardless of whether or not the reasons for their insolubility are supplied in turn by the method of science.

The peculiarity of problems that prove to be scientifically insoluble for reasons which are themselves within the range of science is easily accounted for. We notice in the examples just quoted that the impossibility of scientifically solving any such problem is equivalent to the solution of another scientific problem: for example, the insolubility of the problem concerning the determination of the number π by purely algebraic procedures can be reduced to the fact that the root of any algebraic equation is distinct from π, which is just another algebraic proposition. No wonder, therefore, that the impossibility of scientifically solving the former problem can be established by the same scientific procedures that are effective with regard to the latter, which happens to be a straightforward problem in the "object science" under consideration.

It goes without saying that in the case of scientifically insoluble problems of the empirical kind one has to resort to the variety of scientific methods proper to this field in order to establish the incompetence of scientific method in general to provide the relevant solution. Thus, the impossibility of constructing a *perpetuum mobile* of either kind cannot be demonstratively established by deduction from non-controversial logical premisses; it has to be substantiated by adequate empirical evidence, submitted to logical processes of inductive generalization similar to those applied in other departments of natural science. The insolubility of a problem by empirical methods is therefore based ultimately on empirical evidence. This applies in particular to the puzzling list of unsolvable problems established on the basis of the Quantum Theory of atomic and sub-atomic phenomena, for example, the question of how to improve, beyond certain limits, the accuracy of a simultaneous measurement of two dynamically conjugated quantities in the same dynamical system, such as the position and the momentum of the same particle (Heisenberg's Principle of Uncertainty); the decision on observational grounds as to whether light, or any other physical agency, consists of particles or of waves (duality of waves and particles); the accurate prediction of the individual behaviour of atomic or sub-atomic systems under given initial and boundary conditions (von Neumann's theorem of quantum-theoretical indeterminism), etc. Our only reasons for admitting the impossibility of ever solving these problems coincide exactly with our empirical reasons in support of the basic assumptions of Quantum Theory. It is neither impossible, nor even unlikely, that some day Quantum Theory will come to be replaced by another theory, explaining all the atomic and sub-atomic phenomena accountable for on quantum-theoretical assumptions, and capable of removing, at the same time, the numerous difficulties raised at present by the Quantum Theory in connection with the quantification of electromagnetic fields, and its integration with the (special) Theory of Relativity, etc. Should such a new theory eventually emerge out of the numerous recent attempts, and should it, moreover, not entail the insolubility of the aforementioned empirical problems, then our only reason for condemning these problems as unsolvable would vanish, and we would not be warranted in dismissing them. The difference between mathematical and empirical problems whose insolubility has been established by scientific methods is therefore clear; the impossibility of solving a mathematical problem is demonstratively established by means of mathematical methods, whereas the insolubility of a physical problem is derived from empirical premises, say from the assumptions of Quantum Theory. Therefore, since the verdict of insolubility on empirical problems is derived from the assumptions of a theory which is itself based on empirical evidence, we cannot expect the verdict to be any more final than the admittedly provisional empirical theory used to substantiate it. Nevertheless, we have to acknowledge that the insolubility of empirical and mathematical problems is established by the best scientific methods available in their respective fields.

As to scientific problems held to be unsolvable on non-scientific grounds, they are either relatively isolated, single issues, or consist of a whole group of questions, and constitute therefore a more severe restriction on the range of science. Thus to begin with the first subgroup, we have to acknowledge that several single questions concerned with the subject-matter of science and expressible in terms of contemporary scientific theories (of physics, astronomy, cosmology, applied geometry, biology, etc.) have often been regarded as transcending the reach of science. Questions as to the nature of infinity, as to whether time has a beginning and an end, whether or not matter is indefinitely divisible, whether the origin and nature of life can be accounted for in terms of physico-chemical laws and facts, whether and how the interrelations between the bodily and the mental events in the life-history of a single person can be integrated with the basic laws and assumptions of natural sciences (the "Mind-Body Problem"), whether or not a decision made by a human being is always exactly predictable on the basis of his personality, his background, and his bodily structure (the "Free Will Problem") exemplify such single problems allegedly inaccessible to scientific method. Yet some at least of the problems just listed have been attacked with more or less success by particular sciences—cosmology, Quantum Theory, set theory. It is, of course, out of the question to regard the contributions so far made by the particular sciences to a solution of these problems as having gone beyond the stage of a working hypothesis. Still, the fact is there, and its significance is obvious; since some at least of these single problems, allegedly transcending the potentialities of science, have been actually attacked by scientific method and the result has not been definitively discouraging, they can hardly be cited as demonstrating its inherent incompetence.

It goes without saying, however, that the more sweeping restrictions on the range of science which question its competence for very wide ranges of problems on the basis of non-scientific considerations, carry more weight in our discussion than the relatively short list of single problems just mentioned. Such a wholesale ban on applying scientific method to problems of a given kind may be illustrated by certain positivist attempts at restricting the aims and the potentialities of science. Thus positivistically minded students have often been prepared to grant the competence of scientific method for describing and predicting observable phenomena. Yet, according to their view, science, while perfectly able to say what course the observable phenomena of the past did take and what course the observable phenomena of the future will take, fails entirely to say why this definite course has been or will be taken, in spite of there being an infinite number of other logically possible courses. In brief, science would be able to solve problems concerning the description and the prediction of phenomena, but incapable of coping with problems concerning the explanation of observable phenomena. This is a typical restriction on the range of science, which used to be claimed by positivists of the older schools (from Comte to Mach) in contradistinction to the followers of a "realist" philosophy of science which maintained that the explanation of observable events is within the range of science, along with the description and the prediction of such phenomena.

At present, however, the controversy over the competence of science to cope with explanatory problems has become rather obsolete and has ceased to divide the two rival schools of philosophy of science, for the simple reason that new attempts, promising and partly successful, at elucidating the logic of scientific explanation tend to prove that any scientific method which is capable of solving problems of prediction in a given field is by the same token in a position to explain all the phenomena within this field, and vice versa.[1] In other words, a method is competent for predicting phenomena of a certain kind, if and only if it is also competent to explain such phenomena. Therefore, there is no point any longer in a positivist's emphasis on the predictive function of science, at the expense of its explanatory function, since both are indissolubly interconnected.

Of course, positivist misgivings as to the range of science are not among the most conspicuous. A more sweeping limitation is implied in the view, widely held since the classic analyses of Poincaré, Russell, and Carnap,[2] that science is capable of securing reliable information only about so-called *relational structures* within its subject-matter (or, more precisely, within the "universe of discourse"[3] of the language used by the scientist in expressing his theories), and is not able to get hold either of the individuals who make up this universe of discourse, or of the relations obtaining among these individuals. Let us consider, for example, the concept of spatial distance construed as a relation between material objects or events. According to the "structuralist" or "relationalist" view of the range of science, the relevant theories of applied geometry, physics, astronomy, and cosmology are able to establish the structural properties of the distance relation (i.e., properties which the distance relation shares with any other relation isomorphic[4] with it) without conveying the slightest idea of what physical distance actually is or means. Nor do they indicate the nature of the individuals located in physical space and separated from each other by definite distances,

[1]C. G. Hempel and P. Oppenheim, "The Logic of Explanation," *Philosophy of Science*, vol. 15 (1948); G. Bergmann, *Philosophy of Science* (1957), pp. 75ff.; R. B. Braithwaite, *Scientific Explanation* (1955), pp. 319ff.

[2]H. Poincaré, *La Valeur de la science* (1913); B. Russell, *Introduction to Mathematical Philosophy* (1919); R. Carnap, *Der logische Aufbau der Welt* (1928).

[3]The universe of discourse of a language consists of all the individuals referred to in this language.

[4]Two relations R and S are said to be isomorphic with each other, or to have the same relational structure, if their fields can be bi-uniquely mapped on each other in such a way as to ensure that if, and only if, the relation R obtains among some particular entities then relation S obtains among the objects corresponding to these entities by virtue of this mapping.

although they determine the "structural properties" of the distance relation so efficiently. In other words, the "intrinsic nature" of the distance relation (which is not shared by the whole family of relations isomorphic with it) and of the individuals interconnected by it would essentially transcend scientific method.

This structuralist or relationalist view of Poincaré, Russell, and Carnap concerning the limits of science seems to be supported by a well-known result in mathematical logic due to Lindenbaum and Tarski.[5] Their theorem states, roughly speaking, that any theory which holds true of a given fragment of reality (i.e., a "model" consisting of a subclass of the universe of discourse of the language of science, with a specifiable set of classes and relations of various levels and degrees based on this subclass[6]) will also apply to any other fragment of reality, provided the latter be isomorphic with the former in respect of all the individuals, classes, and relations involved. Thus no theory can be sufficiently specific, according to the Lindenbaum-Tarski Theorem, to apply to a single fragment of reality and to no other. Neither the individuals, nor the classes consisting of these individuals, nor the relations obtaining among the individuals are uniquely determined by the fact that some particular scientific theory applies to them; the same theory will certainly apply to other individuals, classes, and relations provided they add up to a fragment of reality isomorphic with the former fragment (§ 37).

The ability of a scientific theory to restrict the range of its models is even more limited than the Lindenbaum-Tarski Theorem implies. According to their result, one might expect at least that a scientific theory could be made specific enough to determine uniquely the relational structure of its subject-matter. In other words, the optimal strength of a theory would be achieved if any two fragments of reality to which the theory applies were isomorphic with each other. However, a simple argument shows that such strength is actually unattainable, in view of other metamathematical findings of Löwenheim–Skolem and of Tarski. The Löwenheim–Skolem Theorem[7] states that any consistent and appropriately formulated theory has denumerable[8] models. On the

[5]A. Lindenbaum and A. Tarski, "Ueber Beschränktheit der Ausdrucksmittel deduktiver Theorien," *Ergebnisse eines mathematischen Colloquiums* (1936).

[6]Classes of individuals and relations among them are said to be of the first level. The second level is made up of classes of first level classes or relations and of relations which involve first level classes or relations, but no entities of any higher level. Higher levels of classes and relations are defined in a similar way. A class is an entity of degree 1 and a relation among n objects is said to be of the nth degree. R. Carnap, *The Logical Syntax of Language* (1949), pp. 84–7.

[7]Cf. A. Church, *Introduction to Mathematical Logic*, I (1956), pp. 238ff.

[8]A set is said to be denumerable if its elements can be put in a one-one correspondence with the set of all positive integers.

other hand, Tarski[9] has discovered that any theory with denumerable models has also non-denumerable ones. Since no denumerable model is isomorphic with a non-denumerable one, the joint result of these three investigators implies that any consistent and appropriately formulated theory has models that are not isomorphic with each other. Consequently, any such theory fails to determine the relational structure of the fragments of reality to which it applies. This failure is inherent in the scientific method of theory formation.

Another widely held view about the inherent limitations of science credits scientific method with securing relevant and reliable information about the subject-matter of scientific knowledge, but regards this method as inherently incapable of establishing the "real existence" of the objects it is dealing with. For example, although we know a good deal about the atom—its structure, its energy levels, its conditions of stability and disintegration—we still could not possibly infer from all this that the atom does really exist. Similarly, our knowledge of quite a few basic properties of the electron, its mass, electrical charge, spin, its statistical behaviour in large assemblies, etc., would not, on the view under consideration, rule out the possibility that the electron is but a convenient computational device whose function is to facilitate the quantitative description and prevision of large-scale phenomena, said to involve "electrons in action." After all, not a single electron in action has ever been observed. Granted, the basic assumptions concerning the mass, charge, spin, etc. of electrons are so perfectly confirmed by innumerable experiments providing for several independent checks on each assumption, that one cannot help feeling that the statements specifying these fundamental attributes of electrons are true. But, it may be pointed out, no hypothesis asserting the real existence of electrons is to be found in any authoritative treatise, along with the well-confirmed hypotheses about the electrical, mechanical, and statistical attributes of electronic particles. Hence, if somebody were to put forward a hypothesis to the effect that electrons do really exist, the impression might be created that a new *sui generis* hypothesis had been added to those already available. Perhaps one may suspect that this new hypothesis is of interest to philosophers rather than to scientists. This may explain why the existential hypothesis has not occurred so far among the assumptions made by the scientist, who is interested in science proper, not in amplifying it by specifically philosophical hypotheses.

I do not think that this explanation of the absence of a hypothesis concerning the "real existence" of electrons in the entire list of physical

[9]Cf. A. Mostowski, *Logika matematyczna* (1948), p. 360.

assumptions about this particle gives a correct account of the situation. If there is no explicit assumption as to the real existence of electrons in the relevant authoritative monographs, the reason is not its allegedly philosophical nature, which would prevent it from fitting into the framework of a purely scientific theory. The actual reason is much simpler. The assumption concerning the real existence of electrons is an immediate consequence of other assumptions about this particle, a consequence so trivial from the point of view of the scientist that it would hardly make sense to formulate it explicitly. Thus, if the other assumptions about electrons were true, then electrons could not possibly fail to exist. For example, the difference between the shells of a hydrogen atom and a helium atom is that the hydrogen shell consists of a single electron, whereas the helium shell contains two electrons. (I am supposing here that the atoms are not ionized and have their complete shells around their respective nuclei.) If this statement concerning the difference between the hydrogen and the helium shells is true, then electrons must exist: two atomic shells differing from each other by a nonexistent electron would not differ at all. If it is true that electrons are deflected in electric and in magnetic fields, if they bring about a trace visible to the naked eye during their passage through a cloud chamber, if their electric charge is of the order of 10^{-10} electrostatic units and their mass almost 2,000 times smaller than the mass of a hydrogen atom, then electrons must exist. Non-existent electrons could not possibly undergo deflection, produce traces in cloud chambers, or flashes on a scintillation screen. Non-existent electrons could not possibly have either a definite mass or an electrical charge. Thus, if the ordinary, not explicitly "existential" assumptions about electrons are true, the statement asserting that electrons do exist cannot fail to be true as well.

On the other hand, we have to take into account the tripartite classification of scientific statements outlined in a previous section: those which are literally verifiable, those which admit of verification only if suitably reinterpreted, and those which are unverifiable under any admissible interpretation. We have pointed out that statements of the third category are present in any fundamental and well-established theory even though they are not justifiable by the evidence that supports the theory. As a matter of fact, such statements play but an auxiliary part in the theory and will be shown, later on, to be neither true nor false. It may well be that those who deny the scientist's competence to prove the existence of electrons are simply trying to say that existential statements referring to electrons belong in this third category of statements, and more generally, that any statement about the existence of individuals dealt with in scientific theories is a statement of this category. Such a view, however, is untenable for the simple reason that existential statements, being derivable from the most innocuous assumptions about the relevant individuals, could not belong in this third category unless all the remaining statements about the same individuals shared their lot: in other words, if existential statements were unjustifiable by any conceivable observational evidence, regardless of the interpretation put upon them, then the same would hold true of any synthetic (factual) statement within the same theory and the theory would not include a single justifiable factual statement about its subject-matter. This is a sceptical view of scientific knowledge hardly shared by its most cautious analysts.

Of course, it may well turn out on closer inspection that existential statements about electrons or other scientific objects belong in the second category; they may be literally unverifiable, but admit of observational verification provided they be suitably reinterpreted. We shall outline such a reinterpretation of scientific statements about atomic and sub-atomic objects on a later occasion (§18). However, such a need for reinterpretation would then apply not only to existential statements about scientific objects but to any factual statements whatsoever about such objects. Thus, the competence of science for solving existential problems is hardly involved in the query as to the "real existence" of atoms, electrons, the ether, genes, the subconscious, and so many other scientific objects or "constructs": all such objects exist really, if the relevant scientific assumptions are both literally interpretable and empirically verifiable. In particular, the real existence of atoms is tantamount to the empirical and literal verifiability of the atomic theory. More precisely, the real existence of atoms as endowed with specifiable properties is equivalent to the empirical and literal verifiability of those statements of the atomic theory which refer to the properties under consideration.

To sum up: the restrictions on the range of scientific method which are not established in turn by this method, are more difficult to classify and to assess than the scientifically established restrictions. Thus the scientific insolubility of certain single problems has to be qualified by the fact that, recently, some of these problems have been attacked with some success by scientific method. The positivist ban on problems of explanation tends to become obsolete owing to progress made in the logic of scientific explanation. The "structuralist" or "relationalist" limitations of science, which restrict its competence to supplying reliable information about certain properties of certain relations obtaining among certain individuals, on the understanding that both the relations and the individuals are scientifically unknowable, seem to be supported by some results in logic and present a more serious challenge to any analysis aiming at determining the reach of science (cf. § 37). The alleged incompetence of science to solve problems concerning the "real existence" of objects which are otherwise accessible to scientific method seems rather to collapse at the slightest attempt at analysis.

Yet, in spite of all these qualifications concerning scientifically unsolvable problems whose insolubility is asserted on non-scientific grounds, we cannot, of course, rule out the possibility that some of these problems may actually prove to be scientifically unsolvable, while the impossibility of solving them by applying scientific methods cannot in turn be established by scientific methods. We have simply to distinguish, for the time being, within the group of scientifically unsolvable problems, two subgroups, depending upon whether or not the impossibility of ever reaching a scientific solution to such a problem can itself be established by resorting to the method of science.

12   UNSOLVABLE PROBLEMS ABOUT SCIENCE

Let us now consider another group of problems which are still likely to arouse doubts as to the possibility of applying to them the method of science:

1a. How could one prove that whenever the relevant deductive methods are correctly used in constructing basic mathematical theories (for example, elementary arithmetic), these theories will be logically consistent and will never lead to contradictions?

1b. Could one utilize suitable presently unknown quantities ("hidden parameters") at the atomic and sub-atomic level in order to establish laws capable of predicting exactly the result of any future observation made on an atomic or sub-atomic system, provided that sufficient information be available concerning the initial and the boundary conditions of the system?

2a. Is it possible to deduce all presently known laws of nature which govern electro-magnetic and gravitational phenomena from a single set of partial differential equations involving only measurable field quantities? (In other words, is the problem concerning the construction of a unified field theory solvable?)

2b. Is it possible to account for the metamorphoses of elementary particles in terms of present assumptions of the Quantum Theory?

3*a*. Can one deduce the sum-total of presently known biological facts and laws from relatively few independent assumptions, and thus promote biology to the rank of an axiomatic theory?

3*b*. Is it possible to find out to what extent the advance of science in the course of human history was due to the pressure of economic needs?

Most investigators would answer negatively the questions classified under 1: 1*a* has been shown to be unsolvable by Gödel,[10] 1*b*, by von Neumann.[11] The answers to questions 2*a* and 2*b* are likely to vary: quite a few scientists will hesitate as to the possibility of solving 2*a*, in spite of the uniquely great effort made by a uniquely great mind; some may take up a similar attitude towards 2*b* in view of the comparative failure of so many attempts in the last two decades.

Question 3*a* is likely to be answered in the affirmative, since any finite set of facts and laws, including the facts and laws grouped under the heading of biology, can always be axiomatized. Although little headway has been made so far in connection with biology (except for Woodger's interesting attempt at axiomatizing genetics[12]), there is nothing in the nature of the scientific method or of the problem at hand to make the former inherently inapplicable to the latter. Similarly, the facts concerning the interdependence between scientific evolution and economic life may be inadequately known at present and perhaps remain insufficiently known for the foreseeable future, owing to all kinds of extraneous reasons. But the difficulty is still logically accidental and does not prevent the problem from being intrinsically accessible to the scientific method.

It is worth noticing that problems 1*a*, 1*b*, 2*a*, 2*b*, (and, for that matter, 3*a* and 3*b*) differ from the scientifically unsolvable problems dealt with in the preceding section in regard to their subject-matter. Both sections deal with problems related to science. But those of the preceding section are concerned with objects dealt with in other scientific theories, and are therefore themselves formulable within such theories. In contrast, the second group of problems is not within, but rather about science. Thus the admittedly unsolvable problems 1*a* and 1*b* deal with science itself, with its theories, their consistency, the definability of their basic concepts, etc. The person who wonders whether arithmetic is con-

---

[10]K. Gödel, "Ueber formal unentscheidbare Sätze der Principia Mathematica," *Monatshefte für Mathematik und Physik* (1931).

[11]*Ibid.*

[12]J. H. Woodger, *Axiomatic Method in Biology* (1937).

sistent does not deal with numbers, as arithmetic proper does; his subject-matter is the science of arithmetic itself. It matters little that the problem concerning the consistency of mathematics is mostly dealt with by professional mathematicians and by recourse to methods applied within mathematics proper. The problem itself is nevertheless about mathematics, not within it. Similarly, historiography may apply the method of history, but its subject-matter is history itself, not the subject-matter of ordinary history.

We may thus be tempted to distinguish scientifically unsolvable problems pertaining to science proper from those the subject-matter of which is science. This, however, would be incompatible with the position we have taken in discussing the meaning of "science" in chapter Two, where the sciences have been subdivided into object-sciences and meta-sciences of various levels, the meta-sciences dealing with object-sciences, the meta-sciences of the 1st level, etc. The problems listed under 1a-3b are meta-scientific problems concerned with object-sciences and do not themselves belong to any object-science. This does not prevent them from being "scientific" in the broad sense outlined in chapter Two, even if, owing to the comparative vacuum presently prevailing at the higher levels of scientific inquiry, the particular sciences to which some problem would belong may not exist for the time being. We shall have to deal, accordingly, with all scientifically unsolvable problems and their bearing on the limits of science regardless of whether such problems happen to be concerned with the objects themselves, or located on some higher level.

The first impression one gets from a review of these meta-scientific problems 1a-3b is that the situation is basically similar to that of the object-sciences: the range of scientifically solvable meta-scientific problems seems to include some of the relevant questions (e.g., 3a, 3b), but certainly not all of them (e.g., neither 1a nor 1b). More particularly, the subdivision of unsolvable problems according to whether or not the reason for their being scientifically unsolvable is itself located within the range of science, applies to meta-scientific problems as well. Thus the unsolvability of the meta-scientific problems 1a and 1b has been proved by scientific methods. But other meta-scientific problems have often been declared unsolvable without the submission of a scientific proof.

An important example is afforded by the question of whether the basic scientific methods of validation, for example the deductive or the inductive methods, are sufficiently reliable. The deductive method is the most pervasive scientific procedure. A scientist who happens to apply it at some juncture assumes obviously that if his premisses are true and another statement is correctly deduced therefrom (i.e., inferred in accordance with the rules of deductive logic), then the statement so obtained cannot fail to be true. In other words, by the very fact of applying deductive methods, the scientist presupposes that deductive logic "applies to reality," or, to put it in a less grandiloquent way, that this method is absolutely reliable, because any deductive consequence of true premisses is always true. It goes without saying that by preventing the scientist from resorting to this method, we would virtually annihilate science. Yet no scientist has so far succeeded in proving the absolute reliability of the deductive method. As a matter of fact, such an undertaking is of little promise; in order to prove anything one has to apply the deductive method itself, so that a proof of its reliability would be circular. To solve this problem in a satisfactory way, that is without circularity, seems therefore to be impossible. To resume our terminology of chapter Two, no satisfactory theory underlying the deductive method has so far been constructed or can possibly be constructed. One has to take it for granted that the deductive method is reliable and stop worrying about the impossibility of solving the unsolvable problem of constructing a satisfactory reliability proof for it.

Similarly, the reliability of other basic validating methods of science seems to raise important and unsolvable problems. In the empirical field, the scientist applies rules of inductive or probable inference whenever he prefers well-established empirical laws to gratuitous or untenable generalizations, and well-established empirical theories to gratuitous or untenable sets of assumptions. Thus, an empirical universal statement which has proved to apply correctly to all its examined instances, is considered to apply with a reasonable probability to all its remaining instances and therefore probably to constitute an empirical law. The degree of this probability is taken to be conditional on the number and the variety of the examined instances. Similarly, an empirical theory the tested consequences of which have so far turned out to be correct, is considered to be superior to a gratuitous theory with no consequences checked so far, and, of course, to an untenable theory some consequences of which have proved false. This inductive procedure, just described summarily, is obviously as vital for empirical science as deductive procedures are for the whole of science. The empirical scientist who applies inductive procedures must feel that he is somehow justified in doing so, but he is neither interested in nor capable of establishing the reliability of such methods. For inductive procedures are not claimed to be absolutely reliable; anyone who prefers well-confirmed laws and theories to laws and theories that are simply gratuitous may admittedly be mistaken, since laws and theories so far confirmed may nevertheless turn out to be incorrect, whereas gratuitous laws and theories, which have not been submitted until now to any observational test, may nevertheless prove to be correct. Hence, the only advantage of applying inductive methods of probable inference consists in the fact that by so doing one is more likely to obtain correct results than by selecting at random gratuitous laws and theories. Thus, to prove that it is more probable that one will reach correct conclusions by applying inductive procedures of *probable inference*, one would have to resort to the very laws of *probable inference* the validity of which one sets out to establish. The difficulty is exactly similar to the difficulty arising in any attempt at proving the reliability of deductive inference, and, apparently, as insurmountable.

On closer examination, the same circularity turns out to be inevitable in any attempt to prove the reliability of fundamental direct (i.e., non-inferential) methods of validation. It is impossible to establish the reliability of perceptual observation, without presupposing that such observation is reliable—nor is it possible to show that memory is reliable without having recourse, in the proving of reliability, to the reliability of validation based on memory. Thus problems concerning the reliability of all basic scientific procedures seem all of them to be unsolvable.

Similarly, a basic issue in understanding the reach and role of science is connected with a precise determination of what scientific knowledge is about, that is, with defining the subject-matter of science. When literally interpreted, the information offered by various particular sciences deals ostensibly with a very heterogeneous range of entities. In descriptive sciences, such as geography, history, biological taxonomy, the information offered is about directly perceivable common-sense things, including human beings with their psychological states, aptitudes, and traits. In some disciplines, such as psychology and physiology of perception, appearances ("sense data") presented by common-sense objects to human observers are also considered. In the fundamental natural sciences, statements are made about molecules, atoms and sub-atomic particles, electro-magnetic, gravitational, metrical, and nuclear fields. The controversy concerning the "real" subject-matter of science amounts to the question whether this conventional formulation of scientific information in various departments of empirical knowledge is literally interpretable.[13] In other words, does the universe of discourse

---

[13]C. H. Feigl, "Existential Hypotheses: Realistic versus Phenomenalistic Interpretations," *Philosophy of Science*, vol. 17 (1950).

of the language of science include common-sense objects along with sense data and with all kinds of unobservable "scientific objects" or "constructs"? Such a conciliatory attitude, granting eligibility within the universe of discourse of science to all this heterogeneous assemblage of entities, is rather rare. There is a widespread tendency to claim some kind of monopoly for a single group or a combination of a few groups of the foregoing entities. Other entities, ostensibly referred to in the scientific language, are understood as linguistic or computational fictions. Any scientifically valuable information concerned with such fictitious entities is faithfully translatable, it is claimed, into statements concerned exclusively with entities that are granted monopoly by the relevant school of philosophy of science. Thus, a phenomenalist brand of positivism singles out sensory appearances as the only subject-matter of science. Another kind of positivism considers common-sense objects as deserving such a monopoly.

This question concerning the correct ontological interpretation of science, or a determination of the universe of discourse of the scientist's language, has little hope of solution in the view of some outstanding investigators, who feel that scientific knowledge can be adequately expressed in various languages with various universes of discourse. Such alternative linguistic media for science would include, in particular, a "phenomenalistic language" whose basic statements refer always to sensory appearances, a "physicalistic language" with a universe of discourse consisting of common-sense material objects,[14] and a technical scientific language with a universe of discourse made up of "scientific objects" such as molecules, atoms, and elementary particles.[15] The aptitude of these languages with their respective universes of discourse to express adequately the sum-total of scientific knowledge is by no means definitely established at present, in spite of illuminating remarks made by the aforementioned writers about the expressive facilities of the three types of language. It is obvious that if they are justified in dealing with all such languages on a footing of equality, then the problem concerning the subject-matter of science will be inherently unsolvable: each of these languages implies a different answer to this problem, and if there is no ground for choosing one language in preference to another one, there is no definite solution to the problem.

There is another reason that makes one doubt the possibility of solving this problem—a reason not dissimilar to the one responsible for the

[14]R. Carnap, "Testability and Meaning," *Philosophy of Science*, vol. 3 (1936) and vol. 4 (1937).
[15]H. Reichenbach, *Experience and Prediction* (1938).

unsolvability of problems concerning the reliability of various scientific
methods of validation. The fact is that, when we are discussing the universe of discourse of the language of science, we must use a language
with a universe of discourse of its own. The choice of the latter must
necessarily affect the view taken of the former: while discussing the
universe of discourse of the language of science in a meta-language of
our own choosing, we must necessarily locate the scientific universe of
discourse within the universe of discourse of our meta-language, since,
ex hypothesi, we are unable to speak about anything beyond this meta-
linguistic universe of discourse as long as we do use this meta-language.
Hence, if the problem concerning the subject-matter of science can be
discussed at all in various meta-languages with various universes of discourse associated with them—as it apparently can—then there is no
possibility of obtaining a satisfactory solution thereof which would be
independent of our choice of a meta-language.

13   SCIENTIFICALLY UNSOLVABLE PROBLEMS IN PHILOSOPHY

Let us consider finally some philosophical problems and the views
about the possibility of their being solved by scientific methods. It goes
without saying that since we are concentrating our inquiry on the theoretical possibility of attacking these problems by the methods of science,
the fact that scientific solutions of philosophical problems are scarce or
not presently available is not of decisive importance. When solved,
philosophical problems have ordinarily been incorporated in some
special science, for example, in theoretical physics ("natural philosophy"), in cosmology, set theory, psychology, or logic. This historical
tendency to detach from philosophy and to integrate with some particular science any organized body of traditionally philosophical problems which have been successfully dealt with by scientific methods
explains the scarcity of solved problems within the present field of systematic philosophy. It shows at the same time, however, that problems
traditionally allotted to philosophy are not necessarily inaccessible to
scientific method. In any case, regardless of how the scarcity of scientifically solved problems presently allotted to philosophy is accounted
for, there remains the question as to whether the failure of science so
far to yield satisfactory solutions to these problems is due (1) to the
nature of the scientific method, or (2) to the accidental, extraneous
circumstances under which the method has so far been applied to these
problems, or (3), more frequently perhaps, to extra-logical conditions
which so far have altogether prevented the method of science from being
applied to philosophical problems.

Problems traditionally allotted to philosophy can be classified in several ways. As far as the universality of scientific method is concerned, it may suffice here to consider the three classes of problems ordinarily referred to as ontological (or metaphysical), epistemological, and ethical. Ontology is concerned with what there is in the world man lives in; epistemology with what he can know about it; ethics with what he ought to do about it (i.e., with human policies and human values). This tripartite division, convenient for our investigation, implies neither that the whole field of philosophy is exhausted thereby, nor that the three classes of problems are mutually exclusive. The epistemological problems are more closely related to science than those of ontology. We shall start accordingly with a few comments on scientifically unsolvable problems within epistemology.

Scientifically Unsolvable Problems in Epistemology

Some fundamental issues in epistemology are simply generalizations of allegedly unsolvable problems about science. For example, the problem concerning the reliability of scientific validating methods is but a special case of the general epistemological question as to how we can justify our belief in the reliability ("validity") of our basic ways of acquiring knowledge, such as sense perception, introspection, memory, deductive and inductive inference. All these validating ("cognitive") methods are as vitally important in their pre-scientific and extra-scientific use as they are within science proper: man could not have survived until the emergence of science, nor got along in this era where some human decisions are made on the basis of information provided by science, if he had stopped using his sense perceptions, his memory, and his capacity of drawing, however crudely, deductive and inductive inferences. The problem of establishing the reliability of these basic methods is therefore by no means confined to science; it applies to any and every form of human knowledge. The impossibility of solving scientifically the problem of the reliability of any basic validating method within science would entail a fortiori its unsolvability in the wider epistemological setting.

The epistemological issue generally referred to as the Idealist-Realist-Phenomenalist controversy is concerned with the subject-matter of human knowledge: is it always knowledge of introspective data, or of sense data, or of physical objects distinct and independent from sense data? This problem is, of course, a direct generalization of the meta-scientific problem about the subject-matter of science referred to previously; the impossibility of solving the latter would therefore entail the insolubility of the former.

Another classical issue in epistemology is that of empiricism: how could we possibly test and prove the basic empiricist tenet that the scope of human knowledge is determined by what is inferable out of what is humanly observable? This epistemological problem will turn out, once more, to be a direct generalization of one of the basic meta-scientific questions concerning the validity of the Principle of Verifiability within science (cf. Part III).

These three problems are concerned with pervasive features of human knowledge and are therefore allotted to epistemology. The question of interest to us is whether their solution is within or without the inherent range of science. The typical attitudes in this case are more difficult to ascertain than in the two previous cases (§§ 11, 12), where some problems would be declared by most investigators as certainly unsolvable, some others as definitely solvable, and some as of questionable solubility. All one can say in the present case is that certain writers would consider some particular solution to some particular epistemological problem as being either a "philosophical implication" or a "philosophical presupposition" of science, whereas others would view the same epistemological problem as transcending altogether the jurisdiction of science, though possibly solvable by a non-scientific procedure.

In the course of our investigation we shall discuss in some detail the meta-scientific issues of which the last two epistemological problems are generalizations; this may shed some light on the more far-reaching epistemological problems too. As to the Idealist-Realist-Phenomenalist controversy over the subject-matter of scientific and of extra-scientific knowledge, we have already pointed out some reasons for doubting the possibility of a scientifically justifiable solution; we shall nonetheless adduce later on other arguments tending to show that the situation brought about by Quantum Theory, rather unfavourable to phenomenalism, is compatible with an empiricist realism. The findings of atomic physics would thus seem to have at least a bearing[16] on this epistemological problem, though not one supporting phenomenalism, as is often asserted. Finally, the empiricist issue definable as a direct generalization of the meta-scientific problem concerning the validity of the Principle of Verifiability in empirical science will be dealt with in Part III. We shall try to reformulate this principle and to offer a tentative proof thereof. If our approach is correct, the empiricist issue will prove to be accessible to scientific method at a suitable level.

[16]A method may have a bearing upon a problem, though it fails to provide a solution of it, if the method is capable, for example, of refuting premises supporting a particular solution to this problem. The method may thus help one to realize that a given solution to this problem is gratuitous, without either proving or disproving this solution.

The upshot of these highly tentative discussions will be mainly the fact that the prerequisite for any fruitful epistemological (or, for that matter, of any philosophical) investigation is an attempt at clarifying the crucial intervening concepts and at reformulating the problem accordingly, without interpreting it away. Perhaps the hope will thus be strengthened that the whole field of epistemological investigations may come to be integrated with the range of scientific problems, without losing its philosophical status—provided a sufficiently comprehensive formulation of the method of science be laid down, and epistemological problems be reinterpreted as either meta-scientific or generalizations of meta-scientific problems.

Ontological Problems

Philosophical problems concerned with general features of the universe, and not with human knowledge of these, are classified as ontological (metaphysical). Such issues are often held to be unsolvable by scientific methods even by investigators who are less pessimistic in respect to epistemological questions. Yet some at least of the basic ontological problems are just a counterpart of epistemological questions and can be reached by a generalization of object- or meta-scientific problems in the same way as the corresponding epistemological issues. Thus, the metaphysical or ontological issue of Realism versus Idealism is but a counterpart of the corresponding epistemological problem: if the question whether the subject-matter of our knowledge consists of sense data, of introspective data, of physical objects, or of any other specifiable kind of entities is unsolvable within epistemology, then the corresponding ontological question whether reality consists of sense data, or of introspective data, or of any kind of data or entities will be unsolvable as well. For example, if we cannot know whether all our knowledge is of sensory appearances, then the ontological assumption that there is anything else in the universe apart from sensory appearances is gratuitous. Similarly, if the epistemological problem involved in the realist-phenomenalist controversy were scientifically solvable, then the corresponding ontological problem would be susceptible of a solution as well. We shall thus have to associate the ontological controversy over Realism, Idealism, and Phenomenalism with its epistemological counterpart, and, eventually, with the meta-scientific basis of the latter.

Similarly, the question as to whether the real world is governed by chance or by necessity may be interpreted as referring to whether laws can be formulated and established which would enable us to predict exactly the future state of any fragment of reality on the basis of a sufficient knowledge of its present situation and of the boundary conditions.

that will obtain between the present and the future moments under consideration. This, once more, is but a generalization of the meta-scientific problem concerning the possibility of predicting the future state of a material system on the basis of its present state and the relevant boundary conditions. This meta-scientific problem, though meaningful and momentous, has been shown on reasonable grounds to be unsolvable by von Neumann. We would have, then, to conclude that the more general ontological problem of determinism *versus* indeterminism is unsolvable as well.

Generally speaking, although the ontological problems themselves are distinct from epistemological ones, the question concerning the possibility of solving ontological problems is itself an epistemological issue. The hope of making some headway in epistemology by conceiving its problems as generalizations of certain meta-scientific questions is therefore relevant with respect to the applicability of the scientific method to ontological problems as well. It goes without saying that an agreement among the extant solutions of ontological problems is less general than in the case of epistemology. This, however, does not affect our main issue. After all, it is only the solubility of ontological questions by scientific methods, and not these problems themselves, that we are interpreting in terms of meta-scientific problems. This is exactly the procedure we have adopted with those problems of epistemology which differ only by their scope from corresponding meta-scientific problems.

Scientifically Unsolvable Problems in Ethics

Ethical problems are concerned with rules governing human conduct. A complete science of ethics would determine, for example, under what conditions human decisions are right, and which human actions, motives, and characters are good. A considerable number of ethical judgments are thus value-judgments since they involve such crucial value-predicates as good or right. However, the whole field of ethics is hardly subsumable under the category of value-problems. Ethical norms, prescribing what human agents ought to do under specifiable circumstances, are not, ostensibly at least, value-statements.

Admittedly, no comprehensive body of scientifically justifiable solutions to ethical problems (either of the valuational, or of the normative type) is available at present. The question whether the failure of science until now to provide these solutions is accidental or inherent in scientific method is an important aspect of our general problem. In view of the largely controversial boundaries of the ethical field, this is an extremely complex question with which we shall not deal in any detail. At this juncture, it may suffice to point out the types of ethical problems to which scientific methods seem to be inherently inapplicable:

(1) As a result of a concerted attempt at a logical analysis of ethical discourse,[17] it has become increasingly apparent in the last two decades that a considerable part of this discourse does not perform any informative function, but serves rather to express the emotions of the speaker or to induce the listener to take a course of action desired by the speaker. To the extent to which ethical discourse is confined to such non-cognitive tasks and fails to convey information, ethical statements are neither true nor false. Consequently, ethical questions answerable by such statements could not be treated by scientific methods since these methods are essentially validating and serve to ensure that a particular answer to a particular question is reliable, i.e. either probably or certainly true.

(2) The so-called doctrine of ethical relativity asserting that ethical statements are intrinsically neither true nor false because their justification is conditional upon the psychological, social, and historical setting of the investigator has a similar effect upon the range of scientific method. To the extent to which the thesis of ethical relativity is warranted, scientific methods of validation are inapplicable to ethical problems since the relevant answers are not susceptible to validation.

(3) The presence of value-judgments in ethics raises questions similar to those which arise in other fields of valuation.[18] The main objection about the competence of science in this domain centres around the requirement of empirical verifiability of scientifically justifiable statements. It is claimed that the empirical verifiability of any statement consists in its deducibility from purely observational premises and that ethical statements are therefore unverifiable because they involve value-predicates which never occur in such premises. This argument obviously assumes that a statement can be deduced from given premises only if all its constituent terms occur also in these premises. A similar argument against the empirical verifiability of normative ethical statements involving such crucial terms as "ought" or "should" can be based on the fact that these terms occur in no observational premises. We shall touch upon the important question of the empirical verifiability of ethical statements (of either the valuational or the normative type) in connection with our discussion of the Principle of Verifiability (§ 33).

---

[17]Cf. A. J. Ayer, *Language, Truth and Logic* (1946); C. L. Stevenson, *Ethics and Language* (1944); R. M. Hare, *The Language of Morals* (1952); F. Edwards, *The Logic of Ethical Discourse* (1956).

[18]Cf. E. W. Hall, *What is Value? An Essay in Philosophical Analysis* (1952); R. Lepley, *Value: A Cooperative Enquiry* (1949).

(4) A comprehensive theory of human conduct would have to deal with practical problems too. Some of these problems are concerned with single human actions, for example, with whether a particular action A, if taken by a human agent, would bring about his objective O. Other practical problems refer to human policies and involve groups of actions and of objectives. A policy question may ask, for example, what objectives O_i ($i = 1$, 2, etc.) should be chosen and what corresponding actions A_i ($i = 1$, 2, . . .) should be taken in order to achieve a basic objective O. It is easy to see that under certain circumstances such practical problems concerning individual actions or policies are bound to transcend the potentialities of scientific method.

To solve the problem whether an action A would bring about the objective O is tantamount to ascertaining whether the statement describing the action A *implies* the statement that specifies the objective O. Similarly, the solution of a policy-problem would consist in finding out whether every action A_i is describable by a statement which *implies* the statement about the corresponding objective O_i and, in addition, whether the statement referring to the group of all objectives O_i *implies* the statement describing the basic objective O. The decision about these various relations of implication among the statements involved will depend ultimately upon whether these statements are themselves within the reach of scientific method, that is, upon whether they satisfy the requirement of empirical verifiability. Consequently, practical problems concerning human actions and policies will be scientifically unsolvable whenever some or all of the actions and objectives involved are not describable by verifiable statements. For example, the question whether a redistribution of income in a given country would strengthen the electoral chances of a particular political party constitutes a scientifically solvable problem because both the objective and the action involved are describable by verifiable propositions. The somewhat similar question of whether a complete equalization of income would realize the ideal of social justice is not so solvable because the objective is not describable by a verifiable proposition.

14   NATURE OF SCIENTIFICALLY UNSOLVABLE PROBLEMS

The explanations put forward in the preceding sections should clarify in substance the meaning and import of the problem concerning the limits of science. We have seen that to answer this question would amount to delimiting the class of problems that are inherently unsolvable by the method of science. It goes without saying that an attempt at answering this question satisfactorily, that is, on the basis of an adequate body of evidence, depends essentially upon the meanings of the crucial concepts of science, of scientific method, of a scientific solution to a theoretical or a practical problem, of the "impossibility" of reaching such a solution, etc. The attempt requires, in addition to a survey and a critical examination of relevant facts about science, a more thorough clarification of the aforementioned crucial concepts than has been possible in our introductory considerations. The preliminary comments on these concepts may suffice, however, to make the essential contours of the problem visible. They should also make it possible to state in this section, by way of an anticipatory survey, both the steps which will be taken in order to come closer to a satisfactory solution and the kind of solution that our analysis of science will tend to substantiate.

We have seen that to solve a problem by the scientific method is to ascertain that the observational evidence at hand adequately supports the correctness of some particular solution to this problem. This applies both to "pure" and to "applied" problems, both to problems that involve human values and policies and to those that are essentially value-free. The delimitation of the range of scientifically solvable problems which will emerge from our investigation implies that every meaningful practical or theoretical problem can be, in principle, successfully dealt with by the scientific method, even if other methods of dealing with such problems should prove more convenient under special circumstances. In other words, the method of science is inherently universal, in spite of the accidental limitations involved in the extraneous and fortuitous circumstances under which it comes to be applied. There is nothing in the nature of the method of science which would prevent it from being applicable in principle to any problem that could be settled by any method whatsoever. There are no limits inherent in the method of science.

This sweeping statement, though correctly conveying the upshot of our investigation, is dangerously oversimplified and will require several substantial qualifications. The most important qualification relates to the very existence of scientifically unsolvable problems which we have discussed in a preliminary way in the preceding sections of this chapter. We have there subdivided scientifically unsolvable problems into those within science, those about science, and those ordinarily classified as "philosophical." It is obvious that quite apart from the controversial nature of philosophical problems and the possibility of solving them by the scientific method, the demonstrable existence of scientifically unsolvable problems both at the scientific and at the meta-scientific level seems hardly compatible with the aforementioned universality of the scientific method.

In order to realize that the universality of science can nevertheless be maintained on sound grounds and must even be viewed as the most significant feature of the whole situation, we need, first of all, some insight into the nature of unsolvable problems. A vitally important distinction has to be drawn between "decision questions" or "yes or no questions," answerable by a simple "yes" or "no," and "ampliative questions," which can only be answered by a proposition containing, in addition to the data conveyed in the question, some additional item of information. "Is the earth a planet?" is a decision question, the logically admissible answers to which are "yes" and "no," though only the first answer happens to be adequately supported at present by observational evidence. "How many planets are there in our solar system?" is an ampliative question, since a logically admissible answer to this question would have to include, in addition to the data explicitly mentioned in the question, some natural number: 0, 1, 2, etc. "There are 8 planets in the solar system" is a logically admissible answer, whose falsehood seems to be pretty well established by presently available astronomical evidence. "There are 9 planets in the system" is another logically admissible answer whose correctness is now adequately supported by astronomical observational data.

Since the solution of a problem consists in producing a body of evidence which adequately supports the correctness of some logically admissible answer to it, a problem will be unsolvable if it is impossible to produce for it such a body of evidence. It may be unsolvable for two different reasons: (1) if there is no correct answer to it at all; (2) if it is impossible to produce a body of evidence adequately supporting any particular and pertinent answer to it, even if there should be a correct answer among those pertinent. Thus the problem of squaring the circle is unsolvable for the first reason, because none of its logically admissible answers is correct. The problem of determining in what constant ratio R all material objects increase their size every second is unsolvable for the second reason: there could not possibly exist a method of justifying any particular answer to it, for the simple reason that the universal swelling would affect the yardsticks in the same way as the remainder of the material objects and would therefore remain undetected.

The distinction between the two kinds of unsolvability vanishes of course with regard to decision problems. As the only admissible answer in such a case is either "yes" or "no," there is no possibility of all the admissible answers being incorrect or false. If we ask, for example, whether all the material objects double their size every second (and thus include, in the reformulated question, the particular value $\frac{1}{2}$ for the rate of the universal swelling), then the resulting decision problem is unsolvable, because no adequate evidence could possibly be produced for or against an affirmative answer. All in all, we obtain three classes of unsolvable problems:

1. Ampliative problems unsolvable because all their admissible answers are false.

2. Ampliative problems unsolvable because of the impossibility of producing adequate evidence supporting any of their admissible answers.

3. Decision problems unsolvable because of the impossibility of producing an adequate body of evidence in regard to any logically admissible answer.

Now the impossibility of solving an ampliative problem all of whose admissible answers are false can hardly be construed as a limitation of the scientific method. Such a problem raises a question whose correct answer is simply non-existent, and the method of science cannot be blamed for failing to find something that does not exist. In this sense, the failure of science to square the circle, to construct a *perpetuum mobile*, etc. is not an indication of limits inherent in its method, in spite of the fact that such failures cannot be termed accidental; they are inherent in the nature of the problems concerned, which do not admit of any correct answer and are therefore inaccessible to any method, scientific or not. We can take into account the existence of such problems by stating that the inherent universality of science consists in its ability to solve every problem that does admit of a correct answer.

This, however, only takes care of unsolvable ampliative problems of the first kind, by showing that their existence is compatible with a suitably qualified universality of the method of science. The unsolvable ampliative problems of the second kind and the unsolvable decision problems cannot be accounted for in this way, because the failure of science to provide a body of evidence supporting any logically admissible answer to such a problem is compatible, ostensibly at least, with there being a correct and logically admissible answer; such an answer may be discoverable by another, non-scientific procedure, and this would certainly point to inherent limitations of science.

A detailed discussion of the last two classes of unsolvable problems (i.e., of ampliative and decision problems for which the scientific method is incapable of providing a body of evidence in support of any particular and logically admissible answer) will be carried out in Part III of this inquiry. We shall then see that in spite of the undeniable and important difference between the first class of unsolvable problems and the two remaining classes, they do have a common factor which makes the impossibility of solving them by the scientific method compatible with the inherent universality of this method. The ampliative problems of the first class are scientifically unsolvable because all their admissible answers happen to be false. Our investigation will show that the problems of the last two classes are unsolvable because none of their admissible answers is true. In other words, we shall reach the conclusion that if it is impossible to find a body of observational evidence which adequately supports any particular answer to a particular ampliative problem or a particular decision problem, then such ampliative and decision problems do not admit of any true answer at all. In contradistinction to an unsolvable problem of the first kind whose admissible answers are all false, the admissible answers to an unsolvable problem of the second and the third kind are neither true nor false; they simply have no definite truth-value and are *indeterminate* in this sense. Accordingly, we shall call a problem whose admissible answers are neither true nor false an indeterminate unsolvable problem. The main result of the discussions to be carried out in Part III can be stated as follows: If it is impossible to find adequate evidence supporting any admissible answer to a particular ampliative or decision problem by applying any conceivable scientific method, then no admissible answer to this particular problem is either true or false and the problem itself is indeterminate and unsolvable.

This result will go a long way towards establishing the universality of the scientific method, since it states that problems unsolvable by scientific methods are either indeterminate or admit only of false answers. In both cases, the common formula holds that a scientifically unsolvable problem has no admissible true answer or, in other words, it has no solution. And, once more, the existence of scientifically unsolvable problems of the last two kinds described cannot be interpreted as a shortcoming of the scientific method, since in these two cases this method cannot be blamed for failing to discover something that does not exist, namely, a correct answer to the problem under consideration. Thus, we shall have to acknowledge, for example, that it is neither true nor false that every material object doubles its size every second and similarly that whatever the particular number x may be it is neither true nor false that every material object increases its size every second in the ratio $1 : x$. Both the decision problem concerning the doubling of size and the ampliative problem concerning the determination of the ratio of the alleged universal swelling are indeterminate.

The universality of science consists in the fact that problems unsolvable by the scientific method are either indeterminate or answerable by false statements only. In other words, scientifically unsolvable problems have no solution and are therefore unsolvable by any other non-scientific method as well. In this respect, scientifically unsolvable problems are all alike, regardless of whether they are determinate or indeterminate. Nonetheless, the difference between determinate and indeterminate unsolvable problems is important, both for historical and for systematic reasons. The existence of scientifically unsolvable determinate problems is largely non-controversial, whereas the nature of unsolvable indeterminate problems leads us to one of the basic issues in epistemology and philosophy of science, namely the problem of verifiability as a criterion of meaning. This difference cannot be obliterated and we shall have more to say on this score later. But as far as the thesis of the universality of science is concerned, the above explanations of the three kinds of unsolvable problems should prove sufficient.

PART TWO

The Method of Science

CHAPTER ONE

Fact-finding Methods

---

15 DIRECT FACT-FINDING METHODS

IN THIS SECTION we start our critical survey of the basic methods of contemporary science. According to the conclusions reached in Part I, the potentialities and the limits of science depend upon the method of the scientist, rather than on the information at his disposal, or the technical facilities available to him. Hence scientific method, rather than scientific information or scientific technology, can provide a suitable vantage point for investigating our main problem. Our objective is to find out the *inherent* range and the *inherent* limits of science, i.e. the class of problems that are solvable in principle by scientific methods, and the class of problems that are not solvable by such methods. The *accidental* range and limits of science, which reflect the extent to which the potentialities inherent in the nature of science are affected by extraneous (historical, social, psychological) circumstances of the life of science will play but a subsidiary role in the argument.

The present survey of contemporary scientific methods is therefore not an end in itself, but rather a means of determining whether and to what extent the sum-total of all conceivable scientific methods can be viewed as special cases of a single procedure deserving the name of *the* method of science. The inherent range and limits of science can then be identified with the reach of such a procedure. Furthermore, the nature of validating methods to be surveyed in this Part will have to be generalized eventually, in order to provide a framework spacious enough to include all conceivable scientific methods. In other words, we need a critical survey of the basic methods of contemporary science in order to find out a more general procedure that both underlies all of them and is likely to be found in any discoverable scientific method. It will be seen that such a generalization is not completely determined by the supply of presently available scientific methods. An additional rule for selecting a single procedure out of the possible generalizations of present scientific methods will be derived in Part III from the Principle of Verifiability.

The particular objective to be pursued in this Part explains also the special point of view from which the survey of scientific methods will be conducted; that survey will often be in contrast with the usual presentations of these methods. We shall stress mainly the features that are responsible for the range and limits of the various methods and are likely to affect the range and limits of the resulting unique procedure.

It is almost universally agreed that the scientist needs and uses methods that enable him to get hold of particular *facts*, i.e., to answer such questions as whether some particular object had some particular property at a particular time-moment, or bore, at that moment, a particular relation to another particular object. That this piece of iron is now hard and cool, that it was soft a moment ago, that it is twice as long as that piece of iron, are examples of particular statements answering such factual questions. The statements themselves may be said to be *fact-like* for this reason, and to become well-established *facts*, once they are adequately supported by a body of evidence.[1] It is also generally agreed that, although particular facts are indispensable at several stages of the scientist's work, science does not confine itself simply to registering, or to summarizing, predicting, and explaining particular facts. In addition, science contains general laws and theories, which differ from fact-like statements in several important respects. It is essential to our inquiry to investigate the methods used by the scientist for discovering or establishing facts, laws, and theories, and primarily, the methods for acquiring some initial indispensable knowledge of particular facts. What are the basic fact-finding methods of science?

Let us start with "direct," that is, non-inferential fact-finding methods. Such methods are best illustrated by typical questions which fall within their respective ranges. An astronomer may ask his assistant whether a specified star will pass through the telescope before, while, or after his clock strikes three. A psychologist may ask his subject whether he feels tense or relaxed. A physicist may try to ascertain the final reading of a specified instrument. To answer these questions, the astronomer has to refer to his sense-impression of a flicker called "star" and the sounds emitted by the clock, the psychologist to his subject's "introspective

---

[1] A fact, whether well established or not, may be defined simply as a true fact-like statement.

awareness" of a state of tension or relaxation, the physicist to his perception of the scale of the instrument and its pointer. Thus, the experiences of the three observers seem to vary rather considerably, since they range over sensations, perceptions, and introspections. Yet the final result is pretty much the same in all these cases, as far as the solution of a fact-finding problem is concerned: It is these various psychological experiences which suggest the correct answers, or, at any rate, the answers considered to be correct by the scientists. The very fact that a particular answer has been suggested to an observer by his having any of these experiences is admitted as sound supporting evidence for the relevant solution.

In spite of the differences between the psychological experiences referred to as "sensations," "perceptions," and "introspections," which underlie the validating methods in the above examples, there are, of course, several essential features common to all of them. The most important common feature, from our point of view, is that the experiences are used for cognitive purposes, for solving problems. This additional purposive element is recognized in ordinary language, which makes a clear-cut difference between merely perceiving ("seeing," "hearing," etc.) an object, and watching or observing this object. Performing an observation is a planned activity, with a definite objective. What is this objective? The observer tries obviously to answer some question or some set of questions concerning the object under his observation. When he watches or observes something, he does not just stare at it, or merely enjoy the sight of it, without anything else at the back of his mind. The physicist who watches the scale of the instrument tries to make out the mark with which the pointer coincides. The astronomer who looks through the telescope and listens to the clock is anxious to ascertain whether he saw the flickering and heard the clock at the same moment. The psychologist attempts to answer correctly the question concerning his subject's state of tension. The distinctive feature of an *observation*, as compared with the underlying sensation, perception, or introspection, is the desire to answer a definite question by utilizing the relevant experience.

The presence of this question links the process of observation with a particular language; the question is either explicitly formulated in this language or at least formulable in it. The relevance of the linguistic factor at the most fundamental level of inquiry will be commented on later (§ 31). For the time being, let us notice simply that the presence of the question and the desire to answer it transform a psychological experience called sensation, perception, or introspection, as the case may be, into a method of validation, called direct observation. The point of differentiating mere sensation, perception, or introspection from a direct observation is thus to stress the additional conditions which must be fulfilled by certain experiences for them to occur as essential ingredients of problem-solving methods.

It is worth while pointing out that the application of the term "observation" in this context, as implying the cognitive or problem-solving use to which some psychological experiences are being put, is not very felicitous in view of the many and extremely comprehensive meanings presently attached to "observation" in science and its philosophy. We are here primarily interested in direct fact-finding methods, and there is little doubt that, when based on sensory, perceptual, or introspective experiences, such methods may well be subsumed under the genus "observation." When a person tries to make out how he feels, or how a material object looks to him, or the actual nature of this perceived object, he can certainly be said to make an observation. Yet, when not satisfied with the service rendered by his naked eye, he looks through a microscope at the object, if it is quite small, or through a telescope, if it is very far away, he is still said to make an "observation." If he tries to measure the length, or the temperature, or the electrical resistance of an object by making use of the relevant measuring instrument, he is said to "observe" the values of these quantities, possibly in order to compare the values "observed" by him with those predicted. Even if his "observations" require, apart from the use of more or less complicated instrumental equipment, the application of elaborate theories or the carrying out of elaborate computations, he will still be said to have made an "observation." The term "observation" can thus be stretched almost indefinitely, and this is hardly helpful, in view of the fundamental role the concept of observation has come to acquire in the methodological outlook of the whole of science. This term is certainly not illuminating when used in connection with the direct fact-finding methods, based respectively on sensation, perception, and introspection, discussed in this section. We propose, therefore, to classify these three fact-finding methods under the heading of "*direct* observation" and also to keep any further use of the term "observation" within reasonable bounds (§ 18).

We should also point out at the outset that direct observation does not by any means account for all the basic direct fact-finding methods used by the scientist. One of these non-observational methods is so essential that it must at least be mentioned here. Direct fact-finding methods are used to answer not only questions concerning things we presently sense, perceive, or introspect, but also questions referring to particular objects that we have just sensed, perceived, or introspected. In other words, memory is as indispensable a method of discovering particular facts without resorting to inference as is direct observation of present objects. Yet nobody would call memorative validation a case of "observation": we have rather to acknowledge that validation based on memory is a direct fact-finding method of a non-observational type. Its role in the discovery of scientifically significant facts could hardly be exaggerated. For example, with a slight change in the wording of the three afore-mentioned questions solved by direct observation of the astronomer, the psychologist, and the physicist, validation by memory alone would be-come applicable, since observation would no longer be available. The psychologist, for example, would just have to ask his subject "How did you feel a moment ago?" in order to rule out direct observation and make recourse to memory inevitable.

Another common feature of direct observational and memorative methods is their ability to serve purposes of both *discovery* and *veri-fication*. If the astronomer asked whether the star did pass the telescope while the clock was striking three, instead of asking at what time the star passed the telescope, the observational method would have been used for verification instead of for discovery. In other words, an observa-tion is a verification if it is made with a view to checking the correctness of a particular answer to a particular problem. An observation aims at discovery, if, given the particular question, it serves both to find an answer to this question and to produce evidence supporting the cor-rectness of this answer.

Another striking feature of observational and memorative methods is their lack of instrumental equipment. We shall see later that this is not coincidental and is actually connected with the non-inferential nature of these methods. The extremely simple logical structure of problems and solutions pertaining to direct fact-fiinding methods is also remark-able. Only the simplest ("atomic") propositions of either the subject-predicate type or the relational type are susceptible to direct validation; neither "molecular" propositions involving logical connectives ("and," "if," "or," etc.) nor "quantified" statements, which are obtainable from atomic or molecular propositions by inserting quantifying words such as "every" and "some," can be directly validated. I can see that this book is on this shelf, but I could not possibly see (or settle the question by any other direct method) that this book is either on the shelf or in the drawer. The inapplicability of direct fact-finding methods to molecular facts will be explained later. As for quantified statements, it is obvious that no fact-finding method is appropriate for them, since such statements are not fact-like.

In spite of the multifarious similarities among the four basic fact-finding methods, they are separated from each other by fundamental differences. Thus, introspection and sensory observation are generally regarded as absolutely reliable or infallible, in contradistinction to perceptual observation whose reliability is fair, though limited: it is contended that a person cannot be mistaken about how he feels or how some object looks to him, although he might be mistaken as to its real nature. Yet perceptual validation is, in turn, more reliable than memory. It should be added that the absolute reliability of introspection and sensory observation, which has been, since Descartes, a *leit-motiv* of Western thought and a typical expression of man's "quest for certainty," has been challenged recently in lively discussions. We shall see later on, however, that, in spite of its historical and intrinsic interest, the controversy over the degree of reliability of these observational procedures has no bearing upon our central problem. The range and limits of science are not affected by the way the dispute between "fallibilists" and Cartesians is adjudicated: what the scientist may reliably know does not depend upon how much of his knowledge is infallible.

The most important difference between the various methods of direct validation is concerned with the kinds of problems they are fit to solve and the kinds of concepts characteristic of these problems. For example, the question whether the object X looks brown to the observer Y is clearly distinct from the question of whether X is brown. It is not only that Y, depending upon whether he tries to answer the first or the second question, has to take up two different attitudes (referred to as sensory and perceptual observation, respectively). The questions themselves have different meanings, since they are answerable by statements that are not even deducible from each other, let alone synonymous: "X is brown" may be true and "X looks brown to Y" false, and *vice versa*. We shall have to comment later on the terms which occur characteristically in the formulation of problems solvable by sensory, perceptual, introspective, and memorative methods respectively, for example, "looking brown to somebody," "being brown," "liking brown," and "having just looked brown." Such terms fall into four non-overlapping groups with no term in any particular group being either synonymous with, or fully definable by means of, terms included in other groups. Thus, every direct fact-finding method has a specific supply of concepts which, though interconnected with concepts related to other methods, form nevertheless an autonomous group, since their meanings are not fully reducible by definitions to concepts of other groups.

The four groups of terms just mentioned are simply taken over by the scientist from pre-scientific, ordinary usage, and continue to form an important link between science and common sense. They include, in the main, four groups of individual names and of predicates, which may be designated as sensory terms, perceptual terms, introspective terms, and memorative terms, according to the direct fact-finding methods with which they are associated.[2] As the four classes of terms are basically distinct, though logically interrelated, we shall describe all such terms as possessed of a *direct empirical meaning*, since their correct use is determined by the role they play in the context of direct observational and memorative methods; no definitional procedures are necessary.

We should also add that sensory, perceptual, introspective, and memorative terms must not be interpreted in the sense they had in the old-fashioned atomizing psychology. For instance, the sensory terms include not only names of simple qualities, such as "looking red" and "looking round," but also names of complex characteristics involving several simple qualities, for example, "looking like a dog," "looking like a man," provided these expressions serve as predicates in statements susceptible to sensory validation. Similar remarks apply to the remaining classes of terms possessing a direct empirical meaning.

The Range of Direct Fact-Finding Methods

There is little doubt about the fundamental role of observational and memorative procedures in the validation of fact-like statements. Such statements form the core of all empirical sciences, and the methods for validating them are, accordingly, of central importance to empirical science. Yet the precise determination of the inherent ranges of observational and memorative methods is largely controversial.

Thus perception has sometimes been viewed as involving at least subconscious inference, or as not separable from memory, and consequently as non-existent in a pure form.[3] The existence of introspective

[2]Sensory terms are often referred to as "sense-data" terms, or as terms denoting "appearances of material objects." Perceptual terms denote material objects, their observable properties and relations; such terms are also called "ostensive," since they can be explained by pointing out the objects they apply to. Introspective terms are said to refer to "private" experience and are sometimes called "psychological" —an adjective which is unjustifiedly restrictive since the psychologist's vocabulary by no means reduces to such terms. The introspective and sensory terms together are often grouped under the heading "phenomenological" or "phenomenal." The autonomous role of memorative terms is rarely considered.

[3]H. Bergson, *Matière et mémoire* (1896).

validation was flatly denied by A. Comte and the early behaviourists. Accordingly, the problem of relating introspection to sense-perception has played a major part in the evolution of empirical psychology. It has also been contended that pure sensory observation is normally submerged by perceptual activities, which are so much better adjusted to the requirements of practical life that the former materializes only in the aesthetic attitude of the artist and the epistemological analysis of the philosopher. Similarly, the relative importance to be ascribed to all four fact-finding methods, especially to sensory observation as compared with perception, has been variously evaluated by several influential schools of thought. This evaluation is responsible for the idealist-realist-phenomenalist controversy over the subject-matter of scientific knowledge and for the position taken in regard to the question as to whether all scientific knowledge is of phenomena only. This controversy is clearly relevant to our central problem.

At this stage, we shall have to discuss briefly two questions relating to the range of direct fact-finding methods:

(1) Apart from memorative validation, we have distinguished, within science, three observational methods, namely sensory observation, perception, and introspection. Are there no other direct fact-finding methods?

(2) We have acknowledged the direct, non-inferential nature of sensory, perceptual, introspective, and memorative validation. Granted that, psychologically speaking, no inference is involved in solving problems by resorting to any of these methods; but is such a psychological short-cut epistemologically valid? Can knowledge of things perceived or remembered be actually obtained without recourse to inference? This has been denied by several influential schools of thought.

(1) Since we have found that memory of sensed, perceived, or introspected facts is a direct fact-finding method, we can give observations no monopoly in this field. Yet although memorative validation is indispensable to science as well as to common sense, since it is the only means of establishing the bulk of our empirical knowledge, philosophers seem to have become memory-conscious only recently. At present, however, the role we have assigned to memory along with the observational procedures would generally be granted. The point at issue is rather the status of other alleged fact-finding procedures which are neither observational nor memorative: telepathy, extra-sensory perception, clairvoyance, empathy, intuition of values are fact-finding methods not resorted to by the scientists and not taken account of in our discussion. Are problems solvable by such unorthodox fact-finding procedures beyond the reach of science?

We have consciously restricted our discussions to those observational and memorative procedures that are actually used in the direct scientific validation of fact-like statements. We cannot enter into a detailed discussion of the above unorthodox fact-finding methods. The position with respect to them may be summarized as follows: No dogmatic commitment to confine the list of direct fact-finding methods to observational and memorative procedures is either implicit in our account of direct validation or essential to the whole argument. On the contrary, the question concerning the universality of scientific method would not be substantially affected by recourse to the above unorthodox methods of validating fact-like statements. Scientific method, defined in terms of only observational, memorative, and inferential operations is universal regardless of the status of such fact-finding methods as those listed above, for these unorthodox methods, if reliable at all, would theoretically be replacable by the former. Facts that could possibly become known by applying telepathy, or extra-sensory perception, or intuition of values, are also discoverable in principle by resorting to observational and memorative procedures. For example, if telepathy did enable the scientist to ascertain the frame of mind of human beings without observing their bodies, the same psychological information could also be secured in principle by carrying out the relevant observations. Similarly, if extra-sensory perception were able to provide reliable information about unobserved physical events, the observation of these events could yield the same information. And thus our list of direct fact-finding methods seems sufficiently non-committal to avoid any narrow-minded dogmatism.

(2) The doubts often raised in connection with the existence of the "directly given"—doubts which, I take it, are a quaint way of referring to non-inferential fact-finding methods—seem to arise mainly from two confusions. We shall discuss them in respect to sense-perception, the directness of which is most frequently questioned:

(a) It is often held that introspective observation and sensory observation are absolutely reliable. Since they are also the most typical examples of non-inferential validation, these two features of validating methods, namely, absolute reliability and non-inferential nature, come to be confused with each other. This confusion is strengthened by the circumstance that a result achieved by applying a method of limited reliability can often be improved upon, that is, the reliability of the relevant solution can be increased, by utilizing inductive inferential validation as an additional piece of evidence; since inductive inferential validation is admittedly of limited reliability, recourse to it comes to be associated with such limited reliability. Yet, memorative validation is clearly direct and of limited reliability. And once the difference between these two characteristics of validating methods is realized there is no more reason for confusing them, or for doubting the non-inferential nature of perceptual validation. Granted that visual perceptions, for example, are afflicted with illusory and hallucinatory distortions and that, consequently, any one who makes a statement about an object he sees by relying exclusively upon what he sees can commit a mistake and that he might lessen the probability of mistake by providing additional inferential evidence in support of his statement: this does not imply at all that prior to this inferential additional evidence, and while relying exclusively on what he saw, he was unable to validate his statement by simply observing the object this statement is about.

(b) Another reason for questioning the direct nature of observational validation seems to arise from a confusion of observationally validated statements expressed in the object-language of the scientific theory under consideration with statements of the meta-language used to study this object-language. Let us consider any statement susceptible to perceptual validation and formulated, say, in French. "Ceci est vert" can be verified or discovered, as the case may be, by any Frenchman who is not handicapped by some fortuitous circumstances (he must not be colour-blind, his spatio-temporal location in regard to the individual designated by "ceci" must meet certain requirements, etc.). To verify the statement, the Frenchman just needs to have a look at the object and try to answer the question "Ceci est-il vert?" To use his visual observation for purposes of discovery, he has to attempt to answer a different question ("De quelle couleur ceci est-il?"), other things being equal. In neither case is there any necessity for the Frenchman to make inferences. If he is sufficiently trained in using French colour adjectives correctly, he will solve either problem correctly in most cases by having a look at the object and refraining from any inferential operations. And nothing else is required for either question to be solvable by direct observation.

Now, let us assume that a person who speaks English sets himself the problem (formulated in English) of classifying questions (formulated in French) according to whether they are solvable by Frenchmen with or without inference. *This* problem is by no means soluble by mere observation. After all, even the simpler question of the meaning attached to a given word in a foreign language is not answered by pure observation. Suppose now the Englishman comes to consider the possibility of the question "Ceci est-il vert?" being solved by a Frenchman through direct observation: the observational unsolvability of *this* problem formulated in English is quite compatible with the observational solubility of the French problem under consideration. Questions about the solvability of other questions must be differentiated from the latter. But it is rather tempting to identify the two interrelated questions as does C. S. Peirce in his remarks about what he considers to be the observational unsolvability of colour-problems. Once such a confusion has occurred, the legitimate doubts as to the direct solubility of the second, meta-linguistic question are easily and illegitimately transferred to the first question, expressed in the object-language.

These two circumstances may suffice to account for misgivings concerning the non-inferential nature of perceptual validation. There are quite a few other circumstances, of course, which seem to have contributed to blur the line between observation and inference, for example, the disclosure of sense-illusions by inferential procedures.

16   QUANTITATIVE METHODS

Place of Quantitative Methods in Science

Knowledge obtainable by applying direct fact-finding methods is basically pre-scientific in character. Such methods only yield statements that either ascribe perceivable properties to particular perceivable objects, or assert that a perceivable relation obtains among a few perceivable objects. The greatest part of information supplied in pre-scientific exchange between human beings is obviously of this kind. The common stock of knowledge obtained through these direct fact-finding methods, which are both pre-scientific and scientific, illustrates the roots of science in common-sense cognitive procedures. Yet, if science simply followed the ways of handling information implicit in the common-sense outlook, and, more particularly, if science had not invented, in the course of its evolution, new methodological devices, while at the same time abandoning some fundamental common-sense procedures, it would hardly have differentiated itself from its pre-scientific source.

It is obvious that the opposite actually happened. The emancipation of science and its evolutionary metamorphosis into an autonomous social news-agency whose information would eventually dominate man's whole life to an unprecedented extent came about by an interplay of selection and creation. Thus, while taking over the common-sense ways of using sense-observation, memory, and inference for acquiring knowledge (i.e., relevant and reliable information), science abandoned other common-sense procedures of handling information; mere reliance on social or religious authority, mere accordance with a body of traditionally transmitted beliefs, mere adjustment to the interests of a socially influential group, which had played a part in the pre-scientific ways of acquiring knowledge comparable to that of observation and inference, have no validating significance in the scientific way of handling information. Out of the above six major factors discernible in the common-sense cognitive methods, the evolution of science has selected but three: observation, memory, and inference.

Among fact-finding methods *created* by science, the mensurative or quantitative procedures are certainly the most conspicuous. They have often been held responsible for the spectacular advance of scientific knowledge in the last three centuries and, in the popular usage, the word "scientific" has become almost synonymous with "quantitative." This tendency to identify the method of science with measurement is certainly traceable to the extraordinary efficacy of mensurative methods in natural sciences such as physics, astronomy, chemistry, biology. The triumph of Newton's *Mathematical Principles of Natural Philosophy* must have prompted Kant to equate the degree of maturity achieved by any science with its use of mathematical techniques and concepts. A century later, Lord Kelvin declared that we do not know what we are talking about unless we are able to measure it. In our own century, physics has been defined by Campbell as the science of measurement; in conjunction with Lord Rutherford's half-serious saying that only two classes of sciences are distinguishable, physics on the one hand and stamp-collecting on the other, Campbell's metrological definition of physics indicates clearly the importance attached to quantitative methods in science.

The number of sciences whose very names bear testimony to the supremacy of mensurative methods is impressive: the Greek word "μετρεῖν", corresponding to measurement, has often been used to label various sciences. Such names as trigonometry, thermometry, and calorimetry are typical and of long standing. The tendency they exemplify has made considerable headway in the meantime and is very much alive at present. Thus, the application of mathematical, and, more particularly, of statistical procedures to biology has been named biometrics by Galton; his name and the field it refers to are important features of the scientific landscape of our times. Sociometry is a branch of sociology mainly intent upon measuring trends, opinions, attitudes, and similar social phenomena. Econometrics bears a similar relation to economics. Measuring the style of a writer by determining the frequency of specifiable expressions in his writings is the object of stylometry; this measurement has been applied successfully by Campbell and Lutoslawski to the chronology of Plato's dialogues and seems to have settled the conspicuous differences of opinion that used to prevail on this score among the most competent scholars. Moreover it is obvious that there are quite a few sciences dominated by mensurative procedures which do not explicitly refer to measurement in their official designation. Thus, psychophysics, traceable to the publication by T. Fechner of his major work *Elemente der Psychophysik* in 1860, is an attempt to measure accurately the attributes of sensations and their quantitative connections with physical stimuli. Since, then, so many sciences employing measurement as their basic methodological procedure have been extremely successful, it seems natural to assume that other sciences, rather backward at present (especially the social and humanistic studies), could anticipate a comparable success by simply imitating their stress on measurement.

Much can be said in favour of this appreciation of measurement. But a glance at the history of science suffices to convince one that not every major success in science has been due merely to applying quantitative methods, and that measurement is not the only guide to scientific progress. Decisive advances in astronomy and biology were connected with the invention of the telescope and the microscope, respectively, and with the ensuing emergence of vast arrays of astronomical and biological facts, rather than with successful astronomical and biological measurements. Darwin's theory is a qualitative one, regardless of the promising contemporary attempts of R. A. Fisher and his collaborators to substitute a more rigorous quantitative formulation for the basic assumptions of the mutability of species and the agency of natural selection. The very meaning of Kant's identification of scientific maturity with the use of mathematical techniques has changed considerably since the integration of the whole of mathematics with an expanded logic, through the labours of Frege, Russell, and Whitehead. The bearing of this development on Kant's thesis is twofold:

(1) Since mathematics has become a part of pure logic, the extent to which any science has succeeded in mathematicizing its conceptual and inferential apparatus coincides with the extent to which it has come to be pervaded by pure logic. This view of the mathematical component of science is incompatible with Kant's distinction between the analytic nature of logic and the synthetic nature of mathematics and of all the empirical sciences. Further, Kant's position entails the consequence that

a purely empirical science which keeps improving the quality and scope of its mathematical techniques would tend eventually to be absorbed by pure logic. This is untenable.

(2) Mathematics, at present, is no longer a science dealing exclusively with quantities (numbers, functions of numbers, functions of such functions, etc.). Abstract set-theory, abstract group-theory, and various abstract algebras are concerned with general relational structures definable in purely logical terms but having little in common with quantity or number proper. Hence, at present, the ability of an empirical science to utilize mathematical concepts and results, does not imply that quantitative concepts or methods are of any importance in this science.

Thus, on closer inspection, the Kantian over-emphasis on the quantitative and the measurable in assessing the potentialities and the aims of science is hardly defensible. Yet, while qualifying the alleged quantitative monopoly within science, we cannot help but appreciate the overwhelming role and prospects of measurement and of related methods. In this section, I should like to outline some aspects of measurement that are likely to explain the extraordinary power of this method and have, furthermore, a considerable bearing upon our central problem.

Before embarking upon an analysis of quantitative methods and the reasons for their exceptional efficacy, it is worth while pointing out in what respects these methods have proved so effective. In Part I we listed two main theoretical attainments of science: the successful attack by scientific methods on problems of decisive importance to man's over-all outlook; the establishment of a coherent system of scientific facts, laws, and theories which is mainly responsible for the scientist's power to describe, predict, control, and explain man's entire environment, immensely expanded by the very impact of science. In these two basic theoretical attainments of science, quantitative methods can be shown to have played a predominant part. But even this double attainment does not give an adequate idea of the power of these methods. We shall see that they are also responsible for the replacement of human standards of cognitive validity by objective, extra-human criteria and, consequently, for liberating man from the idols of the tribe and the den. The quantitative methods have also contributed decisively to what may be termed the Principle of Intellectual Democracy within science, which stipulates the testability of any proposed scientific fact, law, or theory by any competent (i.e., adequately trained and equipped) investigator as a prerequisite of scientific validity. The discoverer has no priority in this respect, nor is the priority made conditional upon the personal, social, or even the purely scientific standing of the potential investigator. No bit or package of information is to be granted scientific status unless and until it can be tested by any competent investigator.

The surge of science in the seventeenth century and its accelerated progress since that time are not due exclusively to the increasing pressure of economic or strategic needs. Man's theoretical problems have become increasingly complex during this period, but his capacity for coping with them has also kept increasing in spite of the virtually stationary psycho-physical structure of the human race. The explanation is supplied by the quantitative methods of science and the increasingly efficacious tools they have put at the disposal of man.

We are, however, not yet in a position to give a satisfactory account of the multifarious impact of quantitative methods since we have not yet examined some of the basic properties of these methods. Nevertheless, certain advantages and uses of measurement can be seen at a glance, prior to any further analysis, and we might indicate these before turning to measurement itself. Such a starting point may help us to focus our attention on aspects of measurement that are likely to shed some light on its unique efficacy.

Conciseness

The advantages of quantitative methods which will be displayed here are connected with the peculiar nature of the language of mathematics used to express the results of measurement, rather than with measurement proper. That is why they can be set forth at this juncture, prior to an analysis of measurement. These advantages would remain available, in principle, even if the mathematical theories used in empirical science had little in common with either quantity or measurement. As a matter of fact, however, it has to be granted that in the historical evolution of empirical science, the outstanding services rendered by mathematical language have been mainly associated with the mathematics of quantity. Thus, in spite of the theoretical possibility of applying abstract, non-quantitative mathematical theories in the empirical field, it would hardly do justice to the facts of the history of science if the association of the quantitative methods with the advantages now to be discussed, though logically accidental, were disregarded.

The first difference that comes to mind when comparing empirical theories couched in a mathematical language with those formulated in

the vernacular is the extraordinary degree of condensation and conciseness of the former. I do not imply a comparison between mathematical laws of nature and the mentally distorted, awkward, prohibitively
lengthy "paraphrases" which are provided for them in ordinary language,
but refer to the use of mathematics to condense ramified scientific
theories into a few short, clear, unambiguous sentences. The whole
complex science of mechanics can be reduced to the Principle of Least
Action: $\delta \int L dt = 0$ or to similar variational principles. Electro-
magnetism and several other fundamental physical theories can be
equally condensed. Consequently, an accumulation of knowledge which,
in its explicit and developed form, requires between twenty and thirty
sizable volumes, can be literally squeezed into a short list of equations
requiring two or three pages. Never before did man say so much in so
few words. No loss of information is involved, since all the remaining
laws and facts can be derived from this concise presentation by specifiable operations.

Needless to say, such conciseness is not valued because it saves space
but simply because it facilitates the handling of scientific theories. The
role of notational abbreviations in the history of scientific progress is
too familiar to warrant any additional comment. Problems that had
challenged the genius of Apolonius could be solved easily by the use
of Cartesian symbolism; the development of the infinitesimal calculus
by British mathematicians was hampered during an entire century
because Leibnitz's symbolism was shunned. Indeed, the conciseness of
mathematical symbolism is a necessary prerequisite of advanced scientific knowledge for man.

As a matter of fact, the very choice of scientific concepts is often
determined by their anticipated effectiveness in rendering the scientific
laws and theories in which they occur as concise and manageable as
possible. Such concepts cannot be expected to be close to observational
data—and this is the price paid for their effectiveness; electro-magnetic
potentials, entropy, phase spaces are logically remote from sensory,
introspective, and perceptual concepts which are directly connected with
human experience. This gap between scientific concepts and observational data justifies the attempts of investigators such as A. N. Whitehead,[4] B. Russell, and C. D. Broad to bridge it by redefining scientific
concepts in terms of human experience (§35). Such a foundational and
epistemological endeavour can contribute to an understanding of
science, but, as a rule, it does not seem to be required for the scientist's
work.

[4]Cf. E. Nagel, *Sovereign Reason* (1954), sections 10 and 11.

Precision

This advantage is more intimately connected with the use of the language of the mathematics of quantity, though it reaches beyond it as well. Replacing a qualitative statement with a quantitative one generally results in increasing the precision of the former. Thus "X is 20 years of age" is more precise than "X is young"; "X has a temperature of 90 F." is more precise than "X is hot." Now the basic objective of science is providing reliable solutions to relevant problems by formulating statements or theories that pertinently answer these problems and whose truth is guaranteed or nearly guaranteed by available evidence. It is obvious that this basic scientific endeavour will often be frustrated because of the vagueness of the terms occurring in scientific problems and solutions. There is no point in arguing whether a virus is really alive, or in producing any amount of apparently relevant evidence, since "life" is too vague a concept. Nor is it helpful to attempt to decide whether biological evolution was due to the accumulation of accidental "small" changes as long as the vagueness of a "small" change is not removed by appropriate definitions. We shall see, in Part III, that if a statement contains vague terms it will, under certain conditions, be neither true nor false; and if a statement contains terms afflicted with a more restrictive type of vagueness, it will never possess a definite truth-value. Consequently, the scientist's endeavour to solve scientific problems by formulating answers whose truth can be guaranteed by adequate evidence may be frustrated if his answers contain vague terms.

Of course, a scientist who happens to hold a view that is neither true nor false cannot possibly be shown by anyone to be wrong. He plays safe, but purchases this safety at an exorbitant price. As elsewhere in life, the greatest success is obtainable only by running the greatest risk. An investigator who indulges in problems couched in vague terms may be impervious to error; but, by the same token, he has no chance of success. The more vulnerable to an observational test a scientific theory happens to be, the more strength will it derive from a comparison with observational data, should this comparison be favourable. Thus, an astronomical prediction of an eclipse is extremely vulnerable since it would be refuted if the eclipse did not occur at the day, hour, minute, and second that the astronomer predicted. On the other hand, if his vulnerable prediction proves to be true, its value as confirmation of his basic assumptions would increase appreciably. In contrast, the Marxist prediction of an inevitable collapse of the capitalist system with no precise date set for this event is almost invulnerable to observation;

however long the failure of the predicted collapse to materialize continues, there remains always the possibility that a more remote future will bring it about. Yet this virtual invulnerability of the predicted collapse is purchased, once more, at an exorbitant price: the prediction could not possibly confirm the basic assumptions of the whole theory to any appreciable degree. In general, the quantitative nature of a scientific theory makes it extremely vulnerable to an observational test of its consequences. Conversely, the success of the theory in meeting a large number of such tests virtually eliminates any scientific controversy over its correctness.

We must not think that increased reliability is the only advantage obtained by introducing a quantitative, mathematical language, though this is of the utmost cognitive importance because it enables man to come as close as possible to certainty in matters empirical. Apart from being more reliable than their qualitative counterparts in ordinary language, quantitative statements couched in mathematical language convey also a larger amount of information; this is an independent and sometimes startling side-effect. To give a trivial example: if both Mr. X and Mr. Y are said to be "young" we cannot deduce from this bit of information which one is younger. But if we are told that X is twenty and Y is twenty-one, we know not only that they are young, but also the precise relation of their ages. This second advantage is connected with the conciseness of the mathematical language, since the quantitative information happens to say or to imply more than its qualitative counterpart. Hence, in a sense, we come back to an advantage of the mathematical language already discussed. Yet there is a special aspect illustrated in the present case: the conciseness is not due to the formulation of the relevant statement in mathematical terms but to the greater number of consequences derivable from it. Actually, any theory couched in mathematical terms takes advantage of the immensely tighter logical interconnection of mathematical statements as compared with the logical interconnection of statements in ordinary language, that is, the number of deducibility relations obtaining among the latter. The fact that the whole of mathematics is derivable from very few assumptions illustrates this sufficiently. Accordingly, an infinite number of consequences can be deduced from X being twenty and Y being twenty-one, whereas very few inferences can be drawn from both being "young."

Mathematical Language and Correctness of Reasoning

The advantage of mathematical formulations now to be considered is less obvious, though hardly less important. The point is that the very grammar and syntax of this language are so adjusted to expressing the successive phases of inferential operations that these can often be performed almost mechanically, by merely manipulating their linguistic expressions, and can then be checked by a purely visual examination of these expressions. This is why Euler used to say that his pencil was more intelligent than he. What I am now pointing out is to some extent related to the method of constructing formalized axiomatic systems, referred to in §7. Yet the advantage is actually not dependent on the theory in question being formalized, or even axiomatized, and is secured merely by the use of arithmetical and algebraic formulations. A simple example is afforded by the possibility of doing sums or multiplying numbers without having to interpret the actual significance of every step in the operation. In formalized systems this advantage is carried to its maximum, though the complications attending upon complete formalization may offset the gain. But there is no doubt as to the extent to which mathematical formulation facilitates both the carrying out of inferential (logical) operations and the checking of their correctness. Similarly, the replacement of the human computer by a servo-mechanism is entirely based upon the formal parallelism between linguistic structure and logical reasoning.

Mathematical Infinity as a Simplifying Device

It has often been noticed that replacing finite sums by integration, finite differences by differentiation, finite solids by infinite sets of extensionless points leads in scientific practice to considerable simplifications. The geometry of a finite arrangement of small solids is more complicated than the infinitist geometry created by the Greeks. The astronomical theory of planetary systems has been immensely simplified by assuming that they consist of point-like elements. It is true that sometimes this artificial replacement of a finite structure by an infinite and yet more manageable substitute may create new problems, for example, in certain parts of hydrodynamics, where the finite assembly of finite molecules is replaced by a fictitious non-denumerable infinite set of points.

Yet, in spite of such occasional mishaps, which are quite natural in a science that systematically employs a more manageable though fictitious and unobservable infinite in place of an observable finite structure, the success of the infinitist fiction is unquestionable. The reasons for this success are by no means clear and may well vary from case to case. In any event, the job of facilitating the handling of scientific problems by replacing the finite structures they involve by infinite manifolds is entirely to the credit of mathematics. Whether mathematics can be defined as the science of infinity, according to a contemporary suggestion, may be debatable, but there is little doubt that mathematics is the science that has made it profitable to replace the finite with the infinite in virtually every problem.

The infinitist substitutes for empirical problems couched in mathematical language are closely related to the methodological device of replacing actual entities with "idealizations." Geometrical points, curves, and regular solids, physical points, incompressible fluids, electrical surface charges, are a few examples of such "idealizations." They all have an infinitist aspect (the dimensions of a point are said to be "infinitely small," the compressibility of an ideally rigid body "infinitely large," etc.). The value is that while replacing observable objects with infinitist structures, the scientist succeeds in freeing these observable objects from certain complicating features which are a real obstacle in his study, whereas the complications arising out of introducing the infinite are generally concerned with features of the relevant phenomena which play but a minor role in the whole subject-matter. Infinitist idealizations may thus explain to some extent the paradox of simplifying cognitive problems by assuming that they involve fictitious infinite structures.

17   LOGICAL ANALYSIS OF MEASUREMENT

Let us start this attempt at bringing out the main logical features of mensurative methods by some preliminary comments on the crucial concepts involved in measurement. When the height, size, weight, velocity, or duration of a material object is measured, the result is always expressible by the formula "The quantity Q takes on the value Y for the object X," or "The value of the quantity Q for the object X is Y." For instance: "The length in centimetres of X is Y," "The duration in seconds of X is Y," and so on. No explicit reference to the appropriate measuring instrument is contained in such statements, which we shall refer to as *quantitative statements*. In other words, a quantitative statement specifies the value Y of a particular quantity Q for a particular object X. Such a statement can also be interpreted as asserting that a specifiable relation Q obtains between the object X and the number Y. Accordingly, a *quantity* can be viewed as a many-one dyadic relation whose domain and converse domain are made up respectively of individuals and of numbers. The linguistic counterpart of a quantity is therefore a two-place predicate involving an individual and a numerical variable. *Quantitative terms* are predicates of this kind and *quantitative concepts* are the meanings of quantitative terms. From a strictly nominalistic point of view, it would be preferable to identify quantities with quantitative terms, in order to avoid any reference to relations which on this view are fictitious "abstract" entities.

Measurement is a method for solving quantitative problems, i.e., questions which can be answered pertinently and unambiguously by quantitative statements. Consequently, a typical problem solvable by mensurative methods reads: "What is the value of Q for X?" A measurement carried out to solve such a problem is said to be a measurement of Q, performed on X and yielding the result Y.

These tentative explanations of the basic concepts involved in measurement are sufficient to bring out the main features common to measurement and to direct fact-finding methods and also the most outstanding distinctive features of measurement:

(1) Thus, it is obvious that measurement is a fact-finding method, sharing the common characteristics of those fact-finding procedures previously discussed. The object of measuring a quantity Q is to ascertain its value for a particular entity X, i.e., to validate the fact-like statement which specifies the numerical value taken on by the quantity Q for the entity X. In other words, a measurement aims at validating a fact-like statement about a particular relation between a particular individual and a particular number, not at the validation of a law or a theory.

(2) Measurement is a method both of verification and of discovery. In other words, an investigator who carries out the measurement of a particular quantity on a particular object is in a position to indicate the value of this quantity for this object and to produce evidence in support of his evaluation. Thus the method serves both to find out the answer to any relevant quantitative problem and to substantiate the correctness of such an answer. Needless to say, this important advantage places measurement among those methods that are mainly responsible for the continuous and rational advance of scientific knowledge. Verificatory methods are merely the counterpart, within science, of the trial and error procedure. Other things being equal, verificatory procedures cannot compare with methods of discovery in respect to efficiency.

(3) We may add, of course, that observation (sensory, perceptual, and introspective) and memory play almost as decisive a role in measurement as in direct fact-finding methods. One has to observe the object and the instrument and to remember the initial state of the instrument in order to perform a measurement.[5] Accordingly, the range

of objects accessible to quantitative methods includes the four ostensibly distinct ranges of present material objects, present sense-data, and present introspective data, and things that we remember having sensed, perceived, or introspected. Other types of measurable objects which transcend the range of direct fact-finding methods will be discussed in the next section, in connection with indirect observation.

Apart from the common features of mensurative and direct fact-finding methods we have just listed, there are a few extremely important respects in which the two groups of methods differ from each other. It is obvious that these distinctive features of measurement are likely to shed some light on its efficacy. We shall concentrate in the following analysis on three such features.

(1) In contrast to direct methods, mensurative procedures involve always instrumental equipment, which includes at least measuring instruments. One needs a yardstick to measure length, a clock to measure time-intervals, a balance to measure weights, and so on. In some cases, the use of instrumental equipment is coupled with a theory underlying the method. The theory deals with ways of handling the equipment, determines the reliability of the method, and describes how the probability of successfully using the method may be increased by special techniques which prescribe suitable circumstances under which to apply it, in so far as they are not uniquely determined by the definitional statement of the method.

(2) The difference between the quantitative concepts involved in all quantitative problems and their solutions, and the concepts entering problems solvable by direct methods, is profound and far-reaching. Consider ostensive concepts which occur in problems solvable by perceptual observation, say, "being green," or "being round," as compared with quantitative concepts such as "being three yards long" or "weighing 3 lbs." The employment of the former is part and parcel of any correct use of the language, although such handling is neither taught nor controlled by definitions. Ostensive terms are, in this respect, analogues of undefined terms in axiomatic systems. Quantitative concepts, on the other hand, admit and require a definition in ostensive terms. They are, therefore, in this respect epistemologically dependent upon ostensive terms, although the definitions that serve to introduce quantities in terms of ostensive expressions differ in many respects from the usual types of definitions.

⁵This is perhaps the reason for the frequent application of the concept of observation to measurement. In view of the important differences between the two groups of methods, this confusing terminology will be avoided here.

(3) In contradistinction to direct methods, mensurative procedures always involve inference. I can make out by mere observation whether the two ends of a particular object *A* coincide with the two ends of a yardstick. But I cannot *see* that this object *A* has unit-length. To substantiate this quantitative statement, I have first to observe the coincidence of object and yardstick, next to formulate the result of my observations in appropriate "ostensive statements,"[6] and then to *infer* from these fact-like statements the unit-length of the object. The role of inference is indispensable; it will turn out to be due primarily to the presence of quantitative terms in problems solvable by measurement, and to depend also to some extent upon the use of instruments.

The premisses from which the quantitative statement is inferred include, in the above example of measuring length with a yardstick, the observation-statements asserting that the yardstick is superposed on the object and that their ends coincide. But it is obvious that the quantitative statement ascribing unit-length to the object cannot be deductively derived from such observation-statements, for the simple reason that the expression "unit-length" does not occur in these observational premisses. An additional premiss is needed to the effect that any object whose ends coincide with the ends of a yardstick has unit-length. The quantitative statement follows logically from the above observational premisses supplemented by this additional premiss. We shall see, in the second part of this section, that this additional premiss is actually a definition of unit-length.

Consider more generally any quantitative statement inferable from a set of premisses which consists of a few observation-statements about the reading of an instrument corresponding to the quantity *Q* and of a definition of this quantity. Such measurements, called *direct* or *fundamental*, are characterized by the physical fact that the instrument associated with the relevant quantity is attached to the object under consideration and provides a definite reading, and by the logical fact that the validation of the quantitative statement does not require any non-observational premisses apart from the definition of the relevant quantity (the remaining premisses involved in the validation being all observation-statements).

A measurement is said to be *indirect*, if the validated quantitative statement is inferred from a set of premisses which is made up not only of observation-statements about the object measured and the measuring instrument and the definition of the relevant quantity, but also includes

[6]Statements whose terms are ostensive (cf. § 14) are often referred to as "ostensive statements" or as "observation statements."

premisses of a different kind. In the most important cases, an indirect measurement[7] provides for the validation of a quantitative statement by inferring it from premisses that include the results of some previous measurements in addition to observation-statements and the definition of the relevant quantity. A further type of premiss—a statement belonging to a theory and not of the nature admissible in direct or indirect measurement—may also be employed; in this case, the validation of quantitative statements is referred to as a *computation* rather than as an indirect measurement.

The boundary-line between direct, indirect, and computational measurement is not always easy to draw. Its contour will be clarified once the "semi-definitional" function of empirical laws and theories is determined (§§ 20, 25). The above example of measuring the length of an object by superposing a yardstick on it illustrates, of course, a type of direct measurement. All the observation-statements required in this case to validate the relevant quantitative statement refer to the mark on the yardstick with which the end of the object coincides. The only indispensable additional premiss is a definition of length in terms of coincidence and superposition. On the other hand, the measurement of the area of a rectangle is indirect, if it is carried out by first measuring directly the respective lengths of two adjacent sides of the rectangle and then multiplying the results. The diameter of the earth is neither directly nor indirectly measured, but rather computed on the basis of several geodesical observations and theories. I take it here that the unit of length is no longer referred to a terrestrial meridian, but rather conventionally identified with the distance between two ostensively defined scratches on an ostensively defined standard at Sèvres; accordingly, the dimensions of the earth are estimated on the basis of geophysical observations, rather than deduced from the original definition of the standard of length.

Most of the epoch-making measurements reported in the history of science are computations: the determination of the probable size, density, and age of the material universe accessible to scientific observa-

---

[7]Even a direct measurement is, of course, an "indirect validating method" as defined in Part I, chapter Two. The departure from direct methods of validation is, however, rather slight in direct measurement, since such a measurement involves, among its premisses, only the definition of the relevant quantity, in addition to directly validated statements about the observed object and the measuring instrument. This departure becomes more pronounced in indirect measurement, and particularly in computational measurement. In any event, there is no reason for confusing the direct (i.e., non-inferential) type of validation in the general classification of validating methods with a measurement which is indirect in the sense of presupposing results of some other measurements.

tion; the measurements of the speed of light, of the size of atoms and of molecules, of the quantum of action, and of the charges and masses of sub-atomic particles. The same remark applies naturally to Plato's quantitative contention that a just ruler is about 729 times happier than a tyrant.

The preliminary explanations in the first part of this section have brought out the distinctive features and principal types of measurement. We have seen that mensurative methods differ from direct fact-finding methods in three respects: (1) New kinds of ("quantitative") concepts, which do not appear in any problem or solution pertaining to direct methods, occur characteristically in problems and answers proper to measurement. (2) Inferential operations, banned, by definition, from direct validation, play an essential part in measurement. (3) The use of instrumental equipment, unnecessary in principle in direct validation, is part and parcel of any and every measurement. In addition to a preliminary clarification of the above three distinctive features of mensurative methods, we have characterized three main types of mensurative procedures—Direct Measurement, Indirect Measurement, and Computational Measurement. These are differentiated by the kind of premisses required for the validation of the relevant quantitative statement.

This provisional treatment of characteristics and types of measurement will now be supplemented by a few more rigorous considerations which may shed some light on those features of quantitative methods which affect our central problem. The objective is to determine the range of mensurative methods and also to account for their extraordinary efficacy. In order to make some headway towards this goal, we might start with a more rigorous treatment of any of the three distinctive features of measurement: nature of quantitative concepts, role of measuring instruments, and role of inferential operations in mensurative validation. Yet, for pragmatic reasons, it will be best to concentrate on the nature of quantitative concepts almost exclusively, because, as a matter of fact, this will enable us to explain the indispensable role of inference and will clarify, to some extent, the role of measuring instruments as well. I do not imply that the logical devices used by the empirical scientist in creating his quantitative concepts are exclusively responsible for the efficiency of measurement; we have already considered certain reasons for such efficiency quite unrelated to the logical structure of measurement. And we shall find that other reasons, independent of the nature of quantitative concepts, are to be sought for in the field of scientific technology and in the empirical circumstances attending upon the application of quantitative methods in science (§18).

We may anticipate a clarification, rather than an explanation, of the efficacy of measurement when we gain some insight into the nature of quantitative concepts.

Before embarking upon this more rigorous analysis of quantitative concepts, let us notice first that the above provisional explanation of this class of concepts does not amount to a precise definition. One cannot, by definition, identify a quantity with a many-one relation between physical objects and numbers, or define a quantitative predicate by requiring that it involve one individual and one numerical variable. "The pointer on the scale of the instrument X reads Y" is a two-place predicate which meets this requirement, without being a quantitative term. The nature of quantitative terms is not constituted by the number and nature of their arguments. The essential feature is the way in which quantitative terms are directly or indirectly linked by definitional contexts with terms of the ostensive level, without being themselves ostensive. Let us first illustrate the definitional technique involved in the use of quantitative terms by considering an extremely simple quantitative concept, namely, the concept of unit-length.

The term "being one yard long" is not ostensive; we cannot *see* whether a given object perceived by us has such a length. This means that our familiarity with the quantitative term does not consist in our being conditioned to associate a feeling of certainty with the statement "This is one yard long" whenever we check it and point to an object of this length. Yet, while learning the correct use of this adjectival expression, we are taught to apply a yardstick to the object under consideration and to believe or disbelieve the statement, depending upon whether or not each end of the object happens to coincide with the respective end of the yardstick.

The situation is therefore as follows: "The yardstick is superposed on the object A" and "The ends of the object and of the yardstick coincide respectively" are ostensive statements. From these two ostensive statements, and the definitional statement that every object the ends of which coincide with the respective ends of a yardstick is one yard long, it follows logically that the object A is one yard long. The meaning of the term "being one yard long" is obviously fixed to some extent by this definitional context, since, in order to be consistent with this context, the adjective has to be interpreted as applying to every object which is found to coincide with this standard of length.

Similarly, the proposition "B is not one yard long" follows logically from the two ostensive statements, "The yardstick is superposed on B"

and "The ends of the yardstick do not coincide with the ends of B," together with the definitional statement "No object is a yard long if a yardstick is superposed on it without their ends coinciding." Compatibility with this definitional context also characterizes to some extent the meaning of the quantitative term in question, since "being one yard long" must not be applied to any object on which a yardstick is superposed without coincidence ensuing. The two definitional contexts of "unit-length" thus outlined provide a definition of this quantitative concept in terms of the ostensive concepts of superposition (on a yardstick) and of coincidence.

The type of definitional connection with ostensive terms which serves to introduce quantitative concepts can be obtained from this particular example by simply replacing the quantitative concept of unit-length by any arbitrary non-ostensive concept P and by substituting for the two ostensive concepts of superposition and coincidence two arbitrary concepts which we may call the test T and the response R respectively. We have to assume that the test T and the response R either possess a primitive empirical meaning (in the sense of §15) or have been previously defined. The general definitional formula exemplified by the quantitative concept of unit-length can then be put as follows: "By definition, all individuals submitted to the test T with the result R have the property P, whereas no individual submitted to the test T with a different result has the property P" (symbolically: $T(X) \supset [R(X) =_{df} P(X)]$). Here, the property P is the defined concept, whereas test T and response R constitute the defining expression.

There is, so far, no reference to any quantity in the above definitional formula. In these cases, however, where the defined property happens to be a quantity Q measurable by means of an instrument Y, we can very simply specialize the above formula if we make the additional assumption that the instrument Y is gauged directly in accordance with the quantity Q so that the value of the latter will coincide with the reading of the former. We shall then have to say that if the instrument Y has been properly applied to the object X, then, by definition, the value of Q for X is Z if and only if the reading of the instrument Y is Z (symbolically: $T(X, Y) \supset [R(Y, Z) =_{df} Q(X, Z)]$).

The assumption that the measuring instrument Y is gauged directly in accordance with the measured quantity Q is by no means essential. If the reading Z on the scale of Y is connected by an arithmetical function f with the value of Q, instead of being equal to this value, then the above definition becomes simply: "If the instrument Y has been properly applied to the object X, then, by definition, the value of Q for X is $f(Z)$ if and only if the reading of Y is Z." Symbolically: $T(X, Y) \supset [R(X, Z) =_{df} Q(X, f(Z))]$.

The type of definitional context just outlined is used to introduce quantities but is not confined to this task. As a matter of fact, it plays a crucial part in the formation of empirical concepts, both quantitative and qualitative, in the whole of science, and represents the most important definitional technique of empirical science. This was realized and investigated for the first time by Carnap.[8] P. W. Bridgman has covered much of the same ground in his theory of "operational definitions" originally intended to account for the principles of concept-formation within the science of physics and expounded in a non-technical fashion.[9] Sigwart seems to have had an inkling of this definitional technique in the passage of his *Logik* where he refers to what he terms "diagnostic" definitions. But his comments are extremely unsatisfactory and his ideas cannot be considered as a forerunner of Carnap's in this field.[10]

In this study, any definitional context illustrated by the above examples will be called a *conditional definition* or a *definitional criterion* of the new term P in terms of the test T and the response R. The adjective "conditional" is inserted to stress that the definitional equality between the response R and the defined concept P holds only under the condition that the test T has actually been applied. In classical definitions, which will also be referred to as "full-fledged definitions" or as "definitions of the equivalence-type," the equality between the defined term and the defining expression is not conceived of as subject to any empirical condition, which may or may not materialize in any individual case. If a "square" is classically defined as being a rectangle with equal sides, then it is out of the question that any conceivable event may be relevant to a square being such a rectangle. The difference between classical and conditional definitions will prove to have far-reaching consequences for the methodological outlook of the whole of science.

The study of conditional definitions in their bearing upon our central problem will be resumed in this investigation from various points of view, particularly in connection with the "semi-definitional function" of empirical laws and theories and with the need to explore the applicability of conditional definitions of a probabilistic type.

[8]Cf. R. Carnap, "Testability and Meaning," *Philosophy of Science*, vol. 3 (1936).
[9]Cf. *The Logic of Modern Physics* (1927).
[10]Cf. Chr. Sigwart, *Logik* (1924).

Let us turn now to some theoretical advantages of conceiving the link between quantitative concepts and those of the ostensive level as determined either by single conditional definitions or by chains of such definitions. I think that an approach to quantitative methods in terms of conditionally defined quantitative concepts is likely to shed some light on the following important and rather obscure questions: (*a*) How is the indispensable role of inference in quantitative fact-finding methods to be accounted for? (*b*) What part does the instrumental equipment play in measurement? (*c*) How is the inescapable "inaccuracy," or "uncertainty," or "imprecision" of any and every measurement to be accounted for?

(*a*) First, then, the indispensable role of inference in the validation of quantitative statements. We may notice, to begin with, that no difficulty arises in explaining why compound and quantified statements cannot be validated without recourse to inference (cf. §15). I can see that this pen is now on this desk, but I cannot possibly see that this pen is either on the desk or in the drawer. The first statement is "atomic" (without logical constants) and susceptible to direct validation. The second is "molecular" or "compound," since it involves the logical connective "either–or," the presence of which is related in a straightforward and well-understood way to the impossibility of validating the statement without using inference. The whole meaning of logical constants (sentential connectives and quantifiers) is determined by the rules of inference applicable to sentential contexts of such constants. In other words, a person attaches the right meaning to logical constants if, and only if, he is prepared to apply to the sentences in which these constants occur all the rules of inference associated with the constants and characteristic of their correct use and their proper meaning. The interconnection between the meaning of logical constants and the rules of inference governing their sentential contexts has often been asserted, but its definitive clarification is probably traceable to Gentzen's investigations of what he terms "natural inference."[11] Thus, attaching the right meaning to (or using correctly) the logical connective "either–or" consists in being prepared to assert any statement of the form *Q*, whenever "either *P* or *Q*" is deemed true and *P* false. In other words, the meaning attached in ordinary language to logical terms such as the sentential connectives and the quantifiers can be entirely ascertained by finding out what kinds of inferences their correct use requires.

[11]Cf. G. Gentzen, "Untersuchungen über das logische Schliessen," *Mathematische Zeitschrift*, vol. 39 (1934–5).

Atomic sentences are those in which logical constants do not occur: hence, as far as pure logic (i.e., the sum-total of rules which govern the correct use of logical expressions) is concerned, there is no reason why atomic sentences should be governed by any rules of inference. Yet quantitative atomic sentences *are* governed by rules of inference and indeed cannot be validated without resorting to inference. Since logical constants, which could account for the impossibility of validating them without recourse to inference, do not occur in such statements, another reason must be sought. Now the fact that inference can be dispensed with in validating atomic sentences the terms of which are *ostensive* (or otherwise endowed with a primitive empirical meaning) suggests that the only conceivable reason why inferential validation of quantitative atomic statements is indispensable, must relate to *quantitative terms*, which have no primitive empirical meaning and occur characteristically in every quantitative statement. This assumption is confirmed by the fact that the only additional premiss of a non-ostensive type which is required in order to deduce quantitative statements from purely observational premisses is the definition of the quantitative term in question. We may take it, therefore, that the reason for the impossibility of validating a quantitative statement without resorting to inference comes from the presence of quantitative concepts in such a statement.

(*b*) With regard to the role of measuring instruments in mensurative validation, it should be noticed first that the very form of conditional definitions which serve to introduce quantitative concepts provides for such instruments as essential ingredients of any measurement. Conditional definitions of quantitative concepts purport to explain, in terms of a test T and a response R, the meaning of a quantity Q taking on a value Z for an individual X. The special form of conditional definitions of quantitative concepts is significant: the test T is always relational in such cases, "T" denoting the relation ordinarily referred to by saying that the individual Y, called the measuring instrument, has been properly applied to the individual X, called "object of the measurement." Moreover, when quantitative concepts are defined, the response R is denoted by a two-place predicate, which involves an individual and a numerical variable and can therefore be interpreted as the "reading" of an instrument. We thus see that the syntactical nature of test T and response R in conditional definitions of quantitative terms discloses the presence and the role of the measuring instrument.

But there is more to this definition than a formally correct adjustment. We have seen that, in deriving the quantitative statement from the observational premisses concerning the object observed and the measuring instrument, the definition is indispensable as an additional premiss, because no quantitative statement is correctly inferable from purely observational premisses. This role of the conditional definition is more than formal, as can be seen from the following considerations.

If we confined ourselves to describing the observed object, the measuring instrument, and their mutual interrelation in purely ostensive terms, we could dispense with the conditional definition. For example, we could validate by mere observation that the yardstick is superposed on the object and that their ends coincide. In other words, the observation-statements which we would have to validate in order to report on the joint behaviour of the two material objects involved, the observed object and the instrument, do not constitute the whole story. A new aspect of the situation, a new way of looking at the interaction of two material objects is originated by the very act of thought, which creates a new non-ostensive concept, one only conditionally definable in ostensive terms. The possibility of asking new questions about the two material objects therefore arises by the introduction of a quantitative concept. Thus, one of these objects can legitimately be given the name "measuring instrument" because of the conditional definition of a quantity in which the name is involved in a specified way. Accordingly, the nature of quantitative concepts defined by conditional definitions clarifies to some extent the question of how a material object comes to assume the role of a measuring instrument. The important point is the novelty of the questions which can be raised following the introduction of quantitative concepts and which can be answered by using suitable measuring instruments; such questions and their solutions are novel in this precise sense that they are demonstrably untranslatable into questions and solutions in which only the defining expressions occur. It is one of the basic properties of conditional definitions that, in contradistinction to full-fledged definitions of the equivalence type, they do not provide for the possibility of eliminating the terms defined, and replacing them with the corresponding defining expressions (cf. § 24).

(c) Finally, a brief comment should be made on the inescapable inaccuracy of quantitative methods. The problem is studded with technicalities, but has, nonetheless, a definite bearing upon the cognitive reach of science. We shall see, for example, that the boundary line between the scientifically knowable and unknowable (or, alternatively, the tripartite classification of scientific statements referred to in § 3) depends to a large extent upon the inherent inaccuracy of measurement.

One basic remark has to be made concerning the inaccuracy of direct or fundamental measurement. All instruments used in such measurements have a "least count" since their scales are subdivided into smallest units. A measurement of a quantity Q on an object X by means of such an instrument amounts in principle to ascertaining whether the value of Q for X is located between the limits of any particular smallest division of the scale. (I am disregarding here the special techniques which may slightly increase the accuracy of measurement beyond the least count of the scale of the instrument.) Thus, with a measuring rod subdivided into centimetres, the length of an object can be ascertained as lying between N and $N + 1$ centimetres, N being a natural number. Similarly, fundamental time measurement is made in seconds or minutes or hours and the result is always an inequality for the value concerned. The determination of mass by weighing is of the same nature; submultiples of the least count, that is, of the smallest unit available, are simply disregarded.

The epistemological implications of this ostensibly technical fact, that every instrument of fundamental measurement has a least count, will prove far-reaching. The basic theories of contemporary science concentrate on describing, summarizing, predicting, and explaining the results of measurements, and, ultimately, of fundamental measurements. However, these theories express the results of measurements by arithmetical *equalities* in spite of the fact that only arithmetical inequalities provide a straightforward account of the actual results of measurements: instead of stating that the value of the measured quantity is located between N and $N + 1$ least counts of the measuring instrument, the scientist asserts that this value is equal to a fractional or irrational number contained in the interval between N and $N + 1$. How does this replacement of an inequality by an immensely more informative equality, which is neither justified nor justifiable by the use of an instrument involving a least count, come about? Mathematically speaking, the equality is obtained by selecting, out of the non-denumerably infinite set of values within the interval corresponding to the actual result of measurement, a single value to be attributed to the object measured. Yet no unwarranted interpretation of the measurement, no unobtainable precision of the value provided by the mensurative procedure employed, is thereby suggested. The implication is simply that the measurement whose result is stated as settling the value of the quantity in a unique way, is to be qualified with an "inherent inaccuracy," just equal to the least count of the instrument. In other words, instead of the report, for example, that a measurement of the length of an object by means of a rule subdivided into centimetres has yielded the result that this length is between N and $N + 1$ centimetres, we are told that the length is actually N plus ξ cm (where ξ is a real number of the interval

$0 < \xi < 1$, chosen according to certain rules), on the understanding, however, that the whole measurement involves an "inherent inaccuracy" of 1 cm.

The reasons for this roundabout way of stating the actual result of the measurement are numerous, some of them based simply on expediency, some deeply rooted. It goes without saying that equations between numerical or algebraic expressions are considerably easier to handle than inequalities. More important, however, is the fact that a straight-forward account of mensurative results in terms of inequalities would deprive the scientist of the benefit of utilizing the extremely powerful mathematical techniques which comprise the bulk of arithmetic, of algebra, of the infinitesimal calculus, with all their numerous theoretical superstructures (differential equations, calculus of variations, analysis involving infinitely numerous variables, etc.). As a rule, statements occurring in such practically indispensable mathematical theories, clas-sical or modern, intuitive or abstruse, are equations, not inequalities. This may account for the results of measurements being squeezed by the scientist into the pattern of equations.

We must not confuse the inherent inaccuracy of measurement due to the existence of a least count with "systematic corrections" of the mea-sured value conditional upon controllable external circumstances, nor yet with statistical ("random") errors caused by uncontrollable external circumstances. Thus, if the successive measurements of a length which we have good reason to presume constant do not consistently yield the same number of smallest basic units, these discrepancies can often be accounted for by ascertaining, for example, a controllable "systematic" change of temperature in either the object measured, or the instrument, or both. On the other hand, when no such circumstances, external in regard to both the object measured and the measuring instrument, are available to account for the failure of successive measurements con-sistently to yield the same anticipated result, that failure can often be explained by assuming hypothetical random changes in external cir-cumstances; the value of the quantity measured can then be computed by statistical methods, for instance by applying Gauss's Theory of Errors. In each particular case, however, regardless of whether "sys-tematic" or "random" errors have to be corrected, and of whether the value of the quantity measured is known only with an inaccuracy com-patible with these two kinds of external interference with the mensurative process, only a whole number of centimetres—if they constitute the least count—is available and the statistical theory of errors and similar con-siderations aiming at the "reduction of observational data" cannot change anything about the inherent inaccuracy, which is due to the structure of the instrument, not to external circumstances.

We shall therefore have to distinguish carefully the intrinsic inaccuracy of measurement from the accidental errors due to uncontrollable circumstances and the corrections imposed by known external circumstances. This intrinsic inaccuracy is embodied in the very definition of the mensurative operations, and therefore the necessity to compensate for the replacement of the inequalities actually disclosed by the measurement with more manageable equalities has nothing in common with compensation for circumstances external to the object measured and the measuring instrument, whether these circumstances be accidental or not.

Some comments may be added on types of measurement that are not direct before we try to characterize the joint range of all mensural methods. A quantitative statement is validated by direct measurement if it is inferred from a set of premises which consist of observation-statements describing the object measured and the measuring instruments in conjunction with a definitional premiss which explains the meaning of the crucial quantitative concept occurring in this validated statement. The situation is not basically different in the event of an indirect or even of a computational measurement. Thus, when the area of a rectangle is measured indirectly, the proposition specifying the size of the area is deduced from two quantitative statements which specify the length of two adjacent sides, in conjunction with the definition of the area as a product of the length of the sides and the arithmetical premiss stating the value of the product. Thus, instead of one definition, two definitions of quantitative terms are required, and, in addition, an arithmetical premiss. But, basically, the quantitative statement involving indirect measurement is still validated by being shown to follow logically from premises which are either observation-statements about the object measured or the measuring instruments, or definitions of relevant quantitative terms; the arithmetical premiss is redundant, at bottom, in view of the reducibility of arithmetic to logic.

Basically, the same situation prevails in a computational measurement, when a theory or several theories are involved. For, as we shall try to show in some detail later on, the only way in which a theory can contribute to a determination of the referential meaning of a term is to provide a conditional definition for this term. Thus, if the value of a quantity Q for an individual X can be computed by utilizing a theory T, the actual function of T is really to provide a conditional definition of Q applicable to X. Hence, the set of premises which ensure the validation of the statement specifying the value of Q for X consists in this case, too, of observation-statements supplemented by conditional definitions.

It should be added that an indirect measurement is often construed as applying to a quantity which can be measured directly as well. Thus, the length of a side of a triangle would be said to be indirectly measured if it were computed from measurements of some other elements of this triangle, for example, another side and two adjacent angles. The situation is basically the same for both types of indirect measurement because obviously a law connecting the length of the side under consideration with the remaining elements of the triangle is being used; since the service rendered by the law is that of providing a conditional definition (§ 20), the premises which serve to validate the statement specifying the length of an indirectly measured side of a triangle consist still of observation-statements about the object measured and the measuring instruments in conjunction with conditional definitions of relevant quantitative terms.

The fact that the solution of a quantitative problem, that is, the validation of a quantitative statement pertinently answering this problem, consists of observational and inferential operations, is somehow obscured by the conventional way of interpreting the results of measurement in terms of equalities. The value of a continuous quantity, say length, may be asserted to be exactly equal to a real number in the statement reporting on the result of measuring this quantity; but this statement cannot be deduced from the relevant observational and definitional premises because of the inherent inaccuracy of such measurement. Only inequalities can be legitimately inferred from the mensural data. This may cast some doubt upon the range of quantitative methods and upon the possibility of securing by observation and inference the quantitative facts asserted by the scientist. We have seen, however, that this difficulty arises only owing to an incomplete formulation of the mensurative result. If both the value measured and the inherent inaccuracy are stated, the claimed equality is actually the corresponding inequality, and the possibility of validating the equality by a suitable interplay of observation and inference becomes apparent.

To sum up: the range of quantitative methods includes all quantitative problems, that is, questions as to the value of any quantity Q for any object X—on the understanding that quantitative concepts are linked up with the ostensive level by conditional definitions of a specifiable form and that all the operations involved in solving quantitative problems are observational, memorative, or inferential. There are several important aspects of the range of quantitative methods which will have to be discussed in due time. One point, however, can be made at this juncture: quantitative concepts are predicates, not subjects. The only subjects involved in quantitative methods are supplied by direct fact-finding methods. Thus, the universe of discourse of science is not extended by measurement. Only new (viz., quantitative) properties of directly observable individuals are disclosed by measurement. These individuals are accessible to scientific method through direct, qualitative, fact-finding procedures.

18   INDIRECT OBSERVATION

Another fact-finding procedure of fundamental importance is what may be termed *indirect observation*. I have particularly in mind the use of optical instruments such as the microscope and the telescope, enabling the scientist to observe indirectly very small or very remote objects which would otherwise remain inaccessible to him. Of course, instruments are used in measurement as well as in indirect observation. But their role is obviously not the same in both cases. When measurement is carried out, the statement validated by means of the relevant instrument is quantitative. In the case of indirect observation the validated statement is, as a rule, qualitative, for example, that the object observed through the microscope is a microbe of a particular genus. There may be, however, indirect observations in which quantitative statements are validated, for example, when the size of a microbe is measured. This shows simply that these two methods of indirectly validating fact-like statements may sometimes overlap each other, without ceasing to be distinct. When measurement and indirect observation are combined, different contrivances are generally used for validating respectively the qualitative and quantitative statements involved.

It is well known to what an extraordinary extent indirect observation has contributed to broaden man's primitive, pre-scientific environment. Microbes are now indirectly observable through ordinary optical microscopes, molecules and some species of atoms through electronic microscopes, sub-atomic particles (theoretically) through gamma-ray microscopes and similar devices, stars billions of light years away from the earth through telescopes with a diameter which has kept increasing in recent decades. Astronomy, atomic physics, microbiology, and perhaps the better part of biology would not have made the impressive headway they actually have, but for the incessant discovery and improvement of instruments of indirect observation. The numerous devices which enable the human observer to "watch indirectly" single, individual entities and events at the molecular, the atomic, and the sub-atomic levels, are a case in point, and certainly an important one.

Yet the scope of indirect observation is rather controversial. For example, some investigators hesitate to assert that infra-red and ultra-violet light, electrical charge and potential, very low and very high temperatures, have become accessible to indirect observation, even if they have no objection to granting the status of indirect observability to molecules, atoms, and remote stars. Others feel that an atom cannot be said to be "observed" through an electronic microscope, nor can an electron by watching a vapour trail in a cloud-chamber, even if the observation be qualified as "indirect"; the use of the term "observation" in such cases rests, according to these investigators, on a simple ambiguity. There is thus a good deal of discrepancy between the ways in which indirect observational knowledge is construed.

This discrepancy is not merely a matter of terminology. After all, one of the basic philosophical and methodological tendencies of modern science is to confine its investigations to what could be observed. This fundamental empiricist outlook depends obviously upon how "observation" and "observability" are construed, and, more specifically, upon the status of indirect observability. To put it otherwise: according to the modern empiricist outlook, the universe of discourse of the language of science consists of observable individuals. Granted that directly observable individuals, those which are either sensible, or perceivable, or introspectable, do belong to this universe of discourse, what about the status of "scientific objects" which are only indirectly observable?

We thus realize that the universe of discourse of the language of science is involved in the concept of "indirect observation." How should we define this concept in order to remain as close as possible to actual usage and to separate a significant class of scientific procedures? I think that three basic tendencies can be distinguished among the interpretations put on this phrase.

(1) Sometimes the radical view is taken that, literally speaking, only sense-data or sensory appearances of physical objects are "observed" in the genuine sense of the word, whereas all physical entities located above this primitive level involve an inference and should therefore be considered as inferred, rather than as (indirectly) observed objects. Those who hold this view would simply deny that we observe the physical entities inferred in various situations to which the concept of indirect, or even of perceptual observation is usually applied. We do not actually observe stars, since the shining points we see on such occasions cannot be sensibly identified with the relevant celestial bodies, though these points may well be causally related through the agency of the human observer to the stars proper. Similarly, what we actually see

through a microscope is not a microbe, but, once more, an optical sense-datum or a sensory appearance; we may have excellent reasons to account for this appearance by assuming some hypothetical microbe causally interrelated with the human observer, although this causal inter-relation does not entitle us to speak meaningfully of indirectly observing the microbe itself. What we see in a cloud-chamber is a large-scale vapour trail, capable of providing us with some clues about the particles which have crossed the vessel. The particles themselves are neither directly nor indirectly observable, and it therefore makes no sense to claim that we observe indirectly an electron, in spite of the good reasons we have for inferring that the electron has contributed to the formation of the vapour trail in the vessel. Similar views are taken in respect to other directly observable sensory phenomena, usually interpreted as providing for an indirect observation of astronomical, microphysical, or biological entities. Extending the term "observation," even qualified as "indirect," to all such entities is considered an arbitrary and unwarranted change of the meaning usually attached to the term. Accordingly, even the observation of what is usually referred to as a material thing perceivable with the naked eye is objected to as involving an ambiguous use of "observation"; we observe coloured two-dimensional patches, not solid, three-dimensional things, although we may have good reasons for inferring such unobservable things out of the actually observed sensory appearances (for example, a thing called "straight stick immersed in water" out of an observed dark, bent, elongated patch, or the thing called "sun" out of an observed yellow disc against a blue background).

(2) Another tendency is to confine indirect observation to entities of a *thing-like nature* about which qualitative or quantitative statements can be validated by recourse to appropriate observational instruments. According to this terminology, microbes, molecules, atoms, and elementary particles either are or could be indirectly observed, along with distant stars, constellations, and galaxies. In contrast, very low and very high temperatures, infra-red and ultra-violet vibrations, are beyond the reach of indirect observation. The reason for thus discriminating against the second group of physical entities, as far as the status of indirect observability is concerned, is the fact that these physical entities have the nature of properties, relations, processes, etc., but not of things, in contradistinction to the entities of the first group. A microbe or a galaxy is thought of as being a material thing, differing mainly by its size from the ordinary material things of daily life. On the other hand, an ultra-violet vibration is not conceived of as a thing, but rather as a property, somehow analogous to a visible colour.

(3) A fact-like statement can be said to be validated by indirect observation, according to the third view, if it is qualitative (not quantitative), and the only premisses used in deriving the statement are either observation-statements about the instruments made use of, or reports on sensory appearances attending upon their use, or definitions of terms which occur in the validated statement and do not possess a primitive, empirical meaning (such as "microbe," "atom," etc.). In this interpretation, the crucial difference between indirect observation and measurement depends on the nature of the validated statement: it may be qualitative or quantitative, falling respectively within the range of indirect observation or measurement. An equivalent criterion is supplied, on this view, by the nature of the crucial concepts occurring in the validated statement, which may still be either qualitative or quantitative and entail the same classification of the statement. Thus, a microbe is itself indirectly observed through a microscope, on this view; its size is measured, but not observed.

In an assessment of position no. 1 in regard to the scope of observation, it may just be pointed out that we have already twice (§ 15) committed ourselves to rejecting the restriction of this fundamental term to sensory and introspective observation. In the first place, we have tried to show that the frequent doubts which are often voiced in connection with the observational status of sense perception come mainly from the double confusion of identifying the degree of reliability of perceptual validation with its inferential, or non-inferential nature, and of failing to differentiate object-language questions, which are well answerable by perceptual validation without resorting to inference, from related meta-linguistic questions which are not susceptible to direct validation. Secondly, we have stressed the direct, non-inferential nature of memorative validation.

Thus, having already twice rejected the basic tendency of the first position, which grants the monopoly of direct validation to sensory and introspective validation (and is thus in a way somehow reminiscent of eighteenth-century sensationalism), we do not have any reason to admit that the concept of observability should be confined to sensory and introspective data, and withheld from objects which either are directly perceivable or could well be perceived, provided suitable observational instruments, theoretically possible in accordance with well-established laws of nature, could actually be constructed.

Position no. 2, which restricts the status of a genuine physical entity, or, to put it otherwise, membership in the universe of discourse of the language of science, to entities which are conspicuously of a thing-like nature, while relegating properties, relations, and similar "abstract" entities to the realm of linguistic fictions, is obviously a variety of nominalism. Now if, for example, colours or distances between observable objects are viewed as directly observable, there is hardly any purely logical reason for denying (indirect) observability, say to microscopic or sub-microscopic sizes, shapes, and distances. It hardly makes sense to assume that the nominalist ban on abstract entities, such as properties and relations, should depend upon the spatial dimensions of the objects to which these properties and relations are ascribed. If properties and relations among large-scale bodies are epistemologically respectable, those attributed to bodies of atomic or sub-atomic dimensions by well-established scientific theories cannot be treated with Occams razor without promoting size to an undeserved ontological rank. For this reason the tendencies implicit in the second position will be disregarded in this context, in spite of an eagerness to present the substance of our whole argument without presupposing any definite solution to the nominalist-Platonist controversy.

An evaluation of the third position, which puts measurement and indirect observation on a par, as two varieties of indirect fact-finding methods, is much more of a problem, since the undeniable epistemological analogy between measurement and indirect observation is rather difficult to bring out. And yet, according to the view we are now assessing, the question of whether the status of indirect observability is to be granted to objects such as molecules, atoms, sub-atomic particles, agglomerations of astronomical matter, and finally, to properties and relations of such "scientific" objects, hinges upon the analogy between measurement and indirect observation. It seems rather obvious that the three distinctive features of measurement, discussed in some detail in § 17, are attributable to indirect observation as well. Thus, there is no conceivable indirect observation without recourse to appropriate instruments, microscopes, telescopes, Geiger counters, scintillation screens, and so on. Moreover, the concepts of atoms, molecules, elementary particles, which occur characteristically in problems solvable by indirect observation, are non-ostensive, as are the characteristic terms of quantitative problems. The emergence of indispensable inferential operations in every validation by indirect observation is perhaps less obvious, but their existence must be ascertained, since both the basic analogy between measurement and indirect observation and the epistemological status to be allotted to the latter depend upon the parallelism of these two indirect fact-finding methods asserted in the third position. As a matter of fact, we shall have to conclude from our brief analysis that this third position is not tenable either, because the definitional techniques involved in indirect observation do not warrant an extension of the universe of discourse of the language of science beyond the directly observable.

To evaluate this position, we have to determine what indirect observation actually amounts to, in so far as its validating role is concerned. If I look with my naked eye in a definite direction and perceive a rod-like patch, I shall not say that I am seeing a microbe of a specifiable genus. But if a microscope endowed with an adequate magnifying power is properly inserted between my eye and the spot I am looking at, I shall claim that what I am seeing is such a microbe. The sensory data present in the two consecutive experiences may be exactly the same in all their intrinsic characteristics. But the very fact that the second sense-datum originated in connection with the insertion of a microscope between my eye and the relevant region of space constitutes an additional premiss which apparently warrants the completely different conclusion I am now drawing from my unchanged sensory experience. The important point is not the bare fact that the sense-datum is a joint effect of what is going on in the critical spatial region and of the light-beams which reach my eye after having originated in this region (and travelled through the microscope only in the second case). The undeniable physical difference between the physical antecedents of the two consecutive sense-data because of the insertion of a microscope brings about a difference in the resulting effect, that is, the interpretation which I feel entitled to put on the second sensory appearance.

The actual problem is more specific, of course; we are not interested in there being *some* difference between the interpretations which we are inclined or entitled to put on two consecutive similar sense-data, in view of the different conditions under which they originate. The real problem is the nature of the connection between the insertion of a microscope and the precise interpretation of the ensuing sense-datum (i.e., one which construes the sense-datum as the appearance of a microbe of a definite kind). It would not be sensible to accept the vague assertion that the second interpretation should differ from the first in some respect because of the insertion of a microscope—without any precise view of the difference.

"Observing a microbe through a microscope" is logically more complicated than measuring the length of an object with a yardstick, for the simple reason that the entire theory of optics is involved in the former, in contradistinction to fundamental measurement, whose argument consists of a few observation-statements and the definition of the relevant quantity. The proper mensural analogue of microscopic observation would therefore be a computational measurement, rather than a fundamental one. In order to show, nevertheless, the basic analogy between measuring a length with a yardstick and observing a microbe through a microscope, we may oversimplify the actual situation and suppose that "being a microscope of a specified magnifying power" is a non-ostensive concept, conditionally definable by a test T and a response R.[12] This assumption really asserts that in order to ascertain whether a given directly perceivable object is a microscope of a specified magnifying power, it is necessary to validate a few pre-assigned observation-statements, to supplement them by the conditional definition of a microscope, and to draw the relevant conclusion. Once this set of premises is secured in the context of an indirectly observational validation, I have to make sure that the object found to be a microscope has been placed between my eye and a specified region of my environment. Finally, another conditional definition is added to introduce the non-ostensive concept of a microbe of the relevant kind: "If I am looking through a microscope with a specified magnifying power at a particular spatial region, then, by definition, there is a microbe of the genus G in this region, if and only if I notice the appearance of a rod-like elongated patch." Of course, if such a sensory appearance materializes after the insertion of a proper microscope, I shall be entitled to conclude that there is in this region at this moment such a microbe.

Thus, to validate a statement to the effect that a microbe of a given genus is present in a particular region at a particular moment involves, under the simplifying assumptions which we have made (viz., by replacing the theory of optics with a conditional definition of a microscope), the same procedure which is employed in elementary measurement; the non-ostensive statement about the presence of a microbe is validated by being inferred from a set of premises, all of which are either observation-statements, or conditional definitions of the instruments used in this particular case of indirect observation, or a conditional definition of a non-ostensive term occurring in the validated statement. The only appreciable difference between measurement and indirect observation consists in the fact that, whereas the only non-observational premiss in elementary measurement is supplied by the definition of the relevant quantitative concept, there are *two* such non-observational definitional

[12]In this context, it may suffice to stipulate that a microscope is an appropriately shaped object endowed with a magnifying power M if a directly perceivable thing observed through this object looks M times as large as it would if looked at with the naked eye.

premisses in arguments pertaining to indirect observation; neither the instrument (say a microscope with a definite magnifying power) nor the concept occurring characteristically in problems solvable by indirect observation (atom, microbe, electrical charge, etc.) are ostensive but both are introduced into the scientific language by conditional definitions, involving in each case a definite test and a definite response. The difference between the two ways of indirectly discovering scientific facts may seem insignificant, and compatible with the basic analogy between the two methods claimed in position no. 3. Yet, on closer inspection, the analogy between the two methods breaks down at this very point because thing-like concepts (of microbes, molecules, atoms, etc.) are not literally susceptible to conditional definitions in ostensive terms.

The difficulty suggested relates to the inherent inability of conditional definitions to extend the universe of discourse of the language of science. The point is that the variables which occur in the defined and defining expressions of such definitions (that is, in the test T and the response R) range over entities already accessible to the scientist. Consequently, since all conditional definitions start with ostensive terms, regardless of the length of the relevant definitional chain, any variable contained in a conditionally defined expression must range over objects referred to in ostensive terms and over no others. Thus, in the simplest case of a single conditional definition involving a non-relational ostensive test T and a similar response R, the object X has by definition the property P if it was submitted to the test T with the result R. Since both T and R are ostensive concepts applicable to directly perceivable individuals, the defined property P must also apply to such individuals. The same conclusion concerning the defined concept is valid when the test T and the response R are no longer ostensive concepts but are linked up with the ostensive level by a longer chain of conditional definitions: since no link of such a chain involves any extension of the universe of discourse no such extension is involved in the whole chain either. A similar conclusion follows, if, instead of being attributive, test and response are relational.

In our example, the correct conditional definition of a microbe must run: "If the observer X looks through a microscope with a specified magnifying power at the spatial region Y, then, by definition, this region Y contains a microbe of a specified genus if and only if the observer X notices the appearance of a rod-like elongated patch." The directly perceivable individual involved in both the defining test T and the defined concept is the specified region Y. The defined concept is that of "containing a microbe of a specified genus" which is applicable, according to its definition, to directly perceivable spatial regions. It goes without saying that the whole expression "containing a microbe of a specified genus" has to be treated as a single logical unit endowed by the above conditional definition with an empirical meaning. Similarly, such predicates as "containing a molecule of a specified chemical compound," "containing an atom of a specified chemical element," "containing an electron," "containing an electron with a specified velocity v," and so on, are conditionally definable in ostensive terms provided that these predicates be related to directly perceivable spatial regions.

The inability of conditional definitions to extend the universe of discourse of the language of science beyond directly perceivable individuals (or, more generally, the referents of all terms endowed with a primitive empirical meaning) applies *a fortiori* to classical definitions of the equivalence type. Since no other definitional techniques are available in science, there seems to be no possibility, within the framework of scientific concept-formation, of ever transcending the prescientific universe of discourse. This limitation of definitional procedures within science is far-reaching and philosophically significant. Before some of its implications which relate to indirect observation are indicated, it is worth while to point out that the limitation is a logical, not an empirical one. In other words the fact that the scientist is unable by introducing new concepts, to transcend the barrier which we are discussing is not to be explained by a shortcoming in his skill or ingenuity, but by the logic of his concept-formation. To realize this, let us assume that an attempt is made to introduce a new kind of variable ranging over a new type of objects, by using the technique of conditional definitions: If the object X is submitted to a test T by means of the instrument Y, then, by definition, the object X bears the relation S to an object Z if and only if the instrument Y has reacted in the way R. In this case, a new variable, "Z," appears in the definiendum, but not in the definiens. We may, then, have the impression that new variables may appear in the defined expression without occurring in the defining expression and that they may range over a type of entities different from those referred to in the defining test and response.

Yet it is easy to realize that this attempt leads nowhere. A variable which appears only in the defined expression of a conditional definition is obviously observationally irrelevant. Thus, if the conditions of the above definition of the relation S were fulfilled in any single instance, that is, if the test T were applied to any individual X with the result R (or without it), then the defined relation S would obviously obtain (or fail to obtain) between X and any entity comprised by the variable Z. In other words, the only kind of relation which is definable in this way has the peculiar property that whenever an individual bears such a relation to at least one entity of a given type, it also bears the same relation to all the remaining entities of this type. Such relations are of no empirical interest and do not occur in empirical science. They do not form the subject-matter of indirect observation.

The conclusion we have reached does not altogether refute position no. 3 with respect to indirect observation. We cannot say, however, in view of the above argument, that a microbe or an atom is indirectly observed whereas the length of an object is not observed at all but measured—if the implication is that the epistemological status of statements validated either by indirect observation or by measurement is basically the same, since both types of statements are allegedly deducible from ostensive statements supplemented by conditional definitions. In view of later discussions of the Principle of Verifiability, where ostensive statements will be referred to as directly verifiable, whereas statements which follow from directly verifiable statements supplemented by definitions will be termed "indirectly verifiable," one may rephrase position no. 3 by saying that all statements validated either by measurement or by indirect observation are indirectly verifiable. This is not the case if our analyses are correct, since statements literally concerned with microbes, atoms, etc. are not verifiable at all. We have indicated, however, a way of reformulating them which ensures the indirect verifiability of quantitative statements and of those validated by indirect observation, namely, by referring the latter type of statements to directly perceivable individuals and shifting the new concepts characteristically occurring in problems solvable by indirect observation into the *predicates* of such problems and of their solutions. The parallelism between measurement and indirect observation then becomes complete; both methods of validation are concerned with directly perceivable objects, and both serve to disclose non-ostensive properties of such objects and non-ostensive relations among them. The only difference between measurement and indirect observation would relate to concepts occurring characteristically in problems solvable by these two methods respectively; such problems are quantitative in the first case, and qualitative in the second.

It should be noticed that the inability of indirect observation to extend the universe of discourse of science has been substantiated on the assumption that an initial universe of discourse is provided by direct fact-finding methods. The determination of this initial universe of discourse is admittedly controversial, and our contention concerning this limitation of indirect observation is, therefore, compatible with the view expressed on an earlier occasion that the determination of the subject-matter of scientific knowledge, or of the universe of discourse of science, presents an unsolvable problem. The conclusion we have reached in regard to indirect observation amounts to asserting that if the question as to the subject-matter of scientific knowledge had already been answered in some way in regard to direct fact-finding methods, then the universe of discourse involved in this answer could not be extended by indirect observation. In other words, specifically scientific fact-finding methods, such as measurement and indirect observation, have broadened man's pre-scientific environment more by discovering new aspects— either qualitative or quantitative, according to whether indirect observation or measurement is concerned—of objects directly known by man than by revealing totally new objects. This view of the universe of discourse of science does not reflect upon the feeling predominating recently that indirectly observable "scientific" objects are "as real" as directly perceivable common-sense objects. The two types of objects are reliably knowable contents of directly perceivable spatial regions, and, in this context, the "reality" of an entity is simply synonymous with the reliability of our knowledge of it.

The outstanding features of scientific fact-finding methods exemplified by indirect observation and measurement may be summed up as follows:

(1) Although indirect observation and measurement are methods both of discovery and of verification, it does not follow that there is a method for discovering such methods. There is no such super-method at the disposal of science; the perennial dream of discovering it, so eloquently expressed by Bacon, is, and will remain, a dream. In the advance of science, discoveries of methods of discovery or of verification are as essential as are the discoveries of solutions to various scientific problems. However, the discovery of methods of discovery is always obtained by trial and error procedures whereas the very concept of a method of discovery implies that a decision procedure is available for the solution of the relevant problems.

(2) The efficiency of indirect observation and of measurement depends to a large extent upon the fact that they enable the scientist to raise essentially new questions which are yet solvable by applying the age-old procedures of observation and inference. Thus, when a quantitative concept or one pertaining to indirect observation is conditionally defined by an ostensive test and an ostensive response, the questions involving this concept are novel in the precise sense of being demonstrably

untranslatable into questions in which the new term does not occur and only its defining expressions are involved. Similarly, the answers to such questions are essentially novel and, if true, reveal essentially novel aspects of the universe. Yet to solve such novel questions the old observational and inferential procedures are perfectly adequate. The point is that the applicability of the new concept can be decided upon by performing the definitional test, ascertaining the definitional response, and drawing therefrom the relevant conclusion. Since the test and the response are ostensive, and the inference is usually deductive, it is possible to deduce from observations concerning the performance of the test and its outcome whether the new concept applies on any pre-assigned occasion. Thus, by carrying out observations and submitting their results to an appropriate logical processing, the scientist is in a position to discover aspects of the universe which transcend the scope of human perception.

(3) Since the test of a scientific concept is usually relational and consists in connecting the object examined with appropriate testing instruments, the applicability of the new concept depends upon the behaviour of the object and the testing instruments rather than upon the response of the human observer to the object under consideration. This provides for the elimination of human standards of measurement to a very substantial degree, and increases the accuracy of observation and measurement since a highly complex and unstable human organism is replaced by simple, easily controllable instruments specially adjusted to the task under consideration. To use Locke's terminology: As a rule scientific fact-finding methods aim neither at the "primary" nor at the "secondary" qualities of the objects observed, but rather at the "powers" these objects have to influence in specifiable ways other specifiable objects. This shift of emphasis away from primary and secondary qualities which had been of decisive importance in man's pre-scientific knowledge seems to account for much of the scientist's success in exploring the universe.

CHAPTER TWO

Law-finding Methods

---

19 TYPES OF SCIENTIFIC LAWS

THE IDEA THAT the universe is governed by comprehensive and exceptionless laws has impressed the human mind more than any other article of the scientific credo. The supply of universal laws which science has succeeded in discovering constitutes its most significant and least debatable acquisition in the eyes both of educated laymen and of most professional students of science. Yet the procedures the scientist uses to discover and establish scientific laws are more complicated and less understood than those he applies to discover scientific facts. This difference can be explained by several circumstances. Basic fact-finding methods of science are taken over directly from pre-scientific usage with refinements which are negligible in comparison with the features common to scientific and pre-scientific ways of handling factual information. Laws, on the other hand, though present in embryo in pre-scientific knowledge, are, in the main, an acquisition of science. The methods that science had to develop in order to acquire its present extensive knowledge of laws supposed to "govern the universe" were also taken over to some degree from pre-scientific procedures of generalization. But the refinements, the departures, and the extensions have been much greater than in the case of scientific fact-finding methods. The methodology of scientific laws raises, therefore, a host of problems which simply do not arise in connection with the scientific continuation of the pre-scientific quest for facts.

Scientific laws are, further, more complicated logical entities than scientific facts, even than indirectly ascertainable facts. It is one thing to produce an oral or written picture of a fact witnessed or remembered by the observer; it is quite another thing to condense into a single concise statement (i.e., a law) a theoretically infinite number of facts, a negligible fraction of which may have been actually observed in the most favourable cases. We must therefore expect, in surveying critically the procedures applied in discovering and establishing scientific laws, greater difficulties than we encountered while studying fact-finding methods. However, this examination may bring us closer to an understanding of what are generally admitted to be the most characteristic aspects of the scientist's entire procedure.

The first point worthy of consideration is that to ascertain whether a given method for establishing a particular law or group of laws is reliable, we must examine both the *nature* and the *function* of this law or group of laws. Similarly, to determine whether a given set of validating methods is sufficient to establish the sum-total of scientific laws, we need to know just what constitutes such a law. We must therefore postpone our discussion of the scientific methods actually employed to establish laws, until we analyse the main types of laws (§ 19) and their main functions (§ 20), and frame a definition of scientific law (§ 21).

The term "scientific law" will be construed, in this study, as referring to a law established (or capable of being established) in any science whatsoever, either empirical or demonstrative, either natural or humanistic. The phrase "law of nature," on the other hand, will be applied to what J. S. Mill[1] calls "ultimate laws" of the natural sciences, i.e., a selection of laws of these sciences which suffice for deriving all their remaining laws and meet some additional requirements (cf. §§ 19–21). Laws of natural sciences derivable from these ultimate laws are called "derivative"; the ultimate and the derivative laws of natural sciences together add up to the class of "natural laws." Thus, every law of nature is a natural law, but not conversely. All scientific laws are not natural; economic and sociological laws in the empirical field, and laws of pure mathematics in the demonstrative field, are not so classifiable. For instance, Newton's axioms of mechanics exemplify laws of nature. The laws of planetary motions, deducible from these axioms (and a few additional assumptions), are (derivative) natural laws. The sociological law of cultural lag is a scientific law, without being either a law of nature or a natural law.

[1]Mill's classification of laws raises several difficulties to be outlined in the sequel. The distinction between "fundamental" and "derivative" laws, defined by C. G. Hempel and P. Oppenheim ("The Logic of Explanation," *Philosophy of Science*, vol. 15, 1948), renders the same services as Mill's classification, without giving rise to these objections. Cf. J. S. Mill, *A System of Logic* (1874).

There are several important types of scientific laws:

(1) *Conceptually Universal Laws*

We have distinguished fact-like statements from well-established facts. The former attribute, wrongly or rightly, a particular property to a particular individual, or assert a particular relation among a few particular individuals. The latter are selected from the sum-total of fact-like statements in accordance with a criterion based on the existence of an adequate body of evidence. We shall, accordingly, distinguish law-like statements from laws, by recourse to the same criterion; a law-like statement is a well-established law if its truth is adequately supported by available evidence.[2] In other words, a well-established law is a law-like statement known to be true. That this piece of iron expands after having been heated, is a particular or local fact; that every piece of iron would do so under similar circumstances is a law.

Yet, in spite of the basic similarity between the two pairs of concepts, there is an additional difficulty in the case of law-like statements. It is by no means easy to delimit all law-like statements in a satisfactory way, without either distorting the meaning usually attached to the phrase or indulging in a merely lexicographic exercise. The real problem is to provide a definition of the concept of law-like statements which will remain reasonably close to actual usage and shed some light on the considerable role allotted to such statements in the scientific framework since the birth of modern science during the seventeenth century.

One point can be taken for granted: in contradistinction to particular and local facts, laws are "general" or "universal." This much seems to be accepted in all attempts at clarifying the concept of scientific law. The trouble is, however, that this requirement of the universality of scientific laws can be construed in at least three distinct and equally important ways, which refer respectively to the universality of concepts, of quantification, and of spatio-temporal scope. In this section, we shall survey the types of laws which meet these three requirements; the bearing of universality on the definition of scientific law will be discussed in § 21.

Thus, in the first case, a statement is said to be law-like if all its constituent terms are universal, but fact-like if even one of these terms is local or individual. By local terms, I mean proper names of objects and events, such as "the earth," "Napoleon," and "the Second World War," whereas universal terms will be construed as those which, by

---

[2] A law, whether or not well established, can be defined simply as a true law-like statement.

their very nature, are applicable always and everywhere, such as "green," "atom," and "weight." "The earth is a planet" is a particular fact in this sense. On the other hand "Some stars have no satellites" is a law, in so far as the requirement of conceptual universality is concerned.

Yet a proposal to confine the concept of scientific law to statements which are universal in this sense would certainly run counter to many a well-established way of speaking. Thus Kepler's laws of planetary motions refer explicitly to the individual called "sun" and to some of its satellites, but it is rather awkward not to call his discoveries "scientific laws." Similarly, A. Comte's sociological "law of the three stages," which allegedly governs the succession of the religious, the metaphysical, and the scientific (or "positive") stages in the historical development of man, though rather questionable in view of its vagueness and the inadequacy of the support which Comte has derived for it from historical facts, can hardly be denied the status of a law-like statement.

The tendency which is indicated by the requirement of conceptual universality of scientific laws is nevertheless of great importance for the present outlook of science. The fact is that in the most basic sciences, especially in physics, there is an unmistakable trend to get rid of local terms and constants in the formulation of basic principles. This does not mean that all the implications which Eddington, for example, has read into this trend are justified, but its existence is obvious and philosophically significant. Thus, to quote a simple example, the fundamental standards of the ramified system of physical quantities were based until recently upon terrestrial characteristics and were consequently of a local nature; for instance, a second has been, by definition, a definite fraction of the time the earth takes in carrying out a complete revolution around its axis. Since similar considerations were applicable to other standards of physical quantities, the laws of nature involving such standards (inclusive of Newton's laws of mechanics and gravitation and of Maxwell's laws of electro-magnetism) did not meet the requirement of conceptual universality. These theories would have to be classified as systems of fact-like, rather than of law-like statements should conceptual universality be construed as a prerequisite of law-like status. The "universalist" tendency just referred to has, however, changed this situation. The basic assumptions of the aforementioned fundamental physical theories can now be expressed in purely universal terms, since the basic standards of quantities which they involve are so expressible. This is why it seems to be rather in keeping with modern scientific tendencies to stress the importance of conceptual universality, by basing on it the distinction between fundamental and derivative laws. The fundamental laws would, by definition, have to meet the requirement of conceptual universality. Law-like statements which are not fundamental but can be deduced from fundamental laws, may be termed "derivative." It remains to be seen, however, whether this dichotomy accounts for all scientific laws.

(2) *Formally Universal Laws*

Since a law is a general statement and every general statement has a number of "instances" (which are typically, though not necessarily, fact-like), one may wonder whether every law-like statement should claim to be true of *all* its instances. Such a requirement has often been formulated as essential to laws, and the most conspicuous laws are universal in this sense. For example, Einstein's mass-energy convertibility law asserts that the energy content of *every* material system is equal to the product of its mass and the square of the speed of light for *every* possible value of mass and of energy. The law is applicable to every physical system, regardless of its mass and energy content. Hence, all the instances of the law which correspond to particular values of energy and mass and to a particular choice of the system are included. Yet one may feel perhaps that the speed of light is a constant number, not a variable, and that, consequently, only *one* speed is actually involved in the law. The answer to this objection is plain and, in a sense, already alluded to in what has just been said about the universalistic tendency of contemporary science. We have to distinguish between *local constants*, which are simply the local values of spatio-temporal co-ordinates of a single individual (and correspond, in the mathematical language of science, to proper names of individuals in the vernacular) and *universal constants*, such as the speed of light, the magnitude of the quantum of action, or the smallest electrical charge in existence. Universal constants are expressed by a constant, single number, just as local constants are. But their physical significance is completely different, since they refer to any and every space-time region. For example, the speed of light is always and everywhere expressible by the same number of km/sec., provided it is measured from an appropriate "inertial" frame of reference. Thus, universality of quantification implies merely, if properly interpreted, that all *individual variables* are within the scope of a universal quantifier. Non-individual expressions, for instance numbers specifying the speed of light or the quantum of action, may be represented either by a constant value (which would be inadmissible in the case of individual expressions, because of the requirement of conceptual universality) or by a variable operated on by any quantifier, either universal or existential.

As a matter of fact, the requirement of quantificational universality, though ostensibly met by the fundamental laws of nature, determines the appearance rather than the actual meaning of such laws. This requirement is violated by statistical laws, some of which have come to play a decisive role in fundamental theories, for example, in Quantum Theory. A statistical law of the simplest kind specifies the probability p that any object belonging to a class A will also belong to the class B (e.g., the probability that any hydrogen atom that is "excited" at the time t will radiate within the time $t + t_o$). On the now predominant frequency-interpretation of the concept of probability, the meaning of the statement that whatever is A is also B with the probability p can be rendered explicit by the following paraphrase: "If A_n is an appropriately chosen infinite and increasing sequence of finite subsets of A and p_n is the relative frequency of objects in A_n that also belong to B then the sequence of fractions p_n tends towards the limit p as the number n increases indefinitely." Moreover, the convergence of the fractions p_n towards the limit p means that for every positive number ϵ however small there *exists* an integer N such that whenever the integer q exceeds N the fraction p_q differs from p by less than ϵ. Hence, existential quantifiers are indispensable to define the concept of limit, and, in particular, the concept of probability as a limiting frequency.

There is little doubt that existential quantifiers occur also in nonstatistical laws of nature although the presence of such quantifiers is concealed by the equational form given to most scientific laws for reasons of convenience already referred to (§ 16). The point is that the fundamental laws of nature are mostly ordinary or partial differential equations; the concept of derivative which is involved in such laws is also a limiting concept whose definition includes a mixed quantificational pattern.[3] Consequently, whenever a law of nature has the form of a differential equation, the requirement of quantificational universality does not apply to it.

(3) *Laws Universal in respect to Spatio-Temporal Scope*

Finally, a law is often said to be "universal" because it applies to all relevant objects or events in the cosmic space-time, regardless of when

---

[3]The logical structure of limiting concepts, such as the concept of derivative, is explained, for example, in R. Carnap, *Einführung in die Symbolische Logik* (1954).

and where they exist or occur. In contrast, particular or local facts are concerned with a single space-time region, which happens to be occupied by the object or event referred to in the fact-like statement in question. Yet, as a matter of fact, two sciences, physics and chemistry—or perhaps we should say, "the science of physics," in view of the unmistakable tendency of chemistry to be absorbed by physics?—have the monopoly of establishing laws of nature which possess such universality of scope.[4] The laws of astronomy, of geology, of biology, and of human and animal psychology are valid within certain space-time regions, but not in the whole of space-time. I have already referred to Kepler's laws of planetary motions, which apply only to the solar system. Other laws of contemporary astronomy have a spatio-temporal scope vastly exceeding that of Kepler's laws, but finite nevertheless; they certainly fail to cover the whole of space-time. Similar remarks apply to biology. Thus, Darwin's theory is an attempt at reconstructing and explaining a local story which took place on the surface of the earth (or close to it) during less than a billion years. The laws of economics claim validity with regard to human societies, i.e., for a region whose spatial dimensions are surpassed by those of Darwin's theory, and whose temporal dimensions have an order of magnitude of a few thousand years.

The fact that a scientific law applies only within a specifiable region may be due to two circumstances of a completely different nature: in the first place, it may simply happen that entities which by their very nature should be governed by the law under consideration do not occur beyond a specifiable region, so that the law may be said to apply "vacuously" beyond this region. On the other hand, the law may not have a universal scope because entities eligible to be governed by this law which do occur beyond the specifiable region are there precisely at variance with the requirements of the law. Thus, should the statistical interpretation of the law of cosmic increase of entropy (Second Principle of Thermodynamics) be correct and should the entropy keep increasing only within the present period of cosmic history and start decreasing afterwards, then the law would be local in the second sense of the word. For all that is presently known, biological laws may be vacuously universal with regard to their scope. The bulk of physical laws are taken to be non-vacuously universal.

We have thus to distinguish, with respect to spatio-temporal scope, genuinely universal laws from those which are only vacuously universal.

[4]In accordance with his metaphysical views of "cosmic epochs," A. N. Whitehead has denied universal scope even to physico-chemical laws and admitted only regional laws of nature. Cf. his *Science and the Modern World* (1926).

Laws which fail to meet the requirement of (genuine or vacuous) universality of scope, and apply only within a limited spatio-temporal region, may be subdivided in turn into regional and individual laws, depending upon whether they refer to all the individuals within such a region, or to single individuals. Darwin's theory is a set of regional laws; those of Kepler are individual.

(4) Finitist versus Infinitist Laws

The number of instances of a scientific law may be finite or infinite; if infinite, it may be "denumerably" or "non-denumerably" so. Thus, the biological law estimating at about one billion the number of variations needed to lead from the proto-organisms supposedly responsible for the start of life on earth to the present multiplicity of species is a finitist one. Every law concerning human psychology is finitist, since the number of humans is finite. On the other hand, the hypothesis of an infinite space amounts to assuming the geometrical law-like statement to the effect that, for every natural number N, there is an object whose distance from the earth exceeds N units of length. The relevant law would be denumerably infinitist. Those who assume the literal applicability of Euclidean geometry to the cosmic space (even on the assumption that this geometry is reinterpreted so as to deal only with finite regions, for example, in accordance with Whitehead's Method of Extensive Abstraction)[5] and assert accordingly that every finite region includes a non-denumerable class of spatial points are thereby committed to an infinitist law of the second kind.

(5) Statistical versus Causal Laws

I have already mentioned that the simplest statistical laws specify the probability that any element of a class A will also belong to the class B. In general, statistical laws deal with functional relations among the statistical distributions of a set of quantities in a given population. Thus, if the statistical distribution of a given quantity Q in a given population P happens to be uniquely determined by the distributions of a set of other quantities $Q_1, Q_2, \ldots, Q_n$ in the same population, a statistical law of the aforementioned type will state that the distribution of Q in P is a given function of all the distributions of the quantities Q_i ($i = 1, 2, \ldots, n$). If we disregard the unobjectifiability of quantum-theoretical concepts (§ 21) then all the basic laws of this theory are of the statistical kind.

[5] Cf. A. N. Whitehead, *The Concept of Nature* (1920); A. Grünbaum, "Whitehead's Method of Extensive Abstraction," *British Journal for the Philosophy of Science*, vol. 4 (1954).

A typical law of Quantum Theory can be formulated, under the above simplifying assumption, as follows: "If both the distribution of positions and the rate of change of this distribution are given in a population of similar physical entities, then the distribution of any other quantity within the same population can be effectively determined."[6]

Laws which do not deal with populations and the interdependence of statistical distributions in the populations concerned, but rather with single individuals, are often referred to as "causal." Yet this interpretation of "causal laws," though frequent, is hardly felicitous, since it has little in common with the problems of causality and determinism. For example, any statement of the form "Every S is P" would be causal in this sense. We propose, therefore, simply to distinguish statistical from non-statistical laws in the cases just referred to.

Another distinction between statistical and causal laws, more closely connected with the problems of determinism, is involved when a statistical law is said to be causal if it associates a sharp[7] distribution of the determined quantity Q with sharp distributions of the determining quantities Q_i. In this case, the value of Q for any member of the population is uniquely determined by all the corresponding values Q_i, and the latter may be considered to be the "cause" of the former.

(6) Separately versus Contextually Justifiable Laws

One important distinction is connected with the possibility of validating the law under consideration either separately, i.e., without utilizing any other law which is also not separately validatable, or only in conjunction with other laws. For example, Newton's law of equality of action and reaction has been shown by Poincaré to be only contextually capable of validation. The full sense of this distinction will be clarified only after the methods of validating laws and theories have been discussed in some detail. But it may be pointed out at this juncture

---

[6]The mathematically trained reader may be interested in a more precise formulation which involves several technicalities. It runs: "If the distribution $D(p)$ of the spatial position p in a population of similar physical systems (i.e., of systems associated with the same Hamiltonian function H) and the rate of change of this distribution $dD(p)/dt$ are given—$D(p)$ indicating the fraction of the population which occupies the position p—then the average A of any quantity q (such as velocity, momentum, energy, etc.) in this population is determined by the expression $A(q) = \int q^{op} \psi . \overline{\psi} dp$ where q^{op} is the operator corresponding to the quantity q and the state-function ψ is uniquely determined by the distribution $D(p)$ and its rate of change through the Schroedinger equation of the physical systems: $H^{op} \psi = \dot{\psi}$." Cf. H. Reichenbach, Philosophic Foundations of Quantum Mechanics (1946); H. Margenau, The Nature of Physical Reality (1950).

[7]The distribution of a quantity in a given population is said to be sharp if this quantity takes on the same value within the whole population.

that so-called "empirical laws" are just separately justifiable laws. In particular, a law is empirical if all its instances are facts which can be validated by applying the fact-finding methods discussed in §§ 15–18.

Not every quantified statement can be admitted as law-like, if the concept of scientific law is to be of any interest in epistemology and the philosophy of science. "Every electron includes an invisible goblin" is a quantified, but hardly a law-like statement. The source of our hesitation to consider it as possibly eligible for the status of a scientific law is the circumstance that its instances do not represent discoverable facts. Yet the same holds true of statements which we do admit as laws, although they are only contextually justifiable: their instances do not represent discoverable facts either. Thus the distinction between empirical laws, contextually justifiable laws, and quantified statements which are altogether inaccessible to scientific validation is an important aspect of the concept of law, to be discussed in connection with the Principle of Verifiability.

20   FUNCTIONS OF SCIENTIFIC LAWS

The Cognitive Function

Scientific laws are bits of information expressible in single statements, just as are particular facts. "Phosphorus melts at 40 C°" is a law. "This piece of phosphorus has melted once its temperature reached 40 C°" is a fact. If a law-like statement is offered in conjunction with a body of adequately supporting evidence as answering a problem which the scientific community finds interesting (or simply bothering if left unsolved), then this statement supplemented by this piece of evidence is obviously a scientifically admissible solution to a theoretical problem; it meets the same criterion of scientific status as scientifically established facts do. In other words, a scientific law fulfils a cognitive or informational task in science, exactly as well-established facts often do. Some distinctive features of law-like statements, for example, their universality of scope, or of quantification, or of conceptual structure, may (and usually do) permit them to offer information in a more condensed form than statements of facts are able to achieve. As a rule, too, the kind of evidence which substantiates a scientific law will differ considerably from the evidence required by facts, even if the latter are only indirectly ascertainable. A statement specifying the melting point of phosphorus all over the world says more than one which only predicts the melting point of a particular piece of this substance; obviously, the premisses required to validate the latter are insufficient to validate the former. But neither the difference in the amount of information offered, nor in the evidence which is required in either case removes the basic similarity of facts and laws; both enter science if and only if they meet the requirement of offering reliable information.

Let us conclude, therefore, that the first function to be discharged by scientific laws is that of scientifically solving theoretical problems, of constituting scientific knowledge. It is true that, in contrast to facts, scientific laws discharge in addition several other vital functions; they systematize, predict, control, and explain facts and contribute substantially to a definition of the scientist's vocabulary and his conceptual apparatus. That is why the purely cognitive function, which scientific laws share with facts, is often eclipsed by those other functions of laws which facts, by their very nature, cannot discharge. This may account for the frequent unwarranted tendency to ignore the cognitive function of scientific laws.

The Systematizing Function

The function of condensing, systematizing, and making easily accessible a large number of particular facts has often been stressed by philosophers of science in determining the role of scientific laws. E. Mach[8] even contended that to save mental effort, to economize on the labour necessary for pre-scientific man to obtain a satisfactory adjustment to the complex situations he was facing, constitutes the main objective of scientific research and it is attainable because of scientific laws. Thus, the Descartes-Snell law of optical refraction enables one to replace an almost unattainable knowledge of its various instances by a single algebraic formula. This time- and labour-saving quality can be confirmed to a lesser or higher degree with innumerable scientific laws. Yet, the monopoly on virtue granted to this economical attribute of scientific laws by Mach is not thereby justified. The laws of science actually discharge several other important functions within the scientist's activity, and there is hardly any sense in quarrelling about the respective weight of these various functions in the over-all scientific achievement.

The Predictive Function

Moreover, it should be made clear at the outset that this ability to condense, summarize, and systematize is related primarily to facts which have been actually observed and are simply too numerous to be remembered individually. This was the situation of Descartes when con-

[8]Cf. E. Mach, *Erkenntnis und Irrtum* (1905).

fronted with the task of devising a single formula capable of replacing all the particular observations made about the various angles of refraction which correspond respectively to the various angles of incidence. *Predicting* the connection, if any, between future measurements of associated values of these two angles is not automatically made possible by the summary of past observations provided in the law of refraction. Yet prediction is admittedly a task of science at least as vital as that of condensing our knowledge of particular facts obtained in the past by a considerable number of observations.

Indeed, the ability of science to guide our action rests mainly on the predictive power of scientific laws and theories. If I pursue an objective *O* and wonder whether I could bring it about by taking a specifiable course of action *A*, then my practical problem as to what I should do in order to attain my objective *O* hinges upon the prediction: "If I take action *A*, then objective *O* will ensue in due time." Ostensibly, the last statement refers to the connection between a particular action and a particular objective and should therefore be classified as fact-like, rather than as law-like. But on closer inspection, we notice that the fact-like connection becomes available to us only as a special case of the law-like proposition: "Whenever anyone takes action of kind *A*, an event of kind *O*, similar in all the relevant respects to the objective I am after, will follow." Thus, the statement utilized in order to solve a practical problem, namely, to bring about a definite objective, hinges upon the predictive power of a law-like statement.

To put it otherwise, a person who applies some item of information on a given occasion, viz. with a view to bringing about a certain objective, is said to *act on this information*, or on the statement expressing the information. But "to act on a statement *S* with a view to bringing about thereby an objective *O*" means simply to be prompted to take some definite course of action *A* by inferring from the statement *S* that this course of action would be followed by the objective *O*. Thus, acting on a statement amounts to making predictions on the basis of this statement and to being induced by these predictions and the desired event *O* to take a course of action *A*.

The fact that the practical role of science as a guide in action rests upon the predictive power of scientific laws explains why A. Comte conferred upon prediction the rank reserved by E. Mach to economical systematization: "savoir c'est prévoir." Yet, once more, there is little to substantiate the monopoly in the tasks of science which the older positivism has granted to prediction. The most reasonable course seems simply to acknowledge that, in addition to their cognitive function, scientific laws permit both systematization and prediction, without grant-ing either a monopoly as a function of scientific laws.

The Explanatory Function

Furthermore, a fourth function of scientific laws, distinct from the two preceding ones, is undeniable. In regard to past observations, the law is a shorthand formula. In regard to future observations, the law is an instrument of successful prophecy, and, consequently, of practical, reliable, and often indispensable guidance. But what about present observations which we may happen to make without any practical inten-tion at the back of our mind? Neither a summary nor a prediction is required, since we are considering a single presently observed fact: it stands in need neither of a summary, since it is single, nor of prediction, since it is present and known. Are scientific laws of no avail on such occasions? This is hardly the case. We may, and often do wonder why objects change their appearance, for example, when immersed in water. We are then interested in *explaining* the occurrence of a presently observed fact, and the law of refraction is once more helpful. It indicates the basis of the observed change in appearance in a general pattern underlying every passage of light from one medium into another one. This explanatory use to which the law of refraction is put to account for a present fact is distinct from its summarizing function in regard to facts observed in the past and from its predictive function with respect to future facts.

These three functions of laws have been discussed in their application to particular facts; they are also often performed by laws of greater generality for laws of lesser generality. Yet, since the systematization, prediction, and explanation of laws is most frequently attained by means of theories, we shall discuss this trio of functions, as they are applied to laws, in connection with scientific theories.

The Definitional Function

Explaining, predicting, and summarizing are functions performed by laws with regard to facts (and to laws of lesser generality). This does not entirely account for the role of laws in science even if their cognitive function is added. The conceptual apparatus of science is instrumental in discovering and establishing laws in which these concepts occur. But the relationship between laws and concepts is by no means that of uni-lateral dependence. Primarily, of course, concepts are used to facilitate the discovery of laws. Scientific concepts are either taken over from pre-scientific, common-sense usage, or (as is the rule in more advanced departments of science) "introduced," i.e., created by special definitional procedures. In the latter case in particular, it is apparent from a glance at the definitional techniques of the scientist that he tries to form concepts which promise to promote the discovery of laws in which they occur. Thus, the quantitative concepts, for example, are often introduced by definitions because one expects them to be governed by functional, law-like relationships. Should a quantitative concept, introduced in complete accordance with the definitional rules of science, turn out to be sterile because laws including this concept fail to be established, then the concept would in all likelihood be abandoned.

However, the interdependence between laws and concepts goes much further. Concepts are not only introduced to aid the discovery of laws; they are themselves affected in turn by the very laws their introduction has helped to discover. In other words, a concept, that is, the meaning of a term, is primarily circumscribed by the framing of a definition prior to the discovery of laws involving this concept. It turns out, however, especially in the case of concepts introduced by conditional definitions, that the contours of the concept are only partly fixed by this initial definition; although incompletely determined, the concept is nevertheless capable of serving as a guide to discovering laws, which, in turn, will serve to complete its contours and will thus play a role comparable to that of definitions. It may even happen that a term will occur in the context of a theory without having been introduced by any kind of explicit definition, and that the context of the theory nevertheless endows the concept with a meaning in much the same way as ordinary or conditional definitions do. Such concepts, called "theoretical constructs," will be examined in connection with the functions of scientific theories. Conditionally definable concepts (and theoretical constructs, which will be shown later to be a species of the former) represent the main method of concept formation in empirical science. We must therefore consider in some detail the definitional function of scientific laws, in view of its bearing on the technique of conditional definitions.

Let us assume, for example, that a psychologist anxious to reinstate conditional definitions, predominant in more advanced natural sciences, decides to re-define the concept of intelligence. He may declare that, by definition, anyone who is submitted to a specified I.Q. test T and scores R has intelligence to the degree R, while anyone who fails to attain this score is not intelligent to this degree. Taking over a term from ordinary language and endowing it with a new meaning by introducing a definitional criterion for it, is a legitimate procedure in any empirical science. The numerous circumstances essential for its effectiveness are of no concern to us. What is of importance here is that when this term is re-defined in such a way, with the explicit provision that nothing is included in its meaning save what can be found in the definitional criterion, it remains extremely vague, since its applicability is determined only in regard to those individuals to whom the intelligence test has actually been applied. For instance, the question whether Talleyrand was intelligent to any appreciable degree is, thus far, unanswerable, since no psychologist happened to submit him to the test. The psychologist will, however, try to utilize his re-defined concept. He may submit a group of successful politicians to his test and find out that they all make a very high score. He then formulates the law: "Every successful politician is very intelligent." This law is not only established by applying the newly defined term, but also has repercussions on the term itself.

This law has three kinds of instances. (1) There are those from which it has been derived; these instances are facts referring to successful politicians who were submitted to the test and attained a high score. (2) The law predicts that any successful politician would make a high score, if he were submitted to the test. These predictions, when verified by applying the test to successful politicians who were spared during the establishment of the law, form a second group of instances. They are similar in one essential respect to the instances of the first kind: both are factual propositions relating the results either of past or of future applications of the test. (3) The law is supposed to hold for every successful politician, even if he has never been molested by a psychologist. The instances of the law included in this last class are neither reports on past observations, nor predictions of future scores. They are important, nevertheless, because they entitle us, for example, to assert that Talleyrand was very intelligent, since he was a successful politician. Through instances of the third kind, the law has transformed an unanswerable question into an answerable one. The law thus plays, with the third group of its instances, a part comparable to that of a definitional criterion (and, indeed, of any definitional device), since it secures the applicability of the term to instances which otherwise would have been located within the "vagueness-area" of the term (cf. § 29).

The law itself is certainly synthetic, because it summarizes past observations, predicts future observations, and may well be refuted, should some of these future observations reveal the existence of a politician who is both unintelligent and successful. However, since the law entitles us to infer from political success to intelligence in cases where the intelligence test was not applied, there must be some definitional statement justifying this inference; clearly, *logical* deducibility is out of the question. I think the simplest way of accounting for the legitimacy of such inferences is to assume that the very establishment of the law carries with it a new definitional criterion specifying that, by definition, a successful politician who was not submitted to the intelligence test is very intelligent.

More generally, if a concept P is governed at the outset by a definitional criterion involving test T and response R and, by applying this criterion, a law is then established inductively to the effect that a certain property Q (other than R) is accompanied in every individual by the property P, then a new definitional criterion is to be associated with this law. This new criterion stipulates that the applicability of the property Q to any individual not submitted to test T entails by definition the presence of the property P in this individual. For example, since the law asserting that every successful politician is very intelligent has been established by using the I.Q. test, Talleyrand's being very intelligent follows definitionally from his political achievements, on the assumption that he never was submitted to the I.Q. test.

We shall call *semi-definitional* a law whose instances are susceptible to such a tripartite classification. The most important types of semi-definitional laws are the quantitative laws. Suppose, for example, that the quantities P and Q have both been provided with definitional criteria, and that then, by ascertaining the values they take on for any object of a certain class C, it has been established that the values always fulfil a certain functional relationship. In other words, in all the individual instances which have actually been examined, it has turned out that whenever an object X belongs to the category C and the values of P and Q are measured for X, they happen to fulfil the equation $P = f(Q)$. If this equation is then considered as a valid law, instances of it referring to objects for which the values of both quantities have actually been ascertained are summaries of past experiences. The instances concerning objects for which the values of both quantities will be ascertained predict future observations, whereas the instances concerning objects for which only one of the two quantities involved has been measured are semi-definitional and determine the correlated value of the other quantity.

The case is basically similar when, instead of two quantities correlated by a single equation, N quantities are correlated by $N-1$ equations so that when all the quantities but one have been measured for the same object, the value of the unmeasured quantity for this individual can be computed from the equations. The system of equations then performs a semi-definitional role with regard to this quantity and, more generally, with regard to all the quantities involved.

The foregoing analysis shows that in order to account for the ability of a law to make the meaning of its constituent terms more precise, new definitional criteria for these terms have to be associated with the law. That is why, in spite of its empirical nature and its factual implications, the law is termed semi-definitional. However, there is more to be noted than that semi-definitional laws yield new definitional criteria for terms already provided with such criteria. Most statements that are usually considered as conditional definitions should actually be interpreted as semi-definitional laws whose associated criteria perform the definitional task mistakenly ascribed to these statements.

Consider any conditional definition with a relational test, for instance, the statement that if an instrument y has been attached to the object x then, by definition, the quantity Q takes on the value z for the object x if, and only if, the instrument y reads z. This statement implies that if two instruments, y and y', have been attached to the same object x, they will yield the same reading z. For example, if the temperature of an object is defined in terms of the reading of any thermometer containing a given thermometric substance, then one can predict on the basis of this definition that any two thermometers of this type will yield the same reading if applied to the same object. This, however, is a purely factual prediction which does not involve the defined term and may well be refuted by observations if the readings of the two thermometers fail to agree on some occasion. As a matter of fact, it is well known, on the basis of the statistical theory of heat phenomena, that, in a sufficiently long run of simultaneous temperature-measurements, the two thermometers are bound to yield divergent readings in a specifiable fraction of all cases.[9] The possibility of a definition being refuted by observational data is hardly compatible with the nature of a definitional statement, whether conditional or unconditional, since definitions are usually considered analytic, that is, true come what may.

A similar difficulty will clearly arise in connection with any conditional definition involving a relational test, regardless of whether a qualitative or quantitative concept is introduced. In this case, the difficulty will not depend upon whether the measuring instrument involved

[9] We shall try to show later on (cf. § 27) that the clash between the above definition of temperature and statistical thermodynamics not only illustrates the empirical implications of conditional definitions involving relational tests but calls for a substantial, probabilistic modification of the logical structure of conditional definitions.

in the definition is directly gauged to the defined quantity or connected with it by some functional relationship. Since the overwhelming majority of conditional definitions of scientific concepts involve relational tests (cf. our remarks in § 18 about the role of "tertiary qualities" in the conceptual apparatus of science), the difficulty under consideration is rather pervasive.

On closer inspection, the difficulty does not seem to be insurmountable. As a matter of fact, there are several ways of removing it. The simplest solution is to re-formulate slightly any conditional definition with relational tests so as to prevent it from entailing factual consequences, and to re-classify the initial formulation as an empirical law. We may stipulate, for example, that whenever only one instrument y has been applied to the object x then, by definition, the reading y yields the value of the defined quantity for the object x. Since the original definition has now become a law, we are still able to predict that whenever several instruments are applied to measure the value of the defined quantity for the same object, all the instruments will yield the same reading. However, this prediction is now derived from an empirical law, not from a definition. Should the prediction be falsified by observational results, the law would be refuted but the associated conditional definition would not be affected. Accordingly, we shall have to stipulate that, in their usual formulation, conditional definitions with relational tests are actually empirical laws and that only the associated criteria have definitional status.

Let us point out, for the sake of clarity, that the expression "definitional criterion associated with a law," is now being used in a somewhat different sense than previously, since the law is not supposed to provide a new criterion for a term occurring in it, in addition to criteria already available for this term. In the present case, the criterion associated with a law is simply derived from the law by the additional clause that the test involved in the law is applied only once in those cases where the criterion is definitional. Nevertheless, it is still the case that some instances of the law are definitional. Accordingly, we shall use the two related concepts of semi-definitional law and associated definitional criterion, in a broader sense, by extending their applicability to conditional definitions derived in the way just described from semi-definitional laws with relational tests.

Another extension of the concept of semi-definitional law seems to be required whenever several conditional definitions are available for the same concept. It does not matter whether the definitional tests involved are relational or attributive. The need for re-classifying a set of conditional definitions for the same term as a set of semi-definitional laws arises when the tests involved in the original definitions are not mutually exclusive. Suppose, for example, that there are two test-response pairs (T_1,R_1) and (T_2,R_2) for the same concept Q and that T_1 and T_2 are compatible with each other. It then follows from the two conditional definitions that whenever both tests T_1 and T_2 are applied on the same occasion, their responses must agree with each other, that is, either R_1 and R_2 both materialize, or both fail to materialize. This, once more, is a purely factual prediction derived from the two definitions. As previously, the prediction does not involve the defined term Q and is susceptible to an experimental check, in spite of being derivable from purely definitional premisses. For example, if both the deflection of a magnetic needle and the production of an electrolytic effect are considered as definitional criteria for the presence of an electric current, then, by definition, the deflection of the needle must be accompanied by an electrolytic effect; this is an empirical prediction, falsifiable at least in principle. Needless to say, this new difficulty created by the availability of several definitional criteria for the same concept is still pervasive because all important concepts do have several criteria, and for good reasons, since otherwise the concepts would be extremely vague and therefore useless.

This difficulty does not arise, however, when one or both conditional definitions of the same concept are provided by a semi-definitional law because the definitional tests are then mutually exclusive. This fact suggests that whenever several conditional definitions $D_1,D_2, \ldots ,D_n$ are offered for the concept Q, we should consider them a set of semi-definitional laws. The definitional criterion associated with the law D_1 would then read: "If the test T_1 is the only one to have been applied to the object X then, by definition, this object has the property Q if, and only if, the test T_1 has yielded the response R_1." Similarly, in the criterion associated with the law D_i, the test-response pair (T_i,R_i) would be substituted for the pair (T_1,R_1) in the rule we have just formulated. It goes without saying that whenever criteria with a relational test are included in the set, they must previously be subjected to the reformulation already outlined in order to prevent any single relational criterion from having factual consequences.

The definitional criteria associated with the set D of semi-definitional laws have obviously tests that are all mutually exclusive. Consequently, these new criteria do not entail any longer predictions concerning the outcome of a simultaneous application of several tests to the same object. On the other hand, the original statements D_i, now interpreted as laws, do involve tests that are compatible with each other. Hence, if two or more of these tests come to be applied on a single occasion, the agreement of the corresponding responses will follow from these laws, although it is not derivable from the associated definitional criteria. Should experience fail to support the agreement predicted by the laws D_i, some of them would be empirically refuted. This circumstance, however, would have no bearing upon the associated conditional definitions.

I have gone into some detail in discussing these difficulties arising in connection with the technique of conditional definitions because the solution provided by an extended use of semi-definitional laws seems to me both to be in agreement with the actual thinking of the scientist and to shed some new light upon the definitional function of scientific laws. I should like to point out, however, that at least two other simpler solutions could be offered, implying a somewhat different view of the role of conditional definitions. The main drawback of these solutions is that they seem to differ rather considerably from the actual thinking of the scientist.

The first of these alternative views of the role of conditional definitions consists in simply accepting the fact that whenever such definitions involve relational tests or occur in groups centring around the same concept, they do have empirical consequences and may clash with observational results. The point is, that contrary to a widely held view, definitions are not completely arbitrary, even in the logico-mathematical field. It has been pointed out, for example,[10] that in some types of mathematical theories, a definition must be supplemented by a proof of the existence and uniqueness of the concept defined because otherwise the relevant theory would become inconsistent. Consequently, it is not surprising that definitions in the empirical field should also meet certain requirements of consistency, and, more particularly, that these requirements may prove to be empirical, instead of being demonstrable, as in the case of logico-mathematical definitions. Thus the correctness of empirical predictions derivable from a conditional definition would be considered as a necessary requirement for the consistency of the definition. Should this requirement fail to materialize, the definition would have to be abandoned because of the inconsistencies it implies. This would not be interpreted as a proof that the definition has turned out

[10]Cf. S. Lesniewski, "Grundzüge eines neuen Systems der Grundlagen der Mathematik," *Fundamenta Mathematica*, vol. 14 (1929).

to be false; in the same way, in the logico-mathematical field, a definition which fails to meet the consistency requirement is rejected merely because of this failure. Should a definition be rejected because it has allegedly been shown to be false then, by the same token, its negation would turn out to be true and could be incorporated in the theory. This, however, never happens. Consequently, the reason for rejecting a definition (either in the logico-mathematical or in the empirical field) is its failure to meet the requirement of consistency, rather than its falsity.

I realize, of course, that asserting the empirical nature of consistency requirements for definitions in the empirical field implies a considerable departure from the usual views of the status of definitions, and that is why I have rather emphasized the first solution to the problem raised by the derivability of empirical conclusions from conditional definitions. This solution has superior conciliatory power—an advantage unduly neglected in philosophical discussions after Leibnitz. A slightly less conciliatory solution suggested by Bridgman's investigations consists in stipulating that every test-response pair should be considered as defining a distinct concept.[11] Hence, as far as definitions with non-relational tests are concerned, no empirical consequences would follow from any set of such definitions because no two of them define the same concept. The problem raised by conditional definitions with relational tests is not taken care of by this device. The first half of our first solution has to be applied in this case and would thus produce, in conjunction with Bridgman's suggestion, a comprehensive solution of the problem concerning the empirical implications of conditional definitions.

A simple formal implementation of this suggestion would come to the following prescription[12]: whenever a set of conditional definitions (appropriately processed whenever a relational test is involved) $D_1, \ldots, D_n$ is presented for a set of concepts $Q_1, \ldots, Q_n$ and, in scientific practice, the same symbol Q is being used for all these concepts on the understanding that they are all of them coextensive, the definitional set has to be supplemented by a set of empirical laws asserting that any pair of concepts Q_i, Q_j ($i \neq j$) is coextensive. These laws would suffice to obtain all the empirical consequences that can be derived from the conditional definitions under the usual interpretation. Should some of these predictions be falsified by subsequent observations, then the laws concerned

[11]I am here identifying (somewhat inaccurately) conditional definitions with Bridgman's "operational definitions" and am referring to his view that different sets of operations always define different concepts. It is an empirical question, according to Bridgman, whether concepts defined by different operations may be consistently assumed to be coextensive.

[12]I owe this suggestion to Paul Benacerraf.

would have to be abandoned but the status of the definitions would not be altered.

This solution does not require as frequent a re-classification of conditional definitions under the heading of semi-definitional laws as the first solution does (only definitions with relational tests would have to be re-classified). It is less conciliatory, however, because it entails a proliferation of scientific concepts out of proportion with the actual size of the scientist's conceptual apparatus. For example, each definitional criterion for the concept of length would engender a separate concept of length. Since this concept occurs in virtually every physical law, it has as many definitional criteria as there are laws of this type. To say that the actual number of different concepts of length is comparable with the number of physical laws seems hardly illuminating. Moreover, the sum-total of definitional criteria derivable from a given set of laws or from a given theory cannot be effectively determined. Consequently, the empirical equivalences among all the relevant concepts cannot be effectively listed and the use of the same term for a multiplicity of concepts cannot be effectively justified. That is why it seems preferable to re-classify most of the conditional definitions of scientific concepts under the heading of semi-definitional laws and to consider instead the set of associated criteria.

Semi-Definitional Laws and the Problem of Objectifiability

If a concept is introduced by a single conditional definition and no further definitional context is available for it, its areas of precision do not exceed the area of the test involved in this definition; whatever is beyond this area and was never submitted to this test, belongs to the area of vagueness of this term. That is why "dispositional" terms are so vague if not accompanied by semi-definitional laws: he who defines memory as the ability (power, faculty, disposition) to reproduce past perceptions, value as the ability to satisfy needs, beauty as the ability to please (a cultivated onlooker), and ethically right action by its ability to be approved by informed and unbiased arbitrators, attaches extremely vague meanings to the terms under consideration, unless semi-definitional laws are established to provide new criteria for, and decrease the vagueness of, these terms. Without such laws, the conditional definitions offer hardly anything but a modest beginning in the understanding of the term concerned.

Semi-definitional laws are therefore decisive in determining the meaning of conditionally defined terms. As a rule, the initial definitional test and response are relational, and the criterion entitles one to attribute the defined property to a given object only if this object happens to bear to some other entity the relation implied in the definitional test. Once semi-definitional laws become available and provide new definitional criteria for this property, one is in a position to attribute it to a particular object even if this object does not bear to any other object the relation involved in the original definitional test. Thus, a definite length could be attributed only to objects on which a yardstick has been superposed if superposition were the only available criterion of length. However, once other criteria are provided by semi-definitional laws involving the concept of length, a definite length can often be justifiably ascribed to an object regardless of whether it had ever been put in touch with a yardstick. This is what is meant when length is said to be an *objectifiable* concept. It can be ascribed to physical entities *objectively*, that is, independently of whether the investigation implicit in the initial definition of length has been carried out in regard to these entities. Objectifiability of length is thus secured by the laws which govern this quantity and which have been discovered by applying its initial conditional definition.[18]

It is important to realize that according to their content, the laws involving a conditionally defined concept may either provide for, or prevent, its objectification. The most troublesome epistemological feature of quantum theoretical laws consist precisely in their preventing the objectification of the concepts involved. To understand this logical mechanism, let us assume that the same concept P is endowed with two independent criteria involving the test-response pairs (T,R) and (T',R') and that, furthermore, a law comes to be established asserting that every object which meets the condition C will yield response R to test T but will fail to yield response R' if submitted to test T'. It is obvious that if such a law were established, we could not objectify the property P, that is, ascribe it to an object which meets the condition C, regardless of whether any one of the tests T or T' were performed. For we could not explain this objectification by saying that although neither conditional definition is used to prove that the property P applies to a given object, yet we have good reason to assume that if this object were submitted to any test capable of detecting the property P, then the object would respond so as to justify our ascribing to it the property P. In view of the above law, we know in advance that, depending upon whether we apply the test T or T', the object will turn out to exhibit the property P or to lack it.

[18]For example, if the initial criteria of length and temperature are used to establish a law of thermal expansion $L_T = L_0 (1 + c.T)$ correlating the length L_T of an object at the temperature T with its length L_0 at the temperature $0°$ then it is possible to determine the length L_T by measuring only the corresponding temperature, without any recourse to the initial definitional criterion of length.

Thus, in the Stern-Gerlach experiment as interpreted by F. London and E. Bauer,[14] the spatial orientation of a swarm of silver atoms inside a container is shown to be an unobjectifiable concept. For, if a method of investigation is applied to these atoms which is capable of ascertaining whether any of them are oriented in one specifiable direction, then, according to the laws of Quantum Theory, all the atoms in the container will prove to be oriented along this axis, say from south to north. If, however, another investigation, capable of disclosing whether the atoms are oriented in some other direction, is applied to them, then a specifiable non-vanishing fraction of the atomic population will turn out to be oriented along this other axis. It is not the case that a definite orientation (or, rather, re-orientation) of either a single atom or of the whole population would be brought about by applying one or another mode of investigation. The theory simply predicts the outcome of any measurement or experiment which could possibly be performed on the physical system under consideration. Its predictions are, as always, based on universal laws, which in this case, however, happen to prevent the objectification of the concepts they involve. Thus, if neither of two investigations has been carried out, we could not assume that the atoms nevertheless have a definite orientation. For we would then have to assume that they are all of them oriented along one axis, whereas there is also a non-vanishing fraction of the atoms oriented along any other pre-assigned axis. This is plainly impossible. The conclusion is that a definite orientation can be ascribed to the atoms only if some appropriate investigation has actually been carried out, that is, if the atoms have been submitted to some test T and have responded in a specifiable way R.

It is worth noticing at this juncture that there is no inherent logical inconsistency in the situation just described, and that a similar situation could possibly arise with common-sense objects. The fact that the laws which connect the perceivable common-sense objects with their sensory appearances do not prevent the objectifiability of common-sense concepts is a logically contingent feature of our experience. Logically speaking, this experience could be governed by different laws so as to imply the unobjectifiability of common-sense concepts.

A fictitious example of non-objectifiable common-sense concepts may be constructed as follows. Let us assume that a rather moody person X occupies a two-room apartment; we can find out over the telephone whether X is cheerful or depressed, and, by dropping in on him, may ascertain in what room he is then dwelling. We may then discover a

[14]Cf. F. London and E. Bauer, *La Théorie de l'observation en mécanique quantique* (1939), p. 33.

statistical correlation between his mood and his location to the effect that: "If X is cheerful (over the telephone) and is then looked for in Room 1, he can always be found there. But, if cheerful and looked for in Room 2, he can also be found there in a non-vanishing percentage of cases." (We assume, of course, that it is impossible to look for X in the two rooms at the same time, say by employing two investigators.) It is a matter of experience that human beings and other large-scale objects are not governed by laws which would prevent the objectification of concepts referring to them.

Let us conclude that the unobjectifiability of quantum-theoretical concepts is definitely incompatible with the phenomenalistic claim that statements referring to unobserved physical entities are faithfully translatable into equivalent hypothetical statements about the observational results that would ensue if certain specifiable operations were performed. According to the phenomenalistic view of the subject-matter of scientific knowledge, the presence of a table in a dark room at a given moment would be equivalent to the assertion that if an observer entered this room at this moment and lit the room he would perceive the appearance of a table-like object. This alleged equivalence between categorical statements about unobserved common-sense objects and hypothetical statements concerning potential observational results will be touched upon in § 32. At the atomic and sub-atomic level, the equivalence breaks down because of the unobjectifiability of quantum-theoretical concepts. We cannot say that the ascription of a property P to an atomic or sub-atomic object x is equivalent to the prediction that if the definitional test T of the property P were applied to x then a positive response R would ensue: under the same circumstances, we are also in a position to predict that if another definitional test T' of the property P were applied to x the response R' would fail to materialize and x would turn out to lack the property P.

21   ON DEFINING SCIENTIFIC LAWS

What exactly do we mean by a scientific law? We have surveyed, in the two preceding sections, the main types and functions of those statements which are usually granted the status of scientific laws, to prepare for our major objective in the present chapter: a critical survey of scientific methods of discovering and establishing laws. As further preparation, we must consider, on the basis of the preceding discussion, the definition of the concept of scientific law. It must apply to all types of scientific laws listed in § 19 and take into account the main uses to which laws are being put within science, as discussed in § 20. Since the most significant attempts at defining the concept of scientific law have aimed at isolating the types of laws which meet the requirement of *universality*, we shall start with an examination of the definitional value of this requirement.

The main difficulty in any attempt at differentiating law-like from fact-like statements in terms of universality is that the concept of universality is ambiguous. The phrase "universality of statements" was seen to be interpretable in at least three ways, so that there are several definitions of a law-like statement, no two of which are equivalent to each other. (1) A statement can be defined as law-like if it is conceptually universal, that is, all its constituent terms are universal, not local. (2) A statement may be said to be law-like because its spatio-temporal scope is universal; in other words, if the statement applies to any relevant object everywhere and always. (3) A statement may be classified as law-like if it has several instances and is asserted to hold true of all these instances; in other words, if all its quantifiers are universal. Thus, universality of concepts, universality of scope, and universality of quantification are three distinct criteria of law-like status. Moreover, any two, or even all three, of these requirements of universality can be combined in order to define law-like statements. Most of these definitional possibilities have actually been suggested by various authors as a means of circumscribing the meaning of scientific laws. We shall see, however, that these versions of the requirement of universality are all violated by well-known laws of nature.

(1) *Universality of Quantification*

The requirement of universal quantification, to begin with, is disregarded in all those numerous laws whose quantificational pattern is at least partly existential. "There are liquid crystals," "there exists an indivisible unit of electrical charge" are laws of nature which clearly involve existential quantifiers. Laws which clearly combine universal and existential quantification are legion, since, as pointed out in §19, all statistical laws and all law-like statements expressible by differential equations belong in this category. There are also basic laws of nature which are neither statistical nor differential and nonetheless involve a mixed quantificational pattern. For instance, Dalton's Law of Multiple Proportions in chemistry, and Hauy's Law of Rational Indices in crystallography come under this heading. So does the law asserting the existence of atoms of electricity, since it amounts to a statement that the electrical charge of *every* material object is *some* multiple of a specifiable minimum charge.

As a matter of fact, a purely existential quantification would hardly be of value in a scientific law, since such a law would follow from any single fact which happens to instantiate it. The knowledge of this particular fact would make its law-like competitor superfluous. Laws are introduced if they are able to say more, not less, than facts, in approximately the same number of words; if they were less informative than facts, they would hardly be resorted to in science. Actually, purely existential laws are employed primarily in axiomatic presentations of already available theories, since, under such circumstances, the weakest sufficient assumptions are the most desirable. Thus, Veblen, in his classical axiomatic foundation of Euclidean geometry, uses an existential postulate to the effect that "there is at least one point."[15] But an empirical theory, even when exclusively devoted to the search for laws and relegating particular facts to the role of crude, evidential material, is unlikely to care about such existential postulates whose content is more meagre than that of particular facts.

Let us conclude, therefore, that the requirements of universal quantification is certainly incompatible with the actual supply of scientific laws and that even purely existential laws, though scarce, nevertheless exist. Needless to say, while rejecting the requirement of quantificational universality with regard to law-like statements, we do not deny thereby the existence of some quantificational pattern in every law-like statement. Such statements are always quantified, although no formal restriction on their quantificational pattern seems warranted by examining their logical structure.

(2) *Universality of Scope*

With respect to the spatio-temporal scope of validity or applicability, three groups of law-like statements can be distinguished (cf. § 19): (a) *Universal laws* which are applicable to all relevant objects always and everywhere; (b) *Regional laws*, which apply to all relevant objects within, but not without, a limited spatio-temporal region; (c) *Individual laws*, applicable to a single individual or a group of a few individuals. It is obvious that if the requirement of universal scope is accepted, only universal laws fully deserve their law-like status, and neither regional nor individual laws are genuine. Newton's Law of Gravitation is universal in this sense; but biological evolution is governed by regional laws; Kepler's Laws of Planetary Motions have individual scope.

Yet, as already pointed out, it would be an arbitrary fiat to deny law-like status to well-established regional and individual laws. That is

---

[15]Cf. O. Veblen, "A System of Axioms for Geometry," *Trans. Amer. Math. Soc.*, vol. 5 (1904).

why several attempts have been made to reconcile the requirement of universal scope with these two types of laws. One of these attempts consists in partitioning the body of scientific laws in such a way that the dominating role of universal laws is taken into account, while providing some accommodation for laws of a more restricted scope. This is to be achieved by stipulating that a law of limited scope (regional or individual) must always be deducible from universal laws; the latter would then deserve the status of ultimate laws and the former would be only derivative.

Such a defence of universality of scope is hardly satisfactory, however, because, as a matter of fact, regional law-like statements are not always deducible from universal premises. Thus, according to the statistical interpretation of the Second Principle of Thermodynamics, the law asserting an over-all increase of cosmic entropy is valid during certain periods of cosmic history and invalid during the remaining periods. In our own period the law is apparently valid and thus the delusion as to its universality arises. But the regional law of increasing entropy, valid for that space-time region consisting of the whole of space and a considerable time-interval about the present moment, is not deducible from a universal law, since its universal law-like counterpart is false, and hence is not a law. This example shows that if the distinction between ultimate and derivative laws is defined in terms of scope, then the dichotomy "ultimate-derivative" is not exhaustive. A regional law may be neither universal nor derivative. The classification is therefore unable to assure a privileged rank to universal laws. The intention of confining law-like status to ultimate universal laws and those deducible therefrom cannot be carried out.

Another attempt at vindicating the requirement of universality of scope could be made by utilizing the distinction between full-fledged and vacuous universality. In §19 I have referred to vacuously universal laws of astronomy, biology, psychology, and human history, which are useful for limited regions where the relevant objects can be found, but useless outside these regions because of the absence of such objects. Thus, we may grant vacuous universality to Darwin's theory on the assumption that the entities and events governed by the laws of this theory are not to be found beyond a rather restricted spatio-temporal region.

Strictly speaking, two kinds of vacuously universal laws should be distinguished. A law may be prevented from applying beyond a specified region because of the contingent fact that no relevant objects are available outside this region although there is good reason to assume that if such objects were available, the law would apply to them; we shall then say that the law is vacuously universal in the *intrinsic sense*. Thus laws governing the motion of man-made clocks not only hold true of clocks on the earth, but would certainly also apply to non-terrestrial clocks if there were such. On the other hand, a vacuously universal law may be such that there are no grounds for claiming its potential applicability to relevant, though actually non-existent entities beyond a specified region; we shall then say that the law is vacuously universal in an *extrinsic sense*. Thus, the law that the size of any man-made clock never exceeds a specific limit is vacuously universal in the extrinsic sense, because its extra-terrestrial validity is simply due to the lack of relevant objects. There are some reasons for assuming that if the laws of biological evolution are vacuously universal at all, that is, if there is no life outside the earth, then their vacuous universality is at least intrinsic: life everywhere would take the evolutionary course it did take on earth. It goes without saying that vacuous universality in the extrinsic sense is of no interest in connection with the concept of scientific laws. On the other hand, if every scientific law could be shown to be at least vacuously universal in the intrinsic sense as far as its scope is concerned, the requirement of universality could be considered to be fulfilled.

I do not think that this attempt stands up under closer scrutiny, interesting as the concept of vacuous universality may prove to be from other points of view. As just pointed out and illustrated by the statistical interpretation of the Second Principle of Thermodynamics, there are laws which are neither universal nor deducible from universal laws for the simple reason that they are invalid outside a particular region. Hence, such laws are not even vacuously universal, because, though relevant objects are available outside their region of validity, these laws just do not apply to them. Hence regional laws, which are not even vacuously universal, are available, and consequently, this line of defence proves insufficient.

Yet even if we are prepared to grant law-like status to both universal and regional laws, we may feel some qualms about individual laws. Strictly speaking, the statement "Napoleon never seriously contemplated invading Great Britain" is not a particular fact, because of the universal quantifier "never" which stretches over a time-interval and prevents the whole statement from being replaceable by an unquantified, molecular, fact-like statement, under the usual infinitist assumptions concerning the time-variable.

The status of essentially quantified statements about individuals is, of course, arbitrary to some extent. A refusal to grant them law-like status is open to the objection that no essential logical difference is apparent between such statements and those which have regional universality. On the other hand, if we grant them law-like status, we run the risk of blurring the boundary between particular facts and general laws. Actually, none of these difficulties is insuperable because a significant logical difference can be pointed out between non-controversial individual laws and other quantified statements about individuals; such differences may warrant investing some quantified statements about individuals with the majesty of law.

Why do we not hesitate to refer to Kepler's conclusions from Tycho de Brahe's observations as genuine laws in spite of their applicability to but a small number of individuals, while we are rather inclined to consider the quantified statement about Napoleon as not being law-like? The point is that Kepler's laws, though not deducible from the universal laws of mechanics and gravitation, can be derived from these laws provided they are supplemented by a finite number of fact-like premises describing the basic structure of the planetary system, that is, the masses and mutual distances of the sun and its satellites. In other words, Kepler's individual laws are an individual instance of universally valid laws which apply to any and every planetary system; the instantiation rests on the individual data concerning our particular planetary system and is provided by the fact-like premises to be added to the universal laws of mechanics and of gravitation. The quantified statement about Napoleon is not known to be derivable from universal laws supplemented by a few relevant facts about Napoleon, and that is why we are reluctant to grant it law-like status. At bottom, the vague intuition of being law-like which underlies the subdivision of quantified statements about individuals into those which are and those which are not law-like, seems to be a feeling that whenever a quantified statement about individuals instantiates a universal law, it reflects so faithfully and remains so close to this law, that it naturally acquires law-like status by contact, as it were.

To comply with this vague intuition of the law-likeness of some quantified statements about individuals and with our intuitive reluctance to grant law-like status to all quantified statements about individuals, we have tentatively reinstated the distinction between derivative laws governing individuals and ultimate laws of universal scope from which individual laws would have to be derivable. Previously, however, we found that a similar treatment of regional laws is impossible because some of them are not derivable from universal laws. This circumstance suggests that it may be preferable to apply the same procedure to statements about individuals and to grant to these statements law-like status whenever they are essentially quantified, that is, not equivalent to any finite combination of facts and calling for inductive methods of validation instead of the fact-finding methods discussed in chapter Two. The following arguments favour the extension of law-like status to all essentially quantified statements about individuals:

(a) A regional law to the effect that all the individuals in the region R have the property P can always be re-formulated as meaning that the region R has an associated property P' (on the understanding that to have the property P' means to be populated by individuals endowed with the property P). This new formulation is actually required in order to ensure the applicability of inductive logic to the regional law, since, according to the usual assumptions, inductive inferences are valid for statements whose subjects involve local constants but may fail to apply to statements with locally defined predicates.[16] This requirement of inductive logic is met by the changed formulation of regional laws but not by the initial one. The new formulation, however, is hardly distinguishable from a statement about individuals. As a matter of fact, we have found reasons for re-formulating statements about individuals in terms of the spatio-temporal regions they occupy (cf. §18).

(b) Our intuitive reluctance to grant law-like status to quantified statements about individuals whenever these statements are not known to be derivable from universal laws seems to be due, partly at least, to the fact that individuals have come to be so strongly associated with fact-like statements that we are unwilling to consider any statement about individuals as law-like unless its logical proximity to a universal law compels us to do so. This explanation suggests that the whole dichotomy of statements about individuals into those that do and those that do not qualify as law-like is based upon fluctuating linguistic usage and vague psychological associations rather than upon sound logical grounds.

(c) Another source of our reluctance to consider all quantified statements about individuals as law-like is traceable to the circumstance that such statements usually seem to lack a certain flavour of necessity which we have come to associate with other more conspicuous examples of scientific laws. However, we may also feel reluctant to grant law-like status to statements which meet all the three requirements of conceptual, spatio-temporal, and quantificational universality. Consider the statement: "There are no gamma-ray microscopes in the universe." It has

[16]Cf. R. Carnap, "On the Application of Inductive Logic," *Philosophy and Phenomenological Research*, vol. 8 (1947).

certainly universality of concepts and of quantification, and, in all probability, universality of scope as well. The less dignified example "No human male has ever had a hat of more than 10 yards in diameter" also meets the three requirements in a similar fashion, the first two with certainty, the third in all probability (in view of the claim about the dimensions of human headgear being made only in regard to the male population). Yet, in spite of all the three requirements being fulfilled, we feel hesitant about considering the two statements as laws, because they lack the quality of *necessity*, often associated with scientific laws. Similarly, the three-dimensional nature of space is frequently felt to be "accidental" although the corresponding statement about space meets all the three requirements of universality and can hardly be denied law-like status. Actually, this "necessity" of scientific laws, analysed almost incessantly since the time of Hume, has the unpleasant habit of vanishing without notice, at the slightest attempt at logical analysis.[17]

(3) *Requirement of Conceptual Universality*

This requirement is ostensibly incompatible with the proposal to grant, under suitable circumstances, the status of law to quantified statements about individuals. Such statements must necessarily include local constants or individual proper names if the scientific theory is expressed in the vernacular. Similarly, the formulation of a regional law must involve local constants since, in view of the homogeneity of the spatio-temporal continuum, the particular region where the law under consideration is valid can only be specified by using local constants. In other words, the requirement of conceptual universality would automatically entail the universality of scope of all ultimate laws and provide at most for derivative laws of limited scope. We have argued, however, that not all laws of limited scope are derivative. The same argument can serve to refute the requirement of conceptual universality.

Yet, in view of the controversial nature of quantified statements of limited scope, it seems worth while to point out that serious reasons in

[17]During the last two decades, a considerable effort has been made by several authors to rationalize our intuitive reluctance to grant law-like status to some quantified statements of individual, regional, and universal scope. The objective was to explain the feeling of necessity often associated with the concept of law and to lay down a precise criterion capable of separating genuine law-like generalizations from merely accidental regularities even if the latter should be on a cosmic scale. These investigations have shed new light on related problems (e.g., the validity of inductive inference and the adequacy of conditional definitions for the analysis of dispositional concepts) but seem to have failed to bring the basic objective any closer. Cf. N. Goodman, *Fact, Fiction and Forecast* (1955), R. B. Braithwaite, *Scientific Explanation* (1955), W. Kneale, *Probability and Induction* (1949).

support of laws which violate the requirement of conceptual universality can be adduced independently of the attitude one takes up towards such quantified statements.

The requirement of conceptual universality is tantamount to a ban on local constants in scientific laws. What justification can be given for such a ban? We have seen that well-established laws of nature can easily be quoted which contravene it. As in the previous case, two different lines of defence can be devised to protect conceptual universality. These two defensive strategies will be referred to briefly as the distinction between a categorematic and a syncategorematic use of local constants or proper names in scientific laws, and the division of all scientific laws into ultimate and derivative ones, with the intention of reducing laws involving local constants to the latter category.

Let us examine the first line of defence. I have already pointed out that all the quantitative laws of classical science involved local constants or proper names, because the standards of basic quantities, such as length, duration, mass, were defined in local terms. I grant that this is not any longer so, and that there is an unmistakable, successful tendency to get rid of local terms in formulating the basic laws of nature. Yet, one can hardly refuse to apply the term "law" to the classical regularities only because they are formulated in a way involving local terms, either explicitly, or at least implicitly. The circumstance that such local terms have proved dispensable later, when the classical laws of nature have been re-formulated by means of universal terms only, is no real justification for denying the title of laws to the classical regularities in their original formulation. One has to notice that the very possibility of this contemporary "universalistic" re-formulation depends upon an empirical combination of circumstances, not upon the logical nature of the terms involved. If one realizes, moreover, that the basic contentions of, say, Newton and Maxwell would be affected immediately by a ban on local constants in laws of nature and re-classified as fact-like statements, one cannot help concluding that the mere presence of local terms in empirical propositions could not deprive them automatically of law-like status, unless some additional conditions were fulfilled. Some of these conditions refer to inductive validating methods and have already been touched upon.

For the time being, it may suffice to distinguish between statements in which local terms occur as *syncategorematic* components of other constituent expressions and those whose local terms function in a *categorematic* capacity. Thus, the fact that the basic units of physical quantities used to be defined in local terms does not by any means imply that the laws which refer to these quantities are *about* any single individual, because the role these terms play is but syncategorematic. A centimetre may be defined in relation to the individual "earth," but the statement to the effect that the size of an atom is of the order of 10^{-8} cm. is about atoms, not about the earth. On the other hand, Kepler's Laws of Planetary Motions refer explicitly to certain individuals, viz., the sun and its planets; they involve a categorematic use of individual constants.

A merely syncategorematic use of local constants may seem to be more easily reconcilable with the universalist tendency, even if these constants are not replacable by universal terms. For statements involving only such use of local constants can have a quantificational pattern which is universal throughout, and a universal scope of validity as well. On the other hand, a categorematic use of local terms obviously implies that the statement is about single individuals; its specific referential import is a direct consequence of the categorematic role the local signs play in its structure.

Now, the first line of defence of conceptual universality of scientific laws might consist in the claim that local terms, if occurring at all in scientific laws, are able to do so at most in a syncategorematic capacity. Yet, strangely enough, it is precisely this syncategorematic use of local constants that must be submitted to essentially restrictive rules, on grounds of inductive logic (cf. § 22). On the other hand, the categorematic use of such constants is both inductively unobjectionable and essential in regional and individual laws, say, in the geological laws that govern the succession of eras and periods in the history of the earth. There is no possibility of accounting for the presence of local constants in any law of this type by referring to their allegedly syncategorematic function.

Another attempt can be made to save the requirement of conceptual universality by subdividing, once more, all scientific laws into ultimate and derivative laws. This time the ultimate laws would abide by the requirement of conceptual universality, and the derivative laws are those which follow from ultimate laws. Thus a statement would acquire law-like status if it were either conceptually universal or deducible from other conceptually universal statements. For instance, Kepler's laws would owe their law-like status to their deducibility from mechanical and gravitational laws which do not contain a single individual name.

The trouble with this line of defence is exactly the one which arose in the previous case. Of course, regional and individual laws would now have to be derivative in a new sense, that is, deducible from conceptually universal laws. However, since conceptual universality is a more stringent requirement than universality of scope,[18] any quantified statement which can be deduced from conceptually universal premises is also derivable from premises whose scope is universal and would therefore owe its eligibility for law-like rank to its derivability from laws of universal scope. The same consideration would apply to regional laws. We have seen, however, that not all individual and regional laws are so derivable. Accordingly, the requirement of conceptual universality cannot be saved, even when confined to ultimate laws, since the latter do not provide a sufficient basis for deriving all scientific laws.

To sum up: neither conceptual nor quantificational nor spatio-temporal universality is relevant when establishing the law-like status of a statement; nor can such status be confined to statements which are deducible from premises whose universality is either conceptual or quantificational or spatio-temporal. Scientific laws should be defined as essentially quantified statements. According to this definition, the main difference between laws and facts consists in the methods of their validation. Particular facts are established by applying the non-inductive fact-finding methods described in chapter Two. The validation of laws is necessarily inductive.

(4) Some Alternative Definitions of Scientific Laws

The concept of scientific law plays such a prominent role in the whole structure of science that it is no wonder several attempts have been made to formulate a precise definition of it, which would be true to its actual use and also provide for some reliable conclusions to be drawn therefrom about the meaning, nature, and functions of scientific laws. There seems to be rather general agreement among analysts of science that a law-like statement is essentially quantified or general, in contradistinction to a fact, which is a particular unquantified statement about a single individual or a few individuals. The artificial terminology which the older positivists have tried to introduce by designating scientific laws as "general facts," in contrast to genuine "particular facts," did not strike any roots in language. Since we are anxious not to blur the important differences in methods of validation applicable to particular facts and general laws respectively, we have drawn a rather sharp boundary between the two kinds of propositions.

Yet, apart from the consensus about the role played by quantification

[18]If a true statement includes no local constants (and is thus a conceptually universal law), then it does not involve any reference to a single spatio-temporal region either, and is consequently of universal scope; the converse does not necessarily hold, as shown by our previous discussion of local standards of measurement implicit in the laws of nature.

in the logical structure of law-like statements, there is little agreement as to the exact shape of the quantificational patterns characteristic of law-like statements. Thus some writers require that all the quantifiers be universal in such statements (e.g., Hempel and Oppenheim,[19] and Popper[20]) and propose to designate as "theories" those statements, some or all of whose quantifiers are either partly or all of them existential. Burks[21] suggests that a statement be called law-like if all its quantifiers are of the same kind—either all existential or all universal—on the understanding that statements with a different pattern of quantifiers should be called "theories." According to Bergmann,[22] any quantifiers are admissible in a law-like statement, provided that all its terms be "empirical constructs," that is, linked up by a chain of full-fledged or conditional definitions with terms of the ostensive level. This variegated sample of views concerning quantificational patterns admissible in law-like statements could easily be enlarged.

The reasons for the various definitional proposals are rarely explicitly stated. It seems that the main motive is the desire either to come as close as possible to the actual meaning of a "scientific law" or to ensure the possibility of validating scientific laws on observational grounds. For example, if it seems desirable that a law be either finitely provable or finitely disprovable by observational findings then the requirement of quantificational homogeneity will be used to define laws. Bergmann establishes an inferential relation between laws and observational facts by his stipulation that the constituent terms of a law be empirical constructs, and this is why he can afford to be very liberal in regard to the admissible quantificational patterns of scientific laws. On the other hand, the desire to be true to the actual use of the term "law" may be responsible for the requirement of quantificational universality suggested by some writers: the most conspicuous examples of natural laws are universal in regard to quantification, although those which do not meet this requirement are far from negligible. The same motive is likely to be responsible for the ban against local constants and individual names made by several authors. The objections to which all these proposals are open have been discussed in the three previous sections.

An interesting feature of some of these proposals is that scientific laws, by definition, turn out to be a special case of scientific theories:

[19]C. G. Hempel and P. Oppenheim, "The Logic of Explanation," *Philosophy of Science*, vol. 15 (1948).

[20]K. Popper, *Logik der Forschung* (1935).

[21]A. Burks, "Justification in Science," *Proc. Amer. Phil. Ass.*, Eastern Div., vol. 2 (1953).

[22]G. Bergmann, "Outline of an Empiricist Philosophy of Physics," *Amer. Jour. of Physics* (1943).

every law is a theory, though the converse does not hold. We shall see
later that there are serious reasons for separating laws from theories by
a wider gap: theories will prove to be rather deductive systems consisting of laws than single statements which differ from law-like statements
in regard to their quantificational pattern. According to Pap, only
separately provable or disprovable statements can be laws, whereas
other statements of similar logical structure should be considered as
"theories." This amounts to classifying as theories the class of contextually justifiable laws and provides, once more, for too narrow a gap
between laws and theories.

The difference between law-like statements and genuine laws has
been discussed by Hempel. The question is whether laws should be
defined as true law-like statements, or as law-like statements known to
be only probably true, that is, supported by adequate evidence. The
first alternative is in closer agreement with ordinary usage, but raises
several objections: for example, since we can never be absolutely sure
whether a law-like statement is actually true, neither can we be absolutely sure whether any pre-assigned statement is a law, if laws are
defined in terms of truth. The same difficulty obviously applies to the
difference between fact-like statements and actual facts. Since we did
explain the latter difference in terms of truth, it seems reasonable to take
the same course with respect to the former. The difficulty that no statement can be known definitively to be either a fact or a law must not be
overrated, however, for two reasons. In the first place, it does not follow
that logical and epistemological questions concerning the nature, meaning, and validation of laws and of facts depend upon the empirical question of whether a given statement or class of statements fall under the
category of facts or of laws. What is involved in such epistemological
and logical investigations relates to whether a given statement is law-like
or fact-like, and this can be settled without presupposing any empirical
findings. In the second place, any misgivings as to the emergence, in this
context, of the concept of truth, which is unfortunately charged with
a long tradition of speculative vagueness and haziness, are unwarranted,
since modern logic has paved the way towards a precise analysis of
this concept, in accordance with its actual use. The meaning of truth
is of decisive importance for our whole investigation and will be discussed in detail in Part III.

The fact that several investigators have altogether denied the applicability of the concept of truth to law-like statements is easily accounted
for. Thus Ramsey and Schlick have assumed that only sentences meeting
a very stringent requirement of empirical verifiability express genuine

statements and can be labelled "true" or "false."[23] Since infinitist laws of nature do not meet this requirement they would have to be interpreted as recipes for predicting facts rather than as true statements. We shall try to show that this view of scientific laws, which would deny them any genuine cognitive function, is incompatible with an appropriately re-formulated Principle of Verifiability (cf. § 34).

Toulmin[24] seems to have been led to deny the applicability of the concept of truth to laws by his assumption that laws are mere formulae, that is, incomplete statements which do not specify the conditions of their validity. For example, Boyle's law concerning the relation between the volume V and pressure P of a gas at a constant temperature would be identified with the formula $V \cdot P = C$ without any indication being added as to the class of gases to which this formula applies. This is correct as far as it goes since nobody would ascribe a truth-value to an incomplete statement. However, there is no reason for denying truth-value to laws if they are interpreted as complete statements (as laws usually are) with all the conditions of validity either explicitly stated or implied by the meanings of the relevant terms.

22   ON VALIDATING SCIENTIFIC LAWS

Inductive Validation of Infinitist Laws

In the most typical cases, a scientific law is a single quantified statement with an infinite number of instances all of which are fact-like propositions. "This piece of iron expands after having been heated" is a qualitative, directly testable statement. "Every piece of any metal expands when heated" is a law whose class of instances includes the former example. In all these typical cases, even if they are simple enough to ensure that the instances of the law are directly observable facts, a characteristic difficulty in validating the law itself arises in connection with the infinite number of its instances. A body of evidence capable of *guaranteeing* the truth of the law would have to ensure that all its infinitely numerous instances had been checked appropriately and found to be correct. This would require an infinite number of observations, concerned with an infinite number of distinct objects. Clearly, mustering such a comprehensive body of observational evidence is beyond the means of any human being, although no single instance of the law,

[23]F. P. Ramsey, *Foundations of Mathematics* (1931); M. Schlick, "Die Kausalität in der gegenwärtigen Physik," *Die Naturwissenschaften* (1931).
[24]S. Toulmin, *The Philosophy of Science* (1953).

chosen at random, raises any unsurmountable difficulties. Thus, the problem of producing adequate observational evidence in support of infinitist scientific laws raises difficulties not to be compared with those encountered in our survey of fact-finding methods; their solubility seems rather dubious at first glance.

Yet science does not hesitate to incorporate infinitist laws into the knowledge she claims to possess, following in the footsteps of common sense. The classical procedure resorted to by the scientist in this connection is familiar and, admittedly, as indispensable as questionable; if a finite number of instances of an infinitist law have been investigated and all of them happen to be correct, then the law itself is considered as "so far confirmed," or as rendered probable to a degree depending upon the number of examined instances and upon their variety. This is rather surprising, since, the number of all the instances of the law being infinite, any finite number of its confirmed instances would form a negligible fraction of the total and hence fail to affect the final probability conferred upon the law.

If a mathematical proposition that claims applicability to all natural numbers (e.g., the proposition that every even natural number exceeding 4 is the sum of two odd prime numbers) were put forward as a "well-established law," on the ground that a sizable number of its instances had been checked and found correct without any exception, no mathematician would countenance such a claim. From a mathematical point of view, the infinitist mathematical proposition would remain gratuitous and unwarranted, regardless of its confirmed instances, simply because mathematicians can afford to confine themselves to deductive methods of validation. An empirical scientist, however, is not in such a privileged position. He cannot simply dismiss deductively unsolvable problems and must put up with probable or inductive inference as a legitimate and indispensable method of validation, apart from the deductive procedures shared with the mathematicians. This, of course, is not the only methodological difference between the two fields of science. Thus, in contradistinction to the mathematician, the empirical scientist is also compelled to apply methods which involve the use of instruments (microscopes, telescopes, yardsticks, clocks, etc.). Moreover, even though the role of mathematical axioms is duplicated to some extent by directly (say, perceptually) validated facts of empirical science, the analogy between the respective roles of axioms in mathematics and observationally established facts in empirical science is nevertheless misleading, simply because the empirical scientist also has recourse to axioms and similar analytic propositions, independently of, and in addition to, the statements he validates by direct observation. The presence of statements validated by direct observation used primarily to validate, in turn, general laws, is therefore another distinctive feature of empirical science, along with the use of inductive inference and the recourse to instrumental equipment. All three of these features are lacking in the mathematician's methodological equipment.

The question of how a general law with an infinite number of instances can be substantiated by examining a finite selection of these instances is one aspect of the Problem of Induction which has occupied, since Hume, a central place in the philosophy of empirical science. Sometimes, it even seems that this is the only problem of interest in the philosophy of empirical science. We cannot afford to give a detailed survey of the most significant answers which have been put forward so far in connection with this problem and its relevance to the validation of scientific laws. These solutions range from the nihilistic answer, which simply declares the problem spurious, non-existent, or meaningless (and thus, instead of solving it, rather dissolves it by resorting to a brand of logical analysis pretty close to logomachy), to proofs which claim to justify the principle in a demonstrative, quasi-mathematical way, with a host of intermediary positions, illustrated by H. Reichenbach's[25] life-long effort to prove that, if the problem has a solution at all, the inductive inference (from a sample to the whole population) constitutes this unique solution.

We must not think, however, that the whole difficulty raised by the inductive validation of a scientific law by a set of premises consisting of a selection of the fact-like instances of this law, comes from the circumstance that the number of all the instances is infinite. This infinitist nature of so many scientific laws is certainly an aggravating feature, as we shall see before long, but it is not the core of the epistemological problem involved in inductively validating scientific laws. Even if the number of all the instances of a scientific law is finite, there still remains the question of how the investigator, who has succeeded in making sure of the correctness of a sizable sample of the relevant instances, is entitled thereby to consider all the remaining, unexamined instances, and consequently, the law itself, as true in all probability. In other words, in so far as the classical inductive inference from a sample to a whole population is concerned, the finitude of the population does not remove the puzzling nature of the inference.

It is also important to realize that, in spite of the historical interconnection between the Problem of Induction and the Problem of

[25]Cf. H. Reichenbach, *Experience and Prediction* (1938).

Causality or Indeterminism, the two are actually independent of each other. The Problem of Induction, in its most typical though unduly restrictive formulation, runs: What reasons, if any, have we for assuming that the probability of a given law being correct increases in proportion to the number and variety of its tested instances, provided that all the tested instances have proved to be correct? The Problem of Causality runs: What reasons, if any, have we for assuming that every particular fact could be predicted with certainty (i.e., deductively inferred) provided that the investigator has at his disposal the knowledge of the relevant scientific laws and of a suitable number of particular facts related to a time-moment immediately preceding the occurrence of the event described by the fact under consideration? It is needless to add that the "relevant scientific laws" and the required "number of particular facts" would then render the fact in question *predictable* because they would make up a sufficient set of premises for deducing this fact in advance.

In order to demonstrate that the Principle of Induction and the Law of Causality are logically independent of each other, it suffices to point out two cases, each of which involves the correctness of one only of these principles. Thus, in Quantum Theory, the Principle of Induction is as essential as anywhere to substantiate the basic law-like contentions by means of particular facts, but the Law of Causality has been shown to be incompatible with these basic contentions of the theory (von Neumann). On the other hand, even if the Law of Causality were known to be correct, it would simply have no bearing upon the general validity of the Principle of Induction, since the inductive inferences which do not proceed from premises about the past to conclusions about the present or the future, though justified by the Principle of Induction, are not affected at all by the Law of Causality. To put it in a nutshell: predictability of future events, which forms the gist of Causality, is not equivalent to the legitimacy of generalization, the gist of Induction. Historically speaking, this mutual logical independence of the two problems is compatible with Hume's joint discussion of them; as a matter of fact, we often do discover causal laws (i.e., special cases of the Law of Causality) by applying inductive inference from a sample to the whole population. This has nothing to do with the fact that the Law of Causality, even if correct, would fail to substantiate the Principle of Induction, just as the Principle of Induction, even if correct, would not prove the Law of Causality.

This last remark brings up another point about the Principle of Induction which became apparent in recent discussions. It turns out that the validity of the Principle depends essentially upon the conceptual apparatus used in the premises and conclusions of inductive inferences. In other words, should the Principle prove valid with regard to inferences from facts concerning such things as colours, locations, or weights, it may then well prove invalid when applied to similar inferences from premises concerned with sounds, values, etc. Moreover, in the most typical cases, in which the conceptual apparatus consists mainly of a finite set of predicates, it can be proved that, given any set of predicates to which the Principle of Induction is applicable ("projectible predicates"),[26] other sets of predicates can be constructed which do not behave in accordance with the Principle. This makes one question the possibility of solving the Problem of Induction without restricting the relevant inductive inferences to a well-defined conceptual apparatus. Claiming the validity of the Principle with regard to every possible conceptual apparatus is demonstrably false. This leaves open the question of the validity of the Principle with respect to the actual conceptual apparatus of science.

There are thus three essential qualifications in the controversy over the validity of the Principle of Induction and its applicability to the validation of scientific laws. (1) Although the applicability of the Principle to infinitist laws is a puzzling problem, its applicability to finite scientific laws cannot be taken for granted either and raises serious difficulties in its own right. (2) The historical interconnection between the Problem of Induction and the Problem of Causality must not be confused with their logical interrelation. The Principle of Induction is essential in every empirical science, the Law of Causality is not. (3) The validity of the Principle of Induction cannot be sensibly discussed without specifying the conceptual apparatus used in inductive inference.

The problem of the inductive validation of infinitist laws raises an additional, fourth difficulty, superimposed upon the three difficulties mentioned above. Yet, as the basic laws in the fundamental scientific theories are infinitist, we cannot disregard such laws altogether. This would amount to scrapping Newton's mechanics, Maxwell's theory of electro-magnetic phenomena including light, the Quantum Theory of atomic phenomena, etc. If literally interpreted, each basic law of these theories admits an infinite number of instances, and each raises the question of how the validity of an infinitist law can be substantiated to a reasonable degree by verifying a finite sample of its instances selected according to some appropriate rule. I have already referred to the "nihilistic" and the "intermediary" answers to this question. At first

[26]Cf. N. Goodman, "A Query on Confirmation," *Journal of Philosophy* (1946).

glance, the result reached in R. Carnap's theory of inductive inference[27] also seems unsatisfactory. Starting with a general definition of the degree of probability conferred by the premises (or "evidence," expressed in a suitable proposition E) on the conclusion ("hypothesis" H), he shows that, on the assumption that H is a general law and E a sample of instances of H, this probability, as defined by him, increases in proportion to the number and variety of the confirmed instances summarized in E, provided the hypothesis H admits of a *finite* number of instances. The probability, however, conferred upon an infinitist law by any finite number of confirmed instances is always zero in Carnap's inductive or probabilistic logic, which we shall use as a basis for the subsequent discussion.

This result of Carnap's is in a sense disappointing, at least to those who were prepared to view the sum-total of well-established scientific laws as the gist of scientific knowledge and the most valuable information supplied by science. What sense would it make to consider as a piece of reliable information a statement whose probability is zero on the data available to the scientist?

There are two ways out of this infinitist predicament. One may consider, first of all, the probability of any single new instance of an infinitist law being correct, instead of investigating the probability that the law itself is correct. In other words, the inference from a sample of instances of the law to the whole population of instances, that is, the law itself, may be replaced with an inference from the sample of instances to any new instance not included in the sample. It is clear that this so-called instance-probability of a law is different from the probability of the law itself, although, for several theoretical and practical purposes, the former is an adequate substitute for the latter. The point is that, in regard to this instance-probability of an infinitist law, the ordinary procedures of the scientist and the man in the street are justifiable, within the framework of Carnap's inductive logic: the probability that any single new instance of an infinitist law, chosen at random or designated in advance, will prove correct, turns out to be dependent upon the number and variety of the previously verified instances of the law. Yet some of the characteristic functions of scientific laws may seem endangered by this approach. One may wonder, for example, whether the informational function of a scientific law can be performed, should its probability shrink to zero regardless of the data available to the scientist.[28]

Another way out consists in replacing the infinite scope of laws with

[27] Cf. R. Carnap, *Logical Foundations of Probability* (1950).
[28] Cf. §34 and the last paragraph of this section.

a sufficiently wide finite region of validity. It seems rather questionable whether much of the informational and explanatory value of scientific laws would be lost should validity be claimed only for a finite space-time region, say the region of presently accessible galaxies (with a liberal allowance for possible advances in the construction of telescopes) during the last few billion years and with a similar projection into the future? Yet the infinitude of instances of scientific laws comes mainly from the infinitist geometry of space-time underlying these laws. Consequently, even if only a finite portion of space-time were involved, the number of point-instants in this portion would still be infinite. This difficulty may not prove insuperable, since attempts at replacing our infinitist space-time geometry with a finitist one have been made and are far from hopeless. But, so far, it would certainly be premature to admit their feasibility.

The Validation of Special Types of Laws

We have discussed, so far, the validation of infinitist laws. The difficulties arising in this context affect more particularly some important types of laws, for instance (*a*) laws involving mixed quantificational patterns, (*b*) statistical laws, and (*c*) laws of unlimited spatio-temporal scope. Let us comment briefly on how these types of laws are established by inductive methods.

(*a*) The bearing of quantification upon the possibility of validating law-like statements depends upon the number of their instances.[29] If the statement has a finite number of instances, the admissibility of any quantificational pattern (existential, universal, or mixed) does not give rise to any objection as far as the possibility of validation is concerned. If the number of instances is infinite, the law is finitely provable but not finitely disprovable on observational grounds in the event of all its quantifiers being existential, whereas in the event of all quantifiers being universal, the law is finitely disprovable without being finitely provable. If quantifiers of both kinds occur the law is neither finitely provable nor finitely disprovable.

Thus the effect of quantification on the possibility of validation is obvious, whenever infinitist laws are concerned. It is worth while noticing that the bearing of quantification upon validation is entirely due to the concomitant action of the infinitist nature of the law.

---

[29]The instances of a law L with a mixed quantificational pattern can be identified simply with the instances of a quantificationally universal law L' obtained from L by replacing the mixed quantificational pattern with a purely universal one. (Cf. § 30.)

As far as validation proper, that is, the production of adequately supporting observational evidence, is concerned, we may notice that little is gained by recourse to various kinds of quantification in the case of an infinite universe of discourse. Only law-like statements all of whose quantifiers are existential can be validated by a finite number of observational data; yet such laws are deducible from any of their fact-like instances and hardly deserve their status. Laws with mixed quantifiers are no better off than those with universal quantifiers.[30]

(b) The distinction between statistical and non-statistical laws has no immediate bearing on the problem of validation, except through its possible linkage with the problem of infinitude. Since statistical laws deal with whole populations in contradistinction to non-statistical ("causal") laws, where only single individuals are referred to, the whole question of validation hinges upon whether or not the population in question is infinite. This is, as a matter of fact, the case in almost all the important theories. In mathematics infinite populations are preferred, because they are more interesting as an object of study. In the natural sciences the same preference is noticeable, because infinite populations are governed by essentially simpler statistical laws and provide for a manageable approximation to the statistical laws of finite populations, whose mathematical complication is often prohibitive. This is another striking example of the simplifying power of the mathematical concept of infinity in empirical science (cf. § 16).

According to the frequency interpretation of infinitist statistical laws, their quantificational patterns are always mixed (cf. § 19). Such laws are therefore neither finitely provable nor finitely disprovable on observational grounds. This need not prevent them from receiving inductive validation.

(c) On the usual views of space-time, a law applicable always and everywhere would be infinitist. Such universality would therefore affect adversely the validation of the law. If the universe of discourse consists of physical individuals like elementary particles, the number of which is finite according to some investigators (e.g., Eddington), the situation may seem more favourable.

Yet even in the latter case there are serious difficulties. One may doubt, first, whether the finitude of the number of elementary particles is itself an analytic consequence of their respective definitions, or rather a synthetic hypothesis derived from astrophysical data. The latter is

---

[30]As a matter of fact, we shall have to acknowledge (§30) that, regardless of its quantificational pattern, any law is the consequence of at least one suitably delimited class of its instances.

obvious in Eddington's speculations; consequently, according to our concept of universality of scope, the finitude of laws claiming applicability to all elementary particles cannot be maintained. Moreover, Eddington's estimate of the number of particles occupies rather an isolated position in contemporary scientific thought. In Dirac's light theory, there is an infinite number of photons, though most of them seem to enjoy a rather shadowy kind of existence.[31]

Let us conclude, therefore, that universality of scope is obtainable only at the exorbitant price of infinitude when assertions about the whole of cosmic space-time are concerned (e.g., in the theories of Newton and Maxwell), and that, for a universe of discourse consisting of elementary particles, the possibility of achieving such universality at a bargain (finite) price seems rather questionable.

It is worth noticing that the foregoing discussion applies only to what may be termed the general method of *directly* validating an infinitist law by applying inductive processes to its instances. So far, we have not dealt with law-finding procedures of an indirect type, for example, with the extremely important method of validating a law by deducing ("predicting") it from a well-established theory; this type of indirect inductive validation will be touched upon in connection with scientific methods of theory-formation (chapter Three of this Part). Nor have we taken into account the multiplicity of more special law-finding methods such as the method of "confidence-intervals" for validating statistical hypotheses. For reasons indicated on a previous occasion, only the most general validating methods are surveyed in Part II.

It should be pointed out finally that the special difficulty arising in connection with the inductive validation of infinitist laws, namely the fact that the probability conferred upon them by any finite amount of instantial evidence is always zero, will turn out not to affect adversely the range of science. This range will be shown to consist of all statements possessed of a definite truth-value, whereas statements transcending the scientific method will prove to be indeterminate, that is, devoid of any truth-value. Since the truth-values of all the instances of an infinitist universal law are determinate because they can be ascertained in principle by inductive method, it follows that the law itself has also a definite truth-value (cf. § 34). Hence infinitist laws will prove to be inside the range of science even if no non-vanishing probability is ascribable to them on the basis of any conceivable observational evidence.

[31]Cf. P. A. M. Dirac, *The Principles of Quantum Mechanics* (1935).

CHAPTER THREE

Methods of Theory-Formation

---

23 TYPES OF SCIENTIFIC THEORIES

General Remarks

Scientific *facts* and *laws* can be viewed as single items of reliable information. The content of one of these items is expressed by a single statement which formulates the fact or law in question; its reliability rests on a body of supporting evidence which is implied in the concepts of fact and law as distinguished from fact-like and law-like statements, respectively. A scientific *theory* has previously been defined as a potentially infinite set of statements all of which can be deduced from a specified finite set of basic assumptions. If these statements—usually referred to as the "consequences" of the theory—are supported by adequate evidence, we can still view each of them as a bit of reliable information; moreover, we may consider the whole theory, the task of which is to convey all these bits in a condensed form, as a single package of such information. Theories, accordingly, play a cognitive or informational part comparable, in principle, to that of any scientific fact or law; the main difference consists in the amount of information conveyed by facts, laws, and theories. That two particular electrically charged bodies attract each other, is a bit of *fact-like* information. Coulomb's *law* of electrostatic action constitutes a more condensed bit of information. Maxwell's theory of electro-magnetic phenomena offers information in a form even more condensed than Coulomb's law; it contains this law in addition to an infinite number of other laws derivable from the field-equations which constitute the basic assumptions of the theory.

To put it otherwise: The scientific solution to a particular problem may consist either in a single fact-like or law-like statement in conjunction with a body of supporting evidence, or in a system of validated statements adding up to a full-fledged theory. In the latter case, we have to view the whole system of statements which make up the theory under consideration, in conjunction with the relevant evidence, as the solution to a single problem and, hence, as constituting a piece of scientific knowledge, that is, of scientifically relevant and reliable information. Whether scientific information is offered in bits or in packages is rather a technicality. Both bits and packages meet the dual requirement of scientific status. There is no *a priori* reason for discarding packages of information, nor is there any doubt about the essential role which such packages play in scientific method.

Needless to say, a distinction similar to that between fact-like and law-like statements on the one hand, and facts and laws on the other, has to be drawn in regard to theories as well. We might distinguish theory-like systems of propositions from full-fledged theories, that is, propositional systems in conjunction with their validating methods. However, the term "theory," as usually interpreted, should be classified with fact- and law-like statements rather than with facts and laws proper. We shall therefore refer to "well-established theories" or "validated theories" as the counterparts of facts and laws.

Accordingly, we shall have to examine, in this survey of scientific methods, those used especially in order to validate theories, since such methods are as integral a part of any well-established theory as they proved to be an essential ingredient of any scientific fact or law. Owing to the greater logical complexity of scientific theories, their validation proves even more opaque to logical analysis than the inductive substantiation of laws. As a matter of fact, we shall have to acknowledge that very little is known at present about scientific methods of forming theories. Such methods, however, are more likely to provide some insight into the nature of scientific knowledge than those related to scientific facts and laws. For science is not a mechanical assembly of more or less established facts and laws. At bottom, the whole of science is a single organic unit. The particular facts and laws which one can distinguish in what goes at present by the name of science are artificially isolated cells in the scientific organism. The tissues and organs of the living body of science consist of entire scientific theories and of clusters of theories rather than of the cellular components termed facts and laws. There is obviously more probability of obtaining some understanding of how the whole living body of science functions if one concentrates on its organs and tissues rather than on the microscopically accessible cells.

204   THE REACH OF SCIENCE

Before taking up the methods of validating scientific theories, we shall discuss the main types of such theories and their main functions. Classifications of types of theories can be derived from the classifications of their constituent laws, outlined in § 19. If a theory consists of laws of a single type (e.g., of conceptually universal or of causal laws), this common feature is usually attributed to the theory itself. Thus, Newton's "causal" theory of mechanical phenomena and Maxwell's "causal" theory of electro-magnetic phenomena are often opposed to the "statistical" or "statistically interpreted" Quantum Theory of atomic and sub-atomic phenomena, on the understanding that the laws which make up these theories fall under these two headings respectively. We shall not dwell, however, on such derivative classifications of scientific theories but proceed with the discussion of two important divisions which apply to theories alone and have a definite bearing upon their validation.

Axiomatic versus Non-Axiomatic Theories

A scientific theory is always a deductive system,[1] that is, a set of propositions which contains all the logical consequences of any of its finite sub-sets and cannot, therefore, be extended by adjunction of deductively derivable statements. That is why deductive systems are often referred to as propositional systems closed under deductive inference. A deductive system need not be *axiomatizable*, that is, consist of propositions which are all of them deducible from a finite subset of the system: for example, the infinite system of propositions which can be obtained from the formula "The number of electrons is not N" by substituting all the natural numbers for N is clearly not deducible from any finite set of such inequalities and is, therefore, not axiomatizable. Yet scientific theories, as previously defined, are axiomatizable deductive systems: such a theory consists of all the logical consequences of a specified finite set of assumptions and can be deduced from a finite subset of such consequences, since all the assumptions are their own consequences and belong themselves to the theory. Any finite and consistent set of assumptions the consequences of which coincide with a particular theory may be called *an axiomatic basis* of this theory; the choice of some particular axiomatic basis of a given theory is called an axiomatization of this theory. An axiomatizable theory admits usually of an infinite number of axiomatic bases; a particular axiomatization of a given theory is therefore a special representation of it, not to be confused with the theory itself, which remains the same in its various

[1]Cf. A. Tarski, "Grundzüge des Systemenkalküls," *Fundamenta Mathematica* (1935).

axiomatizations. Thus, Maxwell's theory can be axiomatized either by assuming his field-equations or by utilizing assumptions concerning the two electro-magnetic potentials. The difference between these axiomatizations does not affect the unity of the theory itself.

Axiomatizing a theory amounts to constructing an axiomatic system whose axioms, definitions, and theorems add up to this theory. We have commented on the basic structure of a formalized axiomatic system on a previous occasion and have pointed out that such a formalized system is determined by four lists which specify its vocabulary, its grammatical syntax, its supply of axioms, and its rules of inference. Since all non-formalized axiomatic systems can be effectively subjected to formalization, we shall take into account only formalized systems. There is one point to be made in connection with such systems in the empirical field. Apart from intra-systemic definitional rules of inference which entitle one to replace in any theorem a defined expression with its corresponding defining expression, empirical axiomatic theories also involve extra-systemic "interpretive" definitions which may be thought of as separately listed (List No. 5); they explain either intra-systematically undefined terms, or intra-systematically defined terms, or a combination of both.[2] These interpretive definitions, usually of the conditional type, are not couched in the language of the theory under consideration and contain words which do not occur in List No. 1. As a rule, such definitions are expressed in ordinary language supplemented by the vocabulary of some scientific theories; for example, the artificial international language of elementary arithmetic is involved in the interpretive definitions of virtually every axiomatic theory.

Thus, the concepts of electric and magnetic field strength and of electrical charge density which occur in Maxwell's theory are provided with conditional definitions couched in ordinary language with some terms borrowed from the science of mechanics: electrical charge, for example, is conditionally defined in terms of the forces of attraction and repulsion which two bodies with ostensively definable properties (pieces of glass previously submitted to some mechanical treatment, etc.) and separated by a definite distance exert on each other. If Maxwell's theory is axiomatized in the usual way, and electrical charge density and field strength are intra-systemically undefined, the conditional definition just mentioned will happen to explain otherwise undefined terms of the axiomatic theory under consideration. If, on the other hand, the theory

---

[2]H. Margenau discusses a similar distinction between what he terms "constitutive vs. epistemic definitions"; *The Nature of Physical Reality* (1950), pp. 232–6.

is axiomatized by means of potential equations, and the two electro-magnetic potentials are intra-systemically undefined, whereas charge density and field strength are now defined concepts, the interpretive conditional definitions will refer to intra-systemically defined terms. There is thus only a loose connection between intra-systemic and inter-pretive definitions.

Other features of axiomatic theories will be discussed in due time, in connection with the methods of validating theories. In a sense, the clas-sification of scientific theories into axiomatic and non-axiomatic ones is accidental, since every scientific theory is axiomatizable. The fact that only a negligible fraction of well-established scientific theories have so far been axiomatized[3] is logically accidental and hence of little interest in this investigation. The important point is rather that, owing to the extraordinary progress achieved in the last decades in the logical analysis of axiomatic systems, the inherent potentialities and limitations of the axiomatic method, unsuspected by previous generations, have been dis-closed. The Lindenbaum-Tarski theorem about isomorphism, referred to on a previous occasion, is a case in point. Many other illuminating results in regard to the potentialities of the axiomatic method, and, consequently, in regard to any and every scientific theory, have been achieved by a comparative study of axiomatic systems; as a matter of fact, non-empirical axiomatic systems have been and are being produced almost on the assembly line for this very purpose. These axiomatic investigations have also led to several classifications of axiomatic sys-tems, often based on inherent characteristics of the relevant theories, and therefore of interest in this inquiry.[4]

Phenomenological versus Transcendent Theories

This classification of scientific theories, important from both the historical[5] and the epistemological point of view, is based ostensibly on

[3]This includes all major mathematical theories, the most fundamental physical theories (e.g., mechanics, thermodynamics, relativity, Quantum Theory), and some isolated theories of biology, psychology, and economics.

[4]Thus an axiomatic theory is said to be decidable if there is a uniform method for deriving all the consequences that could possibly be shown to follow from the assumptions of the theory; if such a method does not exist, the theory is unde-cidable. In our terminology, the deductive method is a method of discovery when applied in decidable theories and otherwise only verificatory. One of the main results of foundational research in the last two decades (Church, Tarski, Turing) is that only the most rudimentary theories, say, the sentential calculus, are decid-able, and that even elementary arithmetic, for example, is no longer so. This means that in all the major empirical theories, the mathematical formalism of which by far exceeds elementary arithmetic, the validating methods are merely verificatory. Cf. S. C. Kleene, *Introduction to Metamathematics* (1952).

[5]Cf. A. Rey, *La Théorie physique* (1930).

the subject-matter of the basic assumptions underlying the relevant theories. Thus a theory is said to be phenomenological if all its statements, that is, both the basic assumptions and the consequences, are about observable phenomena. For instance, phenomenological thermodynamics is a theory of heat phenomena which derives from three basic assumptions (the "Three Principles of Thermodynamics") all the laws which govern the observable behaviour of observable physical objects involving observable changes in the spatio-temporal distribution of quantities like temperature, pressure, kinetic energy, etc. The three assumptions of the theory are themselves concerned with such observable phenomena. In contrast, statistical thermodynamics is a transcendent theory because it attempts to deduce the axioms and the derivative laws of phenomenological thermodynamics from hypothetical assumptions about such (allegedly) unobservable physical entities as molecules and atoms.

This classification raises serious difficulties because of the ambiguity of the phrase "observable phenomenon." As it stands, the phrase is redundant since a phenomenon is observable by definition. Yet if we drop the reference to "phenomena" and simply stipulate that a phenomenological theory is concerned with *observable* events or processes or objects, there still remains the question of correctly interpreting the concept of observability. We have already dealt with the tendency to stretch the concept of "observation" beyond any reasonable bounds (§ 15); "observability" is even more vague since it involves, in addition, the troublesome concept of possibility. We shall deal with this problem in § 36. For the time being, it may suffice to point out that the classification of scientific theories into phenomenological and transcendent ones cannot be based exclusively upon the subject-matter of the relevant theories, because a theory which deals with observable objects but ascribes to them properties which cannot be ascertained by observation would not be classified as phenomenological. On the other hand, granting phenomenological status to theories which deal exclusively with observable objects, properties, and relations is not unobjectionable either, because the concept of an observable property or relation raises difficulties of its own (cf. § 32) and entails a Platonistic commitment which should be avoided as far as possible (cf. § 31).

It seems preferable to stipulate tentatively that a theory is phenomenological if all its axiomatic and derived statements are susceptible to validation by applying the fact-finding and law-finding methods previously discussed. In other words, a fact-like consequence of a phenomenological theory must be capable of being validated by some direct fact-finding method, or measurement, or indirect observation. A law-

like consequence of a phenomenological theory must be susceptible of
validation by resorting to inductive inference in addition to the fact-
finding methods just mentioned. Transcendent theories would then have
to be defined as involving fact-like or law-like consequences which cannot possibly be validated by the above methods. For instance, the
presence of a fact-like consequence (say, about unobservable particles
in a given spatial region) which cannot be validated by resorting to
either direct or indirect observation or to measurement would mean that
the theory is non-phenomenological.

This tentative formulation of the difference between phenomenological and transcendent theories will be improved upon in due course.
It may suffice, for the time being, to explain the decisive importance of
transcendent theories. Since the assumptions of phenomenological theories can be validated by applying the usual fact- and law-finding methods,
these theories are at bottom superfluous and do not essentially enlarge
the cognitive potentialities of science. Every justifiable statement made
by a phenomenological theory can be derived from its basic assumptions,
and since these assumptions can themselves be validated by the usual
fact- and law-finding methods, the theory can be simply replaced with
the set of its assumptions, that is, with a finite set of scientific facts or
laws. On the other hand, transcendent theories do contain information
which can only be provided in theoretical packages, for the assumptions
of such theories cannot be justified by resorting to any fact-finding or
law-finding methods. Hence, if justifiable at all, they have to be treated
within the context of a whole theory and the theory thus becomes indispensable and not replaceable by any combination of fact- and law-like
statements.

24   SYSTEMS OF INTERDEPENDENT THEORIES

Inferential and Definitional Interdependence of Theories

A scientific theory is rarely used singly. Usually, clusters or systems
of theories are involved in a scientific solution to a theoretical or practical problem. For historical reasons and with a view to facilitating the
division of scientific labour, ramified systems of theories are lumped
together into a single discipline, or a special science. In this investigation, the important question is how the range and limits of science are
affected by such transitions from single theories to systems of theories.
And the first point to be dealt with is a classification of systems of
theories capable of casting some light on the corresponding method of

validation. We shall stress two types of theoretical systems: (1) emergent systems of theories and (2) self-correcting systems of theories. In either case, the nature of the system hinges upon the logical relations among its theories.

The logical relationship between a theory T and another theory T' may involve an *inferential* or a *definitional* dependence of the former upon the latter. Thus, all the theories which make up the science of physics seem to depend inferentially upon one of them, namely, upon the theory of mechanics: physicists feel free to utilize, in arguments pertaining to any extra-mechanical theory, premisses derivable from mechanical axioms, along with those which follow from the specific axioms of the theory in question. Similarly, in chemistry, not only mechanics but thermodynamics and a considerable portion of electro-magnetics are made use of without any qualms; chemical theories are therefore inferentially dependent on these physical theories. In turn, physico-chemical theories are utilized without qualms in the whole of the biological sciences, and, in due course, the latter form the basis of medicine in so far as the aggregate of medical studies now constitutes a genuine science. We thus obtain a whole hierarchy of theories determined by their relations of inferential dependence. On closer inspection, the hierarchy turns out to be representable by a ramified tree rather than by a ladder: thus astronomy, astrophysics, and cosmology depend also on physics, but, in contrast to what happens with biology, this relationship is not mediated by chemistry. Similarly, geology makes use of physical results and, accordingly, bears to the cluster of basic theories which constitute physics a relation similar to those characteristic of chemistry and of astronomy.

An important case of inferential (or evidential) dependency of a theory T upon a theory T' occurs when the theory T' contains a description and explanation of the working of instruments which are indispensable for validating the contentions of T. Thus, as Poincaré once remarked, the whole science of heat phenomena could have been constructed by investigators deprived of thermal sensitivity if they observed and succeeded in establishing the laws governing the expansion of substances under certain conditions, say of a mercury column in a glass vessel. Hence thermodynamics depends inferentially upon mechanical laws governing thermometric substances—so much so that it could have been discovered by investigators either unable to experience the sensations of heat and cold, or unwilling to resort to them in setting up this powerful theory. This aspect of thermodynamics, incidentally, illustrates the structuralist view of the range of science, since it makes it clear that the sum-total of information provided by physical research about heat phenomena could be discovered, understood, and checked by beings deprived altogether of thermal sensations.

Inferential dependence is not the only kind of logical relationship among scientific theories. It may also happen that the basic concepts of a particular scientific theory or of a cluster of theories integrated into a science are defined or definable in terms borrowed from another scientific theory. Thus, electro-magnetic concepts are conditionally definable in terms of mechanical concepts; for example, an electric charge in terms of force. We shall say, accordingly, that the electro-magnetic theory is *definitionally* (not inferentially) dependent upon mechanics. It goes without saying that inferential dependence and definitional dependence of a theory T on a theory T' may, and, as a matter of fact, often do, go together. Thus physico-chemical concepts are utilized in conditionally defining biological concepts, just as physico-chemical facts are utilized as premises for substantiating laws of biology. Yet the two relationships should be in principle distinguished from each other, since they are not always associated.[6]

The inferential or definitional dependence obtaining between two theories may happen to be reciprocal. As a matter of fact, this is perhaps the most frequent case, if we try to consider not merely the core of the relevant theories, but also what P. W. Bridgman has called their background. Thus the concept of a physical body governed by laws of mechanics could certainly not be applied to any concrete situation, unless the elementary chemical concepts required to identify the successive stages of the same body were available. It is unlikely that any headway could be made in an empirical theory without the most elementary notions of psychology and of methodology. Furthermore, the scientist is bound to use a language in order to record and communicate his discoveries, and it is apparent that the efficient use of any language presupposes information encroaching upon several empirical theories. We must therefore not take too literally the watertight division between the various theories and sciences which is implied in our hierarchy of scientific dependence, be it inferential or definitional. Such hierarchical

---

[6]At bottom, definitional dependence of T upon T' entails the inferential dependence of the former upon the latter. If a concept P of T is defined in terms of a test t and a response r, both belonging to T', the laws of T' which govern test t and response r are clearly necessary to handle the concept P within the theory T. We may, nevertheless, differentiate definitional from inferential dependence (of a theory T on a theory T') by stipulating that, in case of a definitional dependence, only those laws of T' be utilized in T which govern the definitional tests and responses associated with the concepts of T.

stratification aims rather at clarifying the logical structure of science than at establishing an actual succession of unilaterally dependent layers in the edifice of science.

The relationship of inferential and definitional dependence between various scientific theories and disciplines has acquired a peculiar significance in the limiting case when the dependence is unilateral and complete and involves both concepts and laws. If *all* the concepts of a given theory T are definable in terms of the concepts of some other theory T' and, at the same time, all the laws of T are rigorously deducible from those of T', the first theory T may be said to be completely *reducible* to the theory T'. On the other hand, if it is impossible either to define all the terms of the theory or group of theories T by means of those of T', or to deduce all the laws of T from those of T', T is said to be *emergent* in regard to T'. The question as to whether or not certain theories or sciences are emergent in regard to certain other theories or sciences has played and is still playing a dominating role in several branches of the philosophy of science. The three sciences which have been most involved in the emergence controversy are psychology, biology, and mathematics. The question whether mathematics is reducible to logic has been discussed for centuries and brought substantially closer to a constructive solution owing to some pioneering ideas of Leibnitz and, mainly, as a result of Frege's life-work. The question whether biology is reducible or emergent in regard to the physico-chemical sciences is known as the controversy between "vitalism" (which sponsors the emergent status of biology) and "mechanism" (proclaiming the definitional and inferential reducibility of biology to the physico-chemical sciences).[7] The question as to the reducibility (or emergence) of psychology in regard to physiology has given rise to the most discussion so far.[8] While either solution is sponsored by certain metaphysical schools of thought, a more cautious temporizing attitude seems to prevail at present among the interested scientists and philosophers.

Needless to say, the three questions—whether or not mathematics is reducible to logic, biology to the physico-chemical sciences, and human psychology to human physiology—offer excellent examples of a metascientific reformulation of perennial philosophical problems. The first question involves the nature and scope of *a priori* knowledge and of the

[7] Cf. M. Schlick, *Naturphilosophie* (1925) and E. Nagel, "Mechanistic Explanation of Organismic Biology," *Philosophy and Phenomenological Research*, vol. 11 (1951).

[8] Cf. C. D. Broad, *Mind and its Place in Nature* (1925); H. Feigl, "The Mind-Body Problem in the Development of Logical Empiricism," *Revue internationale de philosophie*, vol. 4 (1950).

limits of empiricism; the second question bears on the nature of life; the third contains the gist of the mind-body problem.

Self-correcting Systems of Theories

I have already indicated that a theory T may depend on another theory T' in more than one respect: it can happen that some consequences of T could not be deduced from its basic postulates without resorting to T' as a source of additional, indispensable premises. More specifically, the instruments which are indispensable for substantiating T may be described and explained within T'. Furthermore, some of the basic concepts of T may not be definable, even conditionally, without resorting to the basic concepts of T'. If T depends both inferentially and definitionally upon T', we shall say that the theory T *presupposes* the theory T'.

Thus, Einstein's Special Theory of Relativity presupposes the validity of Newton's theory in regard to objects whose speed is neglible in comparison with the speed of light. In a similar sense, Einstein's General Theory of Relativity presupposes his Special Theory of Relativity in regard to very small spatio-temporal regions and to sufficiently weak gravitational fields. The Quantum Theory of atomic and sub-atomic phenomena presupposes the validity of "classical" physical theories of mechanics, electro-magnetics, and thermodynamics under special circumstances, for example, whenever the "quantum of action" can be treated as negligibly small, or when the masses involved are "infinitely large" in comparison with the masses of elementary particles.

The paradoxical nature of the three theoretical systems just referred to consists in the fact that, in each case, the new revolutionary theory T not only presupposes T' because the terms of T would be unintelligible and its claims unjustifiable without resorting to the old theory T': while presupposing the old theory, the new theory also extends the validity of the former to new realms of phenomena and corrects it by replacing some of its tenets with new ones. Thus the new theory T presupposes, extends, and refutes the old theory T'. This seems to be inconsistent. Yet a steady advance towards new theories which presuppose, extend, and refute their predecessors seems to constitute the fundamental pattern of scientific progress. We shall refer to sequences of theories in which each theory presupposes, extends, and refutes its predecessor as *self-correcting theoretical systems.*

The example of classical mechanics may illustrate the gist of the difficulty. Mechanics was developed by Newton into an almost self-contained discipline, which proved, however, in its subsequent development, to be of extreme importance to science as a whole. Newton succeeded in firmly establishing his mechanical theory by showing that it can be reduced to three simple assumptions: the principle of inertia, the principle connecting the force acting on a physical body with the acceleration imparted to that body, and, finally, the principle of equality of action and reaction. In this century, the increasing wealth of new facts to be accounted for has led to successive radical alterations and extensions of Newton's principles. The most revolutionary changes were due to Einstein's Theory or rather Theories of Relativity and to the Quantum Theory of atomic phenomena, often called Quantum Mechanics or Wave Mechanics. All these subsequent changes have been superposed on the original Newtonian edifice without making it either obsolete or dispensable. The objective of the successive modifications has been an adjustment of the Newtonian theory to new realms of phenomena unsuspected by Newton. Within an extremely comprehensive range of phenomena, however, the Newtonian theory remains as adequate and indispensable as ever. In other words, in spite of being both corrected (and to this extent refuted) and extended by the later theories, the classical theory is nevertheless presupposed by these theories even beyond the area where it continues to apply. The new theories would be unable to define their concepts, to interpret their experiments, and to apply their main contentions to the proper realms of phenomena, unless they availed themselves of Newton's unchanged theory for this threefold essential purpose. Thus Newton's mechanics may have lost its monopoly in the physical universe, but there is no question of its losing its indispensable role within the whole of science. Moreover, it seems to us that the very conquest of the pre-scientific, "naïve" world-view by the scientific outlook displays also the three main features of the pattern just outlined: the scientific theories, say of pre-Einsteinian physics, presuppose the common-sense world-picture, because scientific terms are undefinable without recourse to common-sense concepts, scientific facts and laws are not justifiable without taking for granted and utilizing pre-scientific laws and facts, scientific techniques of observation and experiment are unthinkable without rules formulated in ordinary language involving the common-sense world-picture. Yet, in spite of thus presupposing the pre-scientific outlook, science at the same time is incompatible with this outlook and corrects its major mistakes. Scientific information is also more comprehensive than pre-scientific knowledge, since quite a few questions which suggest themselves to a pre-scientific outlook can be answered only by resorting to scientific methods. Thus, classical science presupposes, corrects, and extends the pre-scientific world-picture.

§ 25   FUNCTIONS OF SCIENTIFIC THEORIES

A scientific theory conceived as a potentially infinite set of statements differs considerably in logical structure from scientific facts and laws, both expressible by single propositions. We must not expect, therefore, that the functions assigned to theories will closely resemble those of either facts or laws. Yet, although scientific theories serve several purposes of their own of which neither facts nor laws are capable, any scientific function that can be discharged by a fact or a law can also be discharged by a theory.

We have seen that, in the first place, scientific laws constitute valuable items of reliable information in their own right, and thus fulfil a cognitive or informational function, exactly as facts do. In addition, scientific laws summarize, predict, explain, and control particular facts and thereby perform several functions of which particular facts are incapable. Finally, laws provide an indispensable definitional device of empirical science; the part they thus come to play in scientific concept-formation is also foreign to particular facts. Now the point is that scientific theories do all these jobs of informing, summarizing, predicting, explaining, controlling, and defining exactly as laws do, the main difference being that theories operate, in these various ways, on laws, whereas laws operate primarily on facts. Hence, in so far as the law-like functions of scientific theories are concerned, the situation may be roughly described as follows—Theory: Law = Law: Fact.

As already indicated, the *cognitive function* is essential only in so far as transcendent theories are concerned. A phenomenological theory can be replaced, without any loss of information, with a finite set of laws which constitute an axiomatic basis of the theory. Whatever knowledge such a theory provides can be obtained from the laws contained in its axiomatic basis. Since a finite set of statements is replaceable by a single statement, namely, their logical conjunction, the package of information conveyed by a phenomenological theory does not exceed the bit of information conveyed by such a condensed form of its axiomatic basis. This does not apply, however, to transcendent theories. Since their axiomatic basis cannot be validated by any combination of fact-finding and law-finding methods, there is no possibility of replacing the package of information conveyed by the theory with a bit of information provided by law-like statements. Granted, the impossibility of validating the axiomatic basis of a transcendent theory separately raises the question of how one could substantiate a package of information without validating all the bits it consists of. This will be discussed in detail in

§ 26. With this reservation in mind, we may say that when an infinite set of statements is produced in conjunction with a body of evidence adequately supporting the whole set, the set *is* an item of knowledge, comparable in principle to the knowledge conveyed by a single fact or law. Information contained in a theory is in general infinitely more comprehensive than any information conveyable by a single fact-like or law-like statement; the knowledge provided by transcendent theories can be obtained in no other way.

There are quite a few misgivings connected with the cognitive status of scientific theories. Even if one is prepared to grant such a status to scientific laws, either finitist or infinitist, one may still observe that scientific theories are often held to be neither true nor false and thus to lack cognitive import. This attitude is perhaps partly due to the fact that, historically speaking, scientific theories are more sensitive than scientific laws and facts to the incessant changes of the total observational evidence which dominates the life of science. Accordingly, theories are less enduring, more likely to come and to go, than are laws, let alone facts. That the boundary-line between what any particular theory actually asserts and its merely auxiliary statements is often difficult to draw—actually, the very existence of such a boundary has rarely been take into account in the philosophy of science—may also have contributed to the rather widespread belief that scientific theories have no definite truth-value, and hence do not constitute genuine knowledge. Their function within science would not be cognitive, but rather auxiliary in regard to the body of genuine fact-like or law-like knowledge which science provides. We shall see that all these sceptical judgments concerning the cognitive status of scientific theories, which have been passed by investigators who were not sceptics at all, seem to have been the result of a misinterpretation of the significance of the historical vicissitudes of scientific theories and of failure to distinguish between the core itself of a theory and the body of its auxiliary statements. We shall also have to acknowledge that by relinquishing the construction of theories and confining himself to facts and laws, the scientist would actually impoverish and curtail the scientific method in its cognitive potentialities, since the genuine knowledge provided by scientific theories would then not be available.

Scientific laws are found to summarize the results of the past observations which led to their establishment, to predict the results of future observations, to explain present observations, and to facilitate man's practical activities by enabling him to choose among various courses of action in order to attain a particular objective. Obviously, the sum-

marizing value of a law increases in proportion to the number of independent past observations which led to it; its explanatory value depends
upon the frequency and strangeness of the relevant phenomena; the
predictive and pragmatic value of laws varies according to the number,
strangeness, and importance of the relevant facts. All these feaures
reappear when we make the transition from law to theory. The only
major difference is that a theory is not needed in order to render all the
aforementioned services in regard to facts, since scientific laws take care
of them. Theories are valuable because the laws themselves stand in
need of condensation, prediction, explanation, and practical application:
the laws which govern certain classes of phenomena, for example motion
and rest, are so numerous that it is virtually impossible to handle them
efficiently unless they are condensed in a manageable theory. This
condensation is achieved to a high degree in several advanced theories,
and particularly in mechanics in so far as the laws of motion and rest
are concerned. The *summarizing function* of theories in regard to laws
is extremely valuable and certainly not less effective than the corresponding function of laws with respect to facts.

The *predictive function* of laws is ordinarily linked up with their
ability to yield consequences referring to future facts. On the other hand,
the laws predicted by a theory, that is, unknown prior to their deduction
from it, have no intrinsic reference to the future. The laws of conic
refraction, predicted by Hamilton by means of the undulatory theory of
light, were valid prior to his discovery; the laws concerning the collisions
of electrons and protons which Compton has succeeded in predicting
were valid prior to and independently of their deduction from assumptions of Quantum Theory. These laws are universal and valid throughout
time. A predicted universal law applies to past, present, and future,
exactly as otherwise discovered universal laws do.

Yet there is no basic difference between the temporal aspects of
theories and those of laws in so far as their predictive function is concerned. The predictive power of laws is also utilized to infer their past
unobserved instances, as, for example, when the history of the sun is
reconstructed on the basis of laws of stellar evolution. The proposal of
some authors to distinguish the *predictive* and the *postdictive functions*
of scientific laws did not strike roots in the language. Accordingly, we
shall construe the facts predicted by a law as including all those which
are both new and deducible from it, regardless of whether these facts
refer to past, present, or future; the essential point is that they be new,
that is, unknown prior to the establishment of the law. On such a view,

laws and theories do not differ from one another in regard to the temporal aspect of their predictive function.

We may add that what is predicted by a theory is often a new theory rather than a single law or fact. Thus, Einstein was able to deduce a brand-new mechanics from his relativistic theory of electro-magnetic phenomena. Relativistic mechanics (with its convertibility relation between mass and energy) is thus a theory predicted by the Theory of Relativity, over and above the facts and laws predicted by the latter. It is obvious that, owing to the more complex logical structure of a theory, its predictive power is vastly superior to that of a law, since the latter primarily predicts facts, and, exceptionally, laws, whereas a scientific theory includes within its predictive potentialities facts, laws, and theories.

It is important to realize that when a well-established theory predicts a law, the prediction actually amounts to a validation of the law. The validating function of theories, implicit in their predictive power, is of extreme importance because it enables the scientist to establish laws the instances of which are beyond his experimental or observational control. In particular, fundamental laws of a limiting nature (which are, strictly speaking, vacuously universal because they never apply to actual phenomena) can be validated only indirectly, within the context of a theory. For example, the law of inertia is only vacuously valid since it refers to non-existent objects on which no forces are acting. This law can nevertheless be established within the theory of mechanics by being derived from a non-vacuous law correlating the acceleration imparted to an object with the forces acting upon this object. It follows from the non-vacuous law that the acceleration will vanish if the forces vanish— which is exactly the law of inertia.

Another validating function of theories is discharged whenever a theory strengthens the antecedent probability of the laws derivable from it even if these laws have been used to establish the theory. This is the case because any law deducible from a theory derives additional support from the fact that other well-established laws that are so deducible enhance the reliability of the theory, and, by the same token, of all its consequences. This supplementary validating function of theories referred to as their "consilience" by Whewell, has often been commented upon.[9]

We shall have to recognize at a later stage (§ 34) that the indirect validation of a law by a theory (in terms of prediction or of consilience)

[9]Cf. W. Kneale, *Probability and Induction* (1949), pp. 106. ff.

is conditional upon the possibility in principle of validating this law directly by examining its instances. In this sense, the inherent range of law-finding methods is not enlarged by the use of theories. But this theoretical possibility does not affect the practical indispensability and extraordinary efficiency of methods of indirect validation provided by theories.

Similar remarks apply in principle to the *explanatory function* of theories. Facts are ordinarily explained by recourse to laws; a theory is used to account for a fact only when a law to be deduced from this theory can do the job. Theories are, however, capable of explaining facts, laws, and other theories as well; in other words, new facts, laws, and theories may be accounted for by being derived from the basic assumptions of a given theory. Thus, Maxwell's theory not only explains the *fact* that a stick immersed in water looks bent as well as the *law* of refraction discovered by Descartes: the whole *theory* of geometrical optics is also accounted for in view of its deducibility from the electro-magnetic field-equations.

Let us point out that when facts, laws, and theories are said to be predicted or explained by a given theory provided they be "new," their novelty does not have the same meaning in the predictive and the explanatory context: a fact, for example, is predicted by a theory if it is new in the sense of having been unknown prior to the establishment of the theory, while being deducible from the latter. On the other hand, when a theory explains a fact, the latter is new only in the sense of not having been utilized in the substantiation of the theory, although it might have been established prior to and independently of it. We thus see that the predictive and explanatory powers of a theory are not literally coextensive with each other, since the two concepts of novelty determine two distinct ranges of facts which the theory predicts and explains, respectively. The difference between the two ranges is, however, logically accidental, since it depends upon whether a fact, law, or theory happened to be established prior to a given theory, or utilized in validating the theory. The logically essential feature is the deducibility of the relevant facts, laws, and theories from the theory which either predicts or explains them. Such deducibility is equally important for prediction and for explanation.

I have already referred to the objections raised against the explanatory function of science by the positivist philosophy of science. These objections are probably traceable to the fact that explanatory power had mainly been claimed for transcendent theories, of which positivists disapproved on other (viz., "verificationalist") grounds. That is how the early and the middle positivism came to associate the "metaphysical" (i.e., empirically unverifiable) aspect of empirical science with its explanatory use. Accordingly, the defence of the explanatory function of science has mostly been undertaken by students of science who were rather prepared to countenance its metaphysical import. This situation, however, has considerably changed in the last two decades. Scientific theories have been found able to explain facts, laws, and other theories because the explained items of scientific information meet the dual requirement of being deducible from the explanatory theories without occurring in their supporting evidence. This logical requirement has no ontological strings attached to it and is met by phenomenological theories as well as by transcendent ones. The particular epistemological difficulties raised by any transcendent theory will be discussed later on; they have nothing to do with its explanatory function.

The *definitional function* of theories bears on the concepts they involve, not on laws or facts. What I have in mind here is an analogue of the definitional function of scientific laws. While discussing the latter, we saw that a term introduced by a conditional definition is only partly defined thereby and remains, accordingly, extremely vague. However, the laws in which this term occurs reduce its vagueness by extending its applicability to new realms of phenomena not included in the initial criterion. A law thus contributes to rendering its constituent terms more precise by yielding new definitional criteria for them. Similarly, a scientific theory may provide new definitional criteria for its constituent terms—if, for example, the law-like consequences of the theory are associated with such criteria.

In the most typical theories of mathematical physics, for instance, the basic assumptions are differential equations which establish relations between the spatial and temporal derivatives of measurable quantities. Such derivatives of measurable quantities are not measurable in turn since they involve indefinitely decreasing spatial and temporal intervals which could not possibly be measured, because every mensurative instrument has a definite least count. It is often possible, however, to integrate the fundamental assumptions of such a theory, that is, to deduce from the assumed relations among non-measurable derivatives of measurable quantities other law-like relations among the original measurable quantities. These law-like relations then provide new definitional criteria for the relevant measurable quantities. For instance, in Maxwell's equations of the electro-magnetic field, the non-measurable spatio-temporal derivatives of electric and magnetic field-vectors are connected with the density of electrical charges and their state of motion.

By using these equations for the deduction of law-like relations among the measurable electric and magnetic field-vectors and electric charges, new definitional criteria are obtained for the latter. Similarly the law of the free fall of bodies asserts the constancy of their acceleration, that is, of a non-measurable second derivative with regard to time of the measurable positional co-ordinates of the falling bodies. By deducing, from the differential equation concerning the free fall, relations among the positional co-ordinates, for example, the parabolic shape of the trajectory, one obtains a semi-definitional law interrelating the three co-ordinates which provides for the computation of any one of them whenever the two remaining co-ordinates have been determined.

A theory is thus semi-definitional with regard to a concept it involves if this theory entails laws which are semi-definitional in regard to this concept. In the above examples, the concepts occurring in the theory were endowed with independent definitional criteria, and the theory merely extended their area of precision by providing them with additional criteria associated with some of its consequences. However, the availability of independent definitional criteria for concepts occurring in a theory is by no means a prerequisite for discharge by the theory of a semi-definitional function in regard to these concepts. Even if a concept which occurs in a given theory is not conditionally definable prior to, and independent of, the theory, the theory may still entail law-like consequences which constitute definitional criteria for this concept. Thus, the scalar and the vectorial electro-magnetic potentials which occur in the theory of electro-magnetic phenomena are not directly connected with the ostensive level by tests and responses of their own. Their values can nevertheless be determined in principle in every particular case owing to the fact that the laws governing them within this theory establish such tests and responses in virtue of the functional relationships between these potentials and other independently measurable quantities which these laws entail. Concepts which thus owe their definitional criteria entirely to their theoretical context and lack independent connections with the ostensive level are sometimes referred to as "theoretical constructs." This, however, is not the only meaning of this methodological term.[10]

26   ON VALIDATING SCIENTIFIC THEORIES

Since a scientific theory is not a single statement, but a potentially infinite set of statements, it is natural to expect the validating procedures applicable to theories to differ from those which are used in the case

[10]Cf. H. Margenau, *The Nature of Physical Reality*, pp. 69 ff.

of single statements, and, more specifically, of law-like statements. Yet laws and theories resemble each other in one decisive respect: like a theory, a law may be regarded as a condensation of an infinite number of single statements, that is, of the class of its instances. We have seen that, in the main, the validation of a law is obtained through the validation of some of its instances. And we may expect that, similarly, a theory will be validated if a suitably selected sample of the statements which make it up is validated by the usual methods applicable to single statements.

This analogy between the validating methods applicable to laws and to theories actually obtains, as we shall presently see, but it is no more than an analogy. A theory is established by validating a sample of its *consequences*, whereas with laws, *instances*, not consequences, are validated. Apart from this basic difference, there are several other features which distinguish the validation of laws from that of theories. Some of these differences have far-reaching philosophical implications and bear upon our central problem. It is more convenient, however, to start with an outline of the most typical methods applied by the scientist to the validation of his theories and to postpone an evaluation of the bearing of such methods on our central problem.

The scientist's usual procedure in establishing a full-fledged theory seems to be contained in the following rule: "If all the mutually independent consequences of the theory which have been examined so far by applying conventional fact-finding and law-finding methods have turned out to be correct, then the theory itself may be considered to have been confirmed or rendered probable by this sample of its consequences to a degree depending upon the number and the mutual independence of the consequences included in the sample." However, neither the number nor the independence of consequences of a given theory are definable or ascertainable with a precision comparable to that possible with single law-like statements. The number and mutual independence of instances of a universal law, for example, can be determined, as a rule, without any major difficulties. The instances of such a law can be counted because they originate out of the matrix of a quantified statement by substitution of logically admissible constants for the variables of the matrix. Nor does their mutual independence raise serious difficulties. But the mutual independence of two consequences of any non-trivial theory, for example, of two theorems in Euclidean geometry, may be a challenge to a Hilbert.

The general rule for validating empirical theories may be, nevertheless, rather easily applied when a quantitative theory is concerned. Such a theory, T, often determines in advance the particular values of a given quantity taken on by members of a particular class of physical objects. These particular values computed according to the theory T are then compared with the actually measured or "observed" values for every object of this class. The conformity of the systems of predicted and observed values is then considered as an item of evidence supporting the theory T. Comparing such systems of values is a basic procedure in the validation of any quantitative theory. It is obvious that this procedure is but a special case of the aforementioned general rule concerning the validation of a theory by validating a sample of its consequences: the particular value taken on by the quantity in question for any given individual is specified in a quantitative statement which follows from the theory, and "comparing this predicted value with the actually observed one" amounts simply to validating the quantitative statement which happens to be a consequence of the theory.

In this context, I cannot discuss in detail why an exact conformity of the predicted and observed values is by no means required, and even rather undesirable. A "fair" agreement of the observed and the predicted systems of values is all that is required for validating the theory, or contributing to this end. Part of the explanation can be obtained from previous comments on the measurement of continuous quantities. The extent and validating value of the agreement between the two systems of values are the subject-matter of the "Theory of Errors," a branch of the calculus of probability which exhibits unmistakable traces of Gauss's great mind, in spite of the innumerable modifications and improvements which have originated in the meantime. The "Theory of Statistical Inference," of more recent origin, is also used for settling the problems we are pointing to. The basic idea that underlies the evaluation of the validating force of the two parallel systems of predicted and observed values is the assumption that every observed value depends not only upon the value which the quantity measured actually takes on for the object under consideration, but also upon certain additional factors governed by a random distribution and not taken into account in the theory itself; that is why the theory cannot be blamed entirely for the differences between the values it predicts and those which measurement yields. In spite of the ostensibly arbitrary nature of this assumption, it is well substantiated by innumerable observations of the additional factors and the undeniable over-all success. There is nevertheless a degree of truth in Poincaré's saying that the basic assumptions of the Theory of Errors are never demonstrated by the mathematicians, who believe that only observations and experiments of empirical science are competent to supply the proof; on the other hand, no representative of an empirical science ever attempts to substantiate the Theory of Errors because he assumes that it is in the mathematician's province.

Another typical way of validating scientific theories consists in computing the value of the same quantity for the same object on the basis of the theory under consideration, by applying various mutually independent methods and also a certain amount of independently acquired factual information. If all such values computed on the basis of a given theory are in substantial agreement with one other, the result is considered to confirm this theory, sometimes to a very high degree. This method is extremely effective and represents, once more, a special case of validating a theory by the validation of a sample of its consequences. Since a given quantity has but one value for any given object, it is indeed the consequence of any theory that various mutually independent reliable methods for measuring this value on this theory must yield substantially the same result. As for the efficiency of the method, it suffices to refer to the various independent computations of the age of the universe, the magnitude of the quantum of action, the electron's electrical charge and mass, etc. One cannot help being impressed by the amazing agreement of results achieved by such mutually independent methods.

As a matter of fact, the first procedure for validating quantitative theories must be credited with a similar impressiveness, and for similar reasons, of course. The point is simply that a successful prediction strengthens the probability of the theory responsible for the prediction, in proportion to the improbability of this same prediction considered independently of the theory. If somebody claims prophetic powers and then succeeds in predicting the weather on the following day, this success will slightly confirm his claim (on the assumption that the probability of his claim being correct was not zero before the prediction was made). But the probability of his being a prophet would increase much more if he succeeded, for example, in predicting the exact maximum temperature or the exact time at which it will start raining on the following day. The reason why the latter performance would render the prophetic power of the would-be weatherman much more probable is simply that the probability that the maximum temperature or the beginning of the rain should exactly agree with a temperature or an hour chosen at random is almost negligibly small on the assumption that the predictive powers of the alleged prophet are not resorted to.

At first glance, the validation of scientific theories seems to follow a pattern similar to the inductive validation of a law. A law is validated to a fair degree if a fair sample of its instances has been submitted to an observational test, and all the examined instances have proved correct. Similarly, a theory is considered to be established to a reasonably high degree if a fair sample of its consequences has turned out to be correct. The main difference between methods for validating general laws and for validating theories would thus centre around the distinction between the instances of a general statement and the consequences of a finite set of assumptions. This, of course, is not the only difference between the two methods, but it is important at this stage because it points to their common denominator.

Let us first notice that a statement may be the consequence of another statement without being an instance of it, although the converse is not true. Thus, the relation of instantiation is a special case of the more general logical relationship of entailment or consequence (cf. §31) among statements. This suggests that the method for validating theories should be regarded as the more general procedure since it involves the more general concept of consequence; whereas the inductive validation of laws, which is based on the relation of instantiation, ought to be viewed as a special case of the former. In other words, any set of statements could be validated by validating an appropriate selection of its consequences. If the set of statements consists of all those which are deducible from a finite set of assumptions, the validation involved is that of a theory. If, on the other hand, the set consists of a single statement, and its validated consequences are its instances, an inductive validation of a law takes place. Thus, formally speaking, a general pattern present in the validation of both laws and theories can easily be described. The inductive validation of laws would then be viewed as a special case of establishing theories by validating some of their consequences.

However, this approach is not appropriate, because the inductive validation of laws is now intelligible to some extent, whereas the validation of theories has hardly been explored. It seems preferable, therefore, to attempt to interpret the validation of theories in terms of the inductive validation of laws regardless of the special nature of the latter procedure. In other words, we shall try to extend the inductive method of validation to scientific theories in spite of some crippling implications which such an approach carries with it. Our only reason for so doing is the fact that we are not aware of any other promising approach.

A scientific theory is, by definition, the sum-total of consequences which can be derived by specifiable deductive procedures from a finite set of statements termed the assumptions of the theory. Now, in trying to conceive of validating theories on inductive lines, the crucial question is what fraction of the infinite set of statements which go to make up the whole theory are actually affected by the evidence concerning certain consequences of the theory. As a rule, the theory includes an infinite subset of purely logical and mathematical propositions or of their substitution-instances which obviously are not affected by any conceivable observational evidence, since they are provable *a priori* by purely deductive methods. Apart from this special position of the logico-mathematical constituents of an empirical theory, often referred to as the "formalism" of the theory, there may be other consequences of the theory which stand apart from its formalism, and are nevertheless intrinsically incapable of being affected by any observational evidence produced to support the theory. That this figure would be a square if it were a rectangle with equal sides, need not be supported by observational evidence, because it follows from the definition of a square; the statement is analytic and belongs to the formalism of Euclidean geometry, interpreted as describing observable physical space. That the diagonal of this square is not commensurable with its side is a synthetic consequence of the theory which could not possibly be affected by observational evidence because of the inherent inaccuracy of any and every measurement. Finally, the statement asserting that "if this figure be approximate to a square with a specifiable accuracy and the length of its sides and diagonal be determined with a specifiable accuracy, then the square of the second of these lengths will prove to be the double of the square of the first length with another specifiable accuracy" is a synthetic fact-like consequence of the theory which can be submitted to an observational test and is most likely to be confirmed by such a test.

We may therefore subdivide all the particular or fact-like consequences of any theory into three mutually exclusive classes: (1) analytical or logico-mathematical propositions which pertain to the formalism of the theory and are neither susceptible nor in need of being substantiated by observational evidence; (2) fact-like propositions which could not possibly be substantiated by resorting to the scientific fact-finding methods surveyed earlier in Part II; (3) fact-like consequences of the theory which have been or could possibly be submitted to the aforementioned procedures and were always found to be correct on such examination: in view of later discussions, we may say that the consequences of this category are *empirically verifiable propositions*, in contradistinction to those of categories (1) and (2). (Cf. §§32, 33.)

A similar tripartite classification of the law-like consequences of the theory is obviously determined by the above classification of its fact-like consequences. Those law-like consequences which are susceptible of a logico-mathematical proof belong to the formalism of the theory and are not affected by the relevant observational data. Law-like consequences which could not possibly be validated by applying the law-finding methods surveyed earlier in Part II to verifiable fact-like consequences of the theory are not susceptible of any validation by scientific law-finding methods and resemble, in this respect, the unverifiable fact-like consequences of the theory. Finally, law-like consequences which can be validated by applying law-finding methods to the verifiable fact-like consequences of the theory resemble such fact-like consequences in their amenability to scientific method and may therefore be termed verifiable laws (cf. § 34).

We thus see that all the consequences of a scientific theory can be divided into three exclusive and exhaustive classes: (1) those which make up the *formalism* of the theory; (2) those which could not possibly be submitted to any fact-finding or law-finding methods of validation; (3) those which constitute verifiable facts or laws. Now it seems obvious that if a number of verifiable fact-like or law-like consequences of a theory have been submitted to a test and found to be correct then the only possible conclusion is that the remaining *verifiable* consequences of the theory are likely to be correct as well. No light whatsoever is thrown on the first two categories by examining some consequences of the third category.

Accordingly, in order to replace the validation of a theory by means of its consequences, with the inductive validation of a law-like statement, we have to consider only its consequences of the 3rd category as instances of a quantified statement. To this end, we may substitute for the theory *T*, as formulated in the object-language, a single, law-like statement couched in an appropriate meta-language to the effect that "all the verifiable consequences of the theory *T* are true." It is obvious that when a theory is replaced in this way by a single law-like statement of the meta-language, the validation of it by means of a suitable sample of its consequences would be tantamount to validating the law-like equivalent of the theory in terms of the instances of this equivalent; indeed, all those consequences of the theory which could possibly be validated by applying scientific method are instances of the law-like statement "all the verifiable consequences of *T* are true."

In other words, the empirical validation of theories can be interpreted as a special case of the inductive validation of laws in terms of the instances of the latter, provided every theory couched in the object-language is replaced with an equivalent law-like statement of the meta-language asserting the truth of those consequences of the theory which meet the requirement of verifiability. The greater complication of theories in comparison with laws which appears when both are formulated in the same object-language, is made up for by translation into an appropriate meta-language.

This inductive approach to the validation of theories raises several difficulties which may yet be surmountable. Thus, one may object that any instance of the meta-linguistic equivalent of the theory T is not a single consequence (say, C^1) of T, but rather the statement "C^1 is true." This is correct, but does not reflect upon the meta-linguistic paraphrase of T, since any statement C is equivalent to the meta-linguistic statement "C is true." Hence the replacement of T with a meta-linguistic law of the above kind reduces the validation of T by means of a sample of its consequences to the validation of a sample of instances of a law-like statement: these instances are not literally synonymous with the relevant consequences of T, but logically equivalent to them. For purposes of validation, there is hardly any harm in replacing a statement by another statement logically equivalent to it.

Another difficulty consists in the fact that the meta-linguistic law which asserts the truth of all the verifiable consequences of a theory is not logically equivalent to this theory, but rather to a fragment of it. As just outlined, the class of all the propositions of a theory T consists of the three mutually exclusive classes of all the analytic propositions of T, all the synthetic unverifiable propositions of T, and all the synthetic verifiable propositions of T. The meta-linguistic law, even if correct, would only guarantee the truth of all synthetic and verifiable propositions, but what about the remaining two classes? The analytic ingredients of T are taken care of otherwise, but the synthetic unverifiable consequences of T follow from its assumptions without either occurring in its meta-linguistic paraphrase or being taken care of otherwise. Consequently, the meta-linguistic paraphrase is neither synonymous with, nor even logically equivalent to, the whole theory T.

This second objection is still correct, as it goes. But the point is that major reasons can be adduced against identifying an empirical theory with *all* the propositions which are conventionally construed as pertaining to it. Thus there would be little opposition, I suppose, to considering the formalism of an empirical theory as foreign to it and not being part and parcel of its core: subclass (1) plays an auxiliary part in the set-up of the theory and does not literally belong to it. On the other hand, an attempt will be made later to prove that synthetic unverifiable propositions are neither true nor false, in spite of being logically meaningful, useful, and even indispensable in science. This being so, it would hardly be convenient to construe an empirical theory as including both its verifiable and its unverifiable consequences, even if the latter should prove indispensable (as they often do) in the set-up of the theory; for, if we set such wide limits for the theory, we could not assert that a theory is either true or false, because some of its constituent propositions would be neither. We could not assume that an entire theory can be validated by observational evidence, i.e., shown to be true with a fair probability: the truth of a theory, that is, of a specifiable set of statements, must be identified with all these statements being simultaneously true. This is why it seems preferable to assign to the unverifiable consequences of an empirical theory a purely auxiliary (though possibly indispensable) part in its set-up, similar to the part allotted to the formalism of the theory. In other words, we propose to identify an empirical theory with the sum-total of its empirically verifiable consequences while relegating all its other consequences to the rank of subsidiaries. This identification justifies the scientist who differentiates between true and false theories; a theory which coincides with the verfiable consequences of its basic assumptions can be said to be probably true to the extent to which the sum-total of these verifiable consequences is so. A true theory is one consisting of true statements, and *vice versa*.

The inductive method for validating an empirical theory would thus be confined to its verifiable consequences and leave both the formalism of the theory and its synthetic unverifiable part unaffected. At this price, that is, by sacrificing the analytic and the unverifiable components of an empirical theory, the only difference left between the validation of laws and of theories would consist in the fact that the latter have to be translated into an appropriate meta-language in order to become amenable to inductive treatment, whereas the former need no such translation. The basic procedure would apply to both laws and theories: to establish a universal law, one checks upon a representative sample of its instances and then concludes that what holds true of the sample applies also to the whole population of the instances of the law, that is, to the law itself. When a theory, viewed as a law formulated in the meta-language of science, asserts that all the verifiable statements which follow from a specified consistent and finite set of assumptions are true, there is still the possibility of validating this claim by means of the same procedure: selecting a suitable sample of the population (which consists, in this case, of individuals called "verifiable consequences of the theory") and assuming that the finding concerning the sample (viz., that all the propositions included in the sample are true) applies also to the whole population (i.e., to the sum-total of all the verifiable consequences of the theory itself). This basic analogy between the validation of theories and of universal laws holds for all types of empirical theories, regardless of whether they are phenomenological or transcendent, axiomatized or merely axiomatizable. It holds also regardless of whether we take the traditional view that the findings concerning the sample entitle one to extend their validity to the whole population, or merely claim that the findings concerning the sample apply to any single new individual of the population.

Yet the very analogy between the inductive validation of laws and of theories raises a new difficulty. The applicability of the Principle of Induction (which asserts the validity of the inference from premises concerning a fair sample to the whole population) to a given department of empirical science admittedly depends upon the conceptual apparatus involved and primarily upon the concepts occurring in relevant empirical generalizations. It can be shown, for example, that if the principle does apply in cases where only universal concepts are involved, then it is bound to fail in certain inferences the premises of which include local spatio-temporal constants. The validity of inductive inference is tacitly taken for granted by the scientist who handles the conceptual apparatus of his "object-science"; his success is, so far, the main argument in favour of the Principle of Induction. Yet the validity of this principle at the level of the object-sciences does not guarantee that it will continue to be valid at the next meta-level. In the transition from the object-sciences to the meta-sciences the conceptual apparatus is subject to changes which may be prejudicial to the validity of inductive inference.

To evaluate the force of this objection, we must first ascertain to what extent the conceptual apparatus is changed when a theory formulated in the object-language is translated into a universal law in the meta-language in order to make it amenable to inductive validation. I take it that an empirical theory T is replaceable by the sum-total of the verifiable consequences which can be derived from the assumptions of the theory; such a replacement can be substantiated in terms of the Principle of Verifiability, quite independently of the question concerning the validity of inductive generalization. In other words, the theory T is replaced by the universal law to the effect that every verifiable consequence of T is true. The conceptual apparatus involved in such a universal law includes two basic predicates: (1) that of being a verifiable consequence of the theory T; (2) that of being true. The generalization to be validated by inductive inference runs: Every statement of the language of T which is a verifiable consequence of T is a true statement. Since it is always possible to number all the statements of a given language by using natural numbers as marks, we may simply lay down that the population consists of all the statements (S_1, S_2, S_3, etc.) which could possibly be formed in the language in which T is couched. The meta-linguistic, law-like equivalent of T then runs: No matter what the natural number n happens to be, if Sn is a verifiable consequence of T then Sn is true. This is exactly the form of universal laws which can be inductively inferred from a finite sample of their instances. If the population is finite, any fair sample confers upon this universal law a degree of probability determined by the size of the sample and the variety of its instances. If the population is infinite—as it certainly is in our case since the number of distinct verifiable consequences which follow from any conventional scientific theory can easily be shown to be infinite—the universal law itself cannot be made probable to any finite degree by utilizing a sample of favourable instances, but the probability that a single new verifiable consequence of the theory will prove correct if examined is provided by a representative sample of such consquences.

Accordingly, the new concepts introduced in a meta-linguistic reformulation of a theory are those of the verifiability of statements, of the truth of statements, and of the consequence-relation obtaining between a statement and a set of statements. We shall have to submit all these three concepts to a careful scrutiny and may state now, by way of anticipation, that, on closer analysis, they all prove to be empirical concepts linked up by definitional criteria with the ostensive level, and to involve no local, spatio-temporal constants. The new conceptual apparatus, therefore, resembles the apparatus of the object-sciences in all the characteristics which are relevant to the validity of the Principle of Induction. This does not deny that applying this principle at the meta-linguistic level is an additional assumption, independent of its assumed validity at the level of the object-language of science. We simply have to acknowledge that the logical features are exactly the same in either application of the principle. Furthermore, there is hardly any doubt as to the scientist's readiness to apply the principle at the meta-linguistic level. When he establishes either a single law or a full-fledged theory by utilizing a body of observational evidence and is then challenged as to the legitimacy of his extending the validity of this law or theory to a new case which did not occur in the body of supporting evidence, he would certainly answer: "This formula (law), or this set of assumptions (theory), has often been checked in similar cases with a satisfactory result and, to the best of my knowledge, there is no special circumstance which would make the applicability of the law or the theory to the new case under consideration less likely. Moreover, to the best of my knowledge, there is no other law (or theory) which could be applied instead and which would offer, under the circumstances, a better chance of success than the one I deem well established. Hence, the application of the law or theory to any new case seems warranted by the past record of the law or theory, by a careful scrutiny of the circumstances in the new case, and by the present state of science, which does not offer any more promising law or theory." Thus, the usual way of substantiating the validity of either a scientific law or a scientific theory is actually a meta-scientific one, since the chances of the applicability of a formula or of a set of formulae are estimated on the basis of its past record. In other words, the question as asked by the scientist is not "Will the sun rise tomorrow, just as it did rise so many times in the past?" but rather "To what extent is the hypothesis referring to the future behaviour of the sun substantiated by well-established statements referring to the past?" The difference between the two formulations may seem insignificant when applied to a law all the instances of which can be formulated in the object-language of science. But when applied to scientific theories the difference is important, since it both justifies the inductive validation of scientific theories and necessitates the distinction between the various components of such theories, as outlined above.[10]

27 ON VALIDATING SELF-CORRECTING SYSTEMS OF THEORIES

In § 21 we discussed systems of theories exhibiting a pattern which seems to recur at all the crucial stages of scientific progress. Such a self-correcting system of theories consists of a finite series of theories in which each theory (definitionally and inferentially) presupposes, corrects, and extends the preceding theory. The same threefold relationship of presupposition, correction, and extension may also obtain between whole clusters of theories, instead of between single theories. The most im-

---

[10]It may be noticed that the above analysis of the inductive validation of scientific theories bridges the gap between the so-called "hypothetico-deductive" view of scientific method and the alternative "inductive view." The validation of a transcendent theory couched in an object-language exemplifies the "hypothetico-deductive" view; the reconciliation with the "inductive" approach is obtained by the meta-linguistic paraphrase of the theory in question. In connection with the "hypothetico-deductive" view of scientific method, cf., for example, J. O. Wisdom, *Foundations of Inference in Natural Science* (1952).

portant example of a self-correcting system of clusters of theories is afforded by the sequence of the pre-scientific world-view, the world-view constituted by the fundamental theories of classical science, and the contemporary theories of atomic phenomena and of relativity. This self-correcting sequence of three clusters of theories should be supplemented, I think, by the knowledge we have of sensory appearances or sense-data: the sum-total of validated statements about the appearances presented by objects to the human observer is another theory which is presupposed, corrected, and extended by the common-sense world-view. We thus obtain what might be termed the worlds of sensory appearances, of common-sense things, of pre-atomic scientific objects, and of atomic scientific objects, and this means a sequence of four bodies of knowledge each of which is presupposed, corrected, and extended by the next following body, except the last in the sequence, which may also acquire a successor in due time.

The validation of this self-correcting system of theories, and of other similar systems as well, seems to involve the same devices as those discussed in the preceding section. However, certain specific difficulties arise in connection with the self-correcting function of such systems, which have to be investigated in view of the importance of this mode of forming theories in the life of science. These difficulties are of two kinds: (1) Basic theories sometimes presuppose each other so that it seems to be impossible to validate them without circularity; (2) In a self-correcting system of theories, any non-initial theory both presupposes and refutes the preceding theory: this seems incompatible with a validation of the non-initial theory—it cannot be validated without resorting to the preceding theory which it refutes, and the validation is therefore obtained by deriving the theory to be validated from an untenable theory. This is hardly satisfactory.

I shall deal mainly with the difficulty related to the fact that major theories and clusters of theories both presuppose and refute the same theory, and especially with the problem raised by the sequence of the four comprehensive world-pictures so far evolved within science. The transition from the common-sense world-picture to the pre-atomic scientific world-view has been the object of much debate and construction in the history of philosophy, and, more particularly, in the philosophy of science. The transition between the "world of sense-data or of sensory appearances" on the one hand and the world of common-sense objects on the other has got into the focus of philosophical discussions mainly in the present century, in connection with the very emergence of the concept of the sense-datum and the new emphasis on the problem of perceptual knowledge; a powerful stimulus was also provided by the surge of experimental psychology, both human and animal, since by far the most impressive findings of the new science were concerned with the psychological aspects of perception. The passage from classical to relativistic and atomic science was discussed by philosophizing scientists rather than by professional "pure" philosophers, the main reason for this shift being the technical complexity of the relevant scientific theories, which often required on the part of the philosophers an effort they were reluctant to make. Yet it is worth noticing that the amounts of philosophical labour devoted (either by pure philosophers or by philosophizing scientists) to Relativity and to Quantum Theory are by no means comparable. The Theory of Relativity attained its maximum philosophical influence in the twenties, and this may account for the fact that the most outstanding representatives of pure philosophy did take an active part in discussing its philosophical implications, presuppositions, and limitations: Russell, Whitehead, Bergson, and Cassirer were among them. In contrast, the Theory of Quanta attained its impact in the late thirties. Accordingly, the purely philosophical response of a world already dominated by the spectre of an imminent war was rather weak and certainly not to be compared with the philosophical avalanche let loose by Relativity. The last remark does not apply, perhaps, to one philosophical aspect of Quantum Theory, namely, its bearing on the age-old problem of determinism and free will. But other deeper and admittedly troubling implications of Quantum Theory concerning the relationship between the subject and the object of human knowledge, the controversy over Idealism, Realism, and Phenomenalism, the reign of Probability in the conceptual scheme of our knowledge of nature, the objectifiability of scientific concepts, the applicability of multi-valued non-Aristotelian logics to the physical universe, were discussed, if at all, by the philosophizing scientists rather than by those mainly interested in the epistemology and philosophy of science. There is no doubt that the historical situation is responsible for this development, consequently there is little hope for a decisive change in the foreseeable future.

In this investigation the sequence of four consecutive world-pictures is of interest as a major case of scientific theory-formation. The crucial feature is the mode of transition from the cluster of theories characteristic of a particular world-picture to the theories belonging to the next world-picture. Each stage, characterized by its cluster of theories, bears the threefold relation of presupposition (definitional and inferential), correction, and extension to the immediately preceding stage. Thus the classical world-picture is unintelligible without its common-sense background, since its concepts are undefinable without the common-sense concepts; the classical contentions are not deducible from the classical assumptions, unless use is made of additional premises provided by the pre-scientific world-picture; more particularly, the pre-scientific knowledge of physical objects utilized as observational instruments and mensurative standards is indispensable in the pursuit of classical science. On the other hand, quite a few contentions of the pre-scientific world-picture are either explicitly refuted or declared to be gratuitous and superfluous by the scientific theories; the pre-scientific world-picture is not simply incorporated into the world-view of science but submitted to a careful scrutiny, and then either transformed or found wanting. All sorts of illusory, hallucinatory, mythical, and magical information provided by pre-scientific methods is thrown overboard and either replaced with new, different views justifiable by the method of science, or simply declared unwarranted and not usable. Finally, a host of questions which can be formulated in terms of pre-scientific concepts and which are often extremely puzzling to the human mind (for example, those referring to the explanation of sensory illusions and hallucinations, or to the origin of biological species, or to the probable age of the earth) become answerable, because of the methods and findings of pre-atomic science. The precise determination of how some pre-scientific data are indispensable for the pursuit of science while others are incompatible with it is a difficult problem. As Russell once put it, common sense leads to physics which in turn refutes common sense, and that is why common sense is untenable. This refutation of the pre-scientific world-picture seems irrefutable,[11] but it goes further than Russell might have desired: by liquidating the common-sense world-view, science would dig its own grave since it cannot do without the former.

We shall concentrate our discussion of self-correcting systems of theories around the definitional interdependence of the theories involved. I think that, in the progression from the primitive sensory starting-point towards the advanced stage of atomic science, crucial concepts of any stage are conditionally defined in terms of tests and responses belonging to the preceding stage. Since a conditional definition would be of no use whatsoever without some substantial information concerning the test and the response it involves, one cannot get along in applying the crucial concepts of a given stage without recourse to information about the definitional tests and responses obtainable at the next lower level. Now trouble is bound to occur if, at the new level, the use of a new concept

---

[11] It exemplifies the well-known logical law that any statement which entails its own negation must be false.

leads to the discovery of laws which may be incompatible with the information, indispensable for handling this concept, provided at the lower level. Since such laws are, in turn, associated with new definitional criteria for the same concept, a clash between various definitions of the same term will occur and lead to a serious logical difficulty. If two items of *factual* information clash with each other, no logical scandal is involved, since no internal inconsistency ensues in the body of science; one of these items has simply to be abandoned, and, as a rule, the item pertaining to a higher level will be better substantiated and is therefore likely to eliminate its competitor. However, if a clash of two *definitions* of the same term occurs, this way out is not available because a definition stands in no need of substantiation, its truth being guaranteed by the very meaning of its constituent terms. Hence, the two incompatible definitions enjoy equal cognitive status and another way out of the predicament must be found. This is the root of the difficulty raised by the presence of self-correcting systems of theories within science. As a matter of fact, the scientist copes with the situation by taking a more radical course of action: to cure, he resorts to surgery. The troublesome concept is not simply abandoned, because it remains indispensable; it is rather redefined at the higher level of scientific inquiry so as to avoid inconsistency.

To illustrate this self-correcting technique, let us examine a very elementary case, say that of temperature. Originally, the temperature of a physical object x is conditionally defined in terms of a pre-scientific test and response: "If a thermometer has been applied to the object x for some time, then, by definition, the body x has the temperature T if, and only if, the thermometer reads T." Yet one may wonder whether the temperature of every object necessarily corresponds to the reading of a thermometer which has been applied to it. The question may seem senseless since temperature is defined by the above criterion. Nevertheless well-known phenomena of heat fluctuation cast a doubt upon this inference from the reading of a thermometer to the temperature of the physical object. In view of the laws governing the fluctuation of heat, the statement ascribing a given temperature to a particular body does not follow from the premiss, reporting on the corresponding reading of the thermometer. An inference from the latter to the former presupposes that a state of thermodynamical equilibrium obtains between body and thermometer after their prolonged contact; however, such an equilibrium may fail to occur in this particular case and is bound not to occur in a specifiable fraction of a large number of contacts lasting for a specified time, according to statistical laws of thermodynamic fluctuation. Granted, the probability of the equilibrium failing to materialize on a definite occasion is almost negligible. Yet it does not vanish, and this already suffices to prove that the reading of the thermometer does not entail any statement about the temperature of the body to which the thermometer is applied. As a matter of fact, small fluctuations around the equilibrium are observable under appropriate circumstances, and more considerable departures from the equilibrium are theoretically admissible in any single case. It is even not impossible, though extremely improbable, that a prolonged contact of two bodies of unequal temperature will keep increasing their difference of temperature, the heat continuing to flow from the colder to the hotter body. This is ostensibly incompatible with the second principle of thermodynamics, but actually consistent with the predominating statistical interpretation of it.

Should we conclude that, in view of the impressive body of observational evidence supporting the laws of thermodynamic fluctuations, the conditional definition of temperature—involving the application of a thermometer as test and its reading as response—proves to be false and ought to be abandoned? If put in this way, the question is obviously preposterous. Definitions are neither provable nor disprovable by observational data; they attribute a particular meaning to the term they define and are unconditionally true in virtue of this meaning being actually attached to this term. But the point is that the conditional definition of temperature involving the reading of a thermometer is not the only analytic context in which the term "temperature" occurs, and, consequently, not the only clue as to the correct use of this term. Some empirical laws including the concept of temperature and established on the basis of this initial definitional criterion yield, in turn, new definitional criteria which cannot be disregarded without this term being misconstrued. These new criteria may turn out to be incompatible with the initial definition. Thus, the laws of fluctuation entail that in a nonvanishing percentage of cases, the performance of the original definitional test T in conjunction with the original definitional response R may fail to ensure the applicability of the defined concept P. This is a logical inconsistency between an explicit conditional definition of temperature and a semi-definitional law involving this concept.

How does the scientist deal with this situation? He does not abandon the concept of temperature or its definitional connection with the reading of a thermometer. If he were worried at all about logical difficulties, he could simply declare that, in view of the laws of thermodynamic fluctuations, the concept of temperature needs re-defining in order to avoid inconsistency with the definitional laws which are discovered owing to the term's having been so defined and which subsequently modify its meaning. The re-defining of the concept of temperature has to take into account the fact of fluctuations and may therefore run as follows: "If a thermometer is applied for some time to objects of a sufficiently large class, then, by definition, almost every object of this class has the temperature T if the thermometer applied to it reads T, and, conversely, almost every thermometer applied to such an object endowed with the temperature T will read T." It is obvious that, in the theory of fluctuations, the statistical correspondence between the temperature of the object and the reading of the thermometer is determined with a greater accuracy than in this definition. But, as it stands, this correspondence suffices to show how, by re-defining temperature in a statistical way, one can obtain consistency between the initial conditional definition of temperature and the laws it helps to discover. Thus the connection between response and defined property becomes statistical, without ceasing to be definitional. We shall therefore call *probabilistic criteria* all analytic statements of the following type: "Almost every object which was submitted to test T and yielded thereupon response R has the property P and, conversely, almost every object endowed with the property P, yields the response R if submitted to test T." Every analytic statement of this form can serve as a conditional probabilistic definition of the property P in terms of test T and response R. It is obvious that the usual conditional definitions are a special case of the probabilistic ones; they originate when the connection between response and defined property admits of no exceptions.

Similarly, the connection between sensory and perceptual terms—or, to put it otherwise, the connection between terms describing how things look to a human observer and terms describing how things are—seems to have been originally of the non-probabilistic definitional type: If I look at a given object under usual conditions of distance and illumination, then, by definition, the object is green if it looks green to me. In this conditional definition of a perceptual term by means of sensory terms, the test consists in the object being watched or looked at; the response consists in the sensory appearance then presented by the object to the human observer; and the defined property is the "objective" property attributed to the object within the common-sense outlook. It does not matter at all that the response seems to involve the defined concept: being green is defined in terms of looking green to somebody; the linguistically more complex expression "looking green to me" is epistemologically prior to the simpler linguistic expression "being green." The only trouble is that once the perceptual terms are available and are used for discovering laws of sensory illusions and hallucinations, it turns out that these laws entail "fluctuations" which are incompatible with the initial definitions of such terms. Even if something does look green to me it may still fail to be green because, in view of the laws of sensory illusions and hallucinations, the connection between looking green and being green is not inevitable. Yet the plain man would hardly be induced, by the discovery of illusions and hallucinations, to abandon the definitional connection between how things look to him and how they are. He still feels that the very fact that an object looks green to him is a relevant and important piece of evidence supporting the view that this object is green. He is prepared to admit, however, that such support does not offer an infallible guarantee. I think that the most natural way of accounting for the fact that if an object looks green to me then I feel justified by the very meaning of the term "green" to infer that this object probably is green, is to establish a probabilistic conditional definition which explains the perceptual term by means of its sensory counterpart: "If I watch an object, then, by definition, it has in all probability a definite colour if, and only if, it looks to me as if it had such colour."

Generally speaking, any non-sensory world-picture presupposes the immediately preceding world-picture because its crucial terms are conditionally defined by means of tests and responses which belong to the preceding world-picture. Hence the information obtainable at this lower level about the definitional tests and responses is necessary at the following level, because otherwise the crucial concepts defined by these tests and responses could not be handled with any efficiency. At first, the conditional definitions are non-probabilistic and have to be replaced with probabilistic ones if laws of "fluctuations" involving these concepts come to be discovered. Sense-illusions and hallucinations are such fluctuations affecting the transition from the sensory to the perceptual level; thermodynamic fluctuations illustrate how the need for probabilistic definitions arises in the transition between the common-sense level and pre-atomic science. In all such cases, the inconsistency is removed by a probabilistic re-definition of concepts affected by laws of "fluctuation."

It should be pointed out, finally, that probabilistic re-definition is not the only self-correcting device used to remove the inconsistency implicit in forming systems of theories in which each theory presupposes, refutes, and extends its predecessor. The only vital case where laws of a lower level are used at the following level in spite of being inconsistent with the laws of this level, involves lower level laws concerning definitional tests and responses for higher level concepts. If the validity of the lower level laws for such tests and responses is maintained while their universality is abandoned, the inconsistency may be removed. This procedure is often applied, particularly in the transition between classical and atomic or relativistic science. Thus, although the pre-relativistic laws governing clocks and yardsticks are abandoned in special relativity, pre-relativistic physics with all its laws concerning standards of length and time is used without qualms whenever the speed of physical objects involved is negligible in regard to the speed of light. Similarly, the laws of pre-atomic physics which govern the behaviour of essential measuring and observational devices are assumed in Quantum Theory to be applicable with sufficient accuracy to the relatively large physical objects made use of in all such instruments. Hence the inconsistency between the classical laws implied in the use of indispensable instruments and the new laws is removed by restricting the originally universal classical laws to the rank of merely regional laws. No probabilistic re-definition of the classical concepts would do in this case, since the new laws are not merely asserting fluctuations incompatible with the universal validity of the classical laws; the gap is wider in this case and can be bridged only by abandoning the claim for universal validity of the old laws.

PART THREE

Universality of Science

CHAPTER ONE

Verifiability and Truth in Science

Objective and Logical Ranges of Science

We have seen in our previous discussions that the *epistemological* range of science, that is, the sum-total of problems to which scientific method could possibly be applied, has to be distinguished from what may be called the *logical* and the *objective* ranges of science. The objective range of science is determined by the largest possible group of objects which are knowable by scientific method (the universe of discourse of science). The logical range of science depends upon the supply of scientific concepts, the logical complexity of scientific statements, and the nature of inferential procedures applied by the scientist. The objective and the logical ranges of science, though distinct from its epistemological range, have clearly a bearing upon it. One may wonder, therefore, whether our central problem concerning the epistemological range of science could be solved by determining its objective or its logical range. A simple argument shows why this is not feasible.

The inadequacy of the objective range can be seen from the following consideration. An object is accessible to scientific method whenever this method provides some reliable information about this object; this does not necessarily prevent other problems concerning the same object from being unsolvable by any scientific procedure. The size of the earth can be determined with sufficient precision by computational measurement; the earth is therefore within the universe of discourse of science. But the question as to whether the diameter of the earth is commensurable with the diameter of any other planet is scientifically unsolvable. Similarly, the question of whether various human individuals have exactly similar

sensations of colour is beyond the reach of scientific method, whereas
similar questions related, for example, to the statistical distribution of
colour-blindness in human populations have been scientifically solved
(cf. § 1). Of course, in both cases the inapplicability of scientific method
can be accounted for by specific features of these problems, for example,
the role of the concept of incommensurability in the first question and
that of intersubjective similarity of sensations in the second. But, even
if this explanation of the failure of scientific method to cope with such
questions be correct, it does not deny this failure. Thus, there is little
doubt that the range of scientifically solvable problems is not determined
by the range of objects referred to in these problems.

One should also observe that the extremely controversial nature of
the boundaries of the scientific universe of discourse makes this approach
unsuitable, irrespective of the difficulty just referred to. As long as there
is no solution to the problem concerning the translatability of scientific
information referring ostensibly to a very heterogeneous assembly of
entities (common-sense material objects, sensory appearances, constructs of the natural and social sciences such as atoms, molecules,
elementary particles, fields, civilizations, literary trends, personalities,
etc.) into statements about a more homogeneous set of objects, there
remains the possibility that the body of scientific knowledge will eventually be shown to involve a substantially reduced universe of discourse,
less comprehensive than the apparent universe of discourse of a literally
interpreted science. Moreover, there are some reasons for assuming that
a determination of the universe of discourse of science constitutes an
inherently unsolvable problem (cf. § 12). If such were the case, the
bearing of the objective range of science upon its epistemological range
would remain permanently controversial.

Furthermore, we have already called attention to the fact that apart
from a determination of the basic universe of discourse, there still remains the choice between a nominalist and a Platonist interpretation
thereof, either by restricting the range of science to the concrete individuals of the universe under consideration, or by claiming that this set
of concrete individuals is but a starting point for a whole infinite
hierarchy of scientifically knowable abstract entities such as properties
and classes of these individuals, relations among them, properties of
these properties, etc. There is also the question concerning what I shall
call *elusive entities* (cf. § 30); although such entities are often referred
to in the usual formulations of scientific theories, their status is very
much a matter for debate, quite independently of the nominalist-
Platonist controversy. These reasons make imperative an attempt to

characterize the reach of science independently of its universe of discourse.

On the other hand, a determination of the *logical* structure of scientific facts, laws, and theories would still fail to answer our main question. For such logical characteristics of scientific information, though extremely important in connection with our main problem, provide by themselves answers to the following questions only: What is the basic undefined vocabulary of the language of science? What are the ways of forming sentences out of the elements of this vocabulary? Which sentences, out of all those correctly formed, are to be taken for granted without further ado, simply because the correct use of the language so requires? Given any set of sentences in the language of science, which other sentences follow from this set according to the rules of inference obtaining in this language? Which sentences can be said to derive at least a specifiable degree of probability from this initial set of sentences? Of course, most of the logical questions of this kind concerning the language of science are far from being solved at present; there are also reasons for doubting the possibility of constructing a single consistent language capable of expressing all that is scientifically knowable. However, even if a whole hierarchy of languages used by the scientist in order to formulate all his theories were investigated from all possible angles, and the fundamental logical questions concerning the hierarchy were answered, the range of problems that could be formulated in these languages and solved by applying scientific methods would not be determined thereby: the entire logical structure referred to in such questions may be contained in a single scientific theory and could not possibly account for the cognitive content of *all* the scientific theories which are well established at present, let alone for those which will, or could possibly be established.

Expressing the Reach of Science in Terms of Verifiability

To determine the epistemological range of science, we shall have to utilize our survey of scientific methods. It is true that only methods actually used have been surveyed. However, all problems which are solvable by any conceivable scientific method might be obtained from those to which some presently available method has been successfully applied, by appropriately replacing the *actual* circumstances under which the method came to be used by some *potential* or *possible* circumstances. This passage from actuality to potentiality can be effected in several ways, leading to different determinations of the range of science; it raises quite a few other difficulties as well. Still, the approach seems promising, since the epistemological range of science takes in problems to which scientific method could *possibly* be applied: it is natural enough that the troublesome concept of possibility, present in the very formulation of the problem, should reappear in attempts to solve it. The multiplicity of transitions from actual to possible methods will be restricted by applying the Principle of Verifiability.

We have thus to realize what the actual application of the scientific method amounts to and subsequently to ascertain which actual features of such an application are replaceable by potential ones. Of course, some features will have to remain unchanged. Since our aim is to determine the entire range of science, we cannot restrict ourselves to its methods of discovery: we shall have to consider a problem to be solvable in principle by scientific methods, even if the relevant methods are merely heuristic or verificatory (cf. § 10). In other words, we have to extricate the basic features of the situation which obtains whenever a particular problem P is considered to have been solved by a scientific method, in view of the fact that this method has enabled the investigator to ascertain that a particular answer A to this problem is correct; we do not, however, assume that the answer A has been suggested to the investigator by the method, or even uniquely determined by it. All that matters in successfully applying a scientific method to the problem P is that the correctness of a particular solution A to this problem (which the investigator may have obtained by any procedure whatsoever) has been ascertained by applying this method.

For example, the solution to the question as to the forces which any two bodies of known masses exercise on one another was obtained by Newton, when he succeeded in proving, on the astronomical data available to him, that all such data would be accounted for if the mutual attraction of any two masses were equal to the product of their masses divided by the square of their distance. Whether or not any pomological experience suggested such an answer to Newton is entirely immaterial.

Now the basic features of this situation are obviously as follows: the astronomical data available to Newton were ultimately based upon observations made by Tycho de Brahe; they constituted the body of relevant evidence. And the process by which Newton solved the problem consisted actually in his showing that this observational evidence, if properly utilized (I mean adequately expressed in observation-statements and used in conjunction with other premises provided by pure mathematics), entails the correctness of his answer to his problem—or, rather, renders the correctness of his answer to his problem fairly probable.

Tentatively, we have assumed (cf. § 10) that applying the scientific method to any problem always consists in showing the bearing of the available observational evidence upon the correctness of some solution to this problem. As a matter of fact this is the case in all the basic scientific methods we have surveyed. Making a direct or indirect observation, performing a measurement, validating a universal law, establishing a scientific theory by drawing consequences from it, and comparing them with observational data, always involves the two basic operations: utilizing either extant or newly obtained observational data and then submitting these data to some suitable logical processing, in order to ascertain the bearing of such observational evidence on the correctness of some solution to the problem under consideration.

We may therefore widen the concept of the application of scientific method to a particular problem by replacing the actual features in the foregoing description with potential ones. In other words, we shall tentatively assume that a problem is solvable in principle by scientific methods, if *some* conceivable amount of observational evidence submitted to *some* conceivable logical processing would have *some* bearing upon the correctness of *some* solution to this problem. Conversely, a problem would be unsolvable in principle by scientific methods if no conceivable observational evidence submitted to any conceivable logical processing would have any bearing whatsoever upon the correctness of any of its solutions. We may call such problems whose logically admissible solutions are all of them completely impregnable to *any* combined attack of observational evidence and logical processing, scientifically unsolvable. Propositions which provide logically admissible answers to such problems are termed unverifiable propositions. The task of determining the epistemological range of science would thus hinge upon an appropriate delimitation of the class of verifiable propositions. In particular, the universality of science would be established by showing that all problems solvable by extant scientific methods have verifiable solutions, whereas problems answerable only by unverifiable propositions have no solution.

Of course, by so reformulating our problem, we have merely substituted one wording for another. Actually, the new wording is rather vague since it includes such phrases as "some conceivable observational evidence," "some conceivable logical processing of a body of evidence," "some specifiable bearing of a logically processed body of evidence on the correctness of some particular solution." In our subsequent investigations, we shall try to define clearly all these vague expressions. Reformulating the problem concerning the epistemological range of science in terms of verifiability will then prove to provide two advantages: once the concept of verifiability is appropriately defined, it can be proved that problems answerable by unverifiable propositions have no solution and are therefore unsolvable by any method, be it scientific or not; it can also be shown that all problems which can be solved by actual or discoverable scientific methods admit of verifiable answers. The Principle of Verifiability would thus determine both the range and the limits of science.

Private versus Public Verifiability

For the time being, let us supplement this account of scientific method in terms of observational evidence and inferential processing (or, alternatively, in terms of verifiability, which involves both observation and inference) with a few comments on the role of the investigator. Observational evidence consists, obviously, of reports on observations made by some human observer, or a group of such observers. The verifiable propositions which answer any cognitive problems must be verifiable by some investigators or a group of investigators. To what extent is our problem affected by the part played by observers and investigators? How should their part be taken account of in formulating the Principle of Verifiability?

It is important to realize that the distinction between observational evidence and logical processing implied in any satisfactory solution to any definite problem is not uniquely determined by the nature of the problem but varies with the investigator. In other words, the boundary between "observational evidence" and "logical processing" in any validating procedure will depend, in general, upon who applied the method to the problem and also upon the circumstances under which the method came to be applied. Thus, in order to ascertain whether the weather is fine right now, I have only to look through the window. There is no logical processing involved at all; direct observation only is required, because I can see whether the sky is blue and the sun shining. To answer the same question, a blind person would have to make inquiries and use reports of other persons endowed with normal sight; some logical processing on his part would be necessary for him to obtain a reliable answer. I myself would be unable to answer reliably a similar question concerning today's weather if it had been put to me a month ago, or a month from today. Thus, the evidence and logical processing required for solving the same question by persons who differ in their perceptual equipment or by persons who are faced with this problem in different space-time regions will vary considerably, although the problem itself and the correct answer to it remain unchanged.

It goes without saying that this variability of the boundary between evidence and logical processing in the application of a scientific method to some particular problem by several investigators has to be taken into account when the determination of the range of scientific methods is formulated in terms of verification. The Principle of Verifiability asserts that only verifiable propositions have a definite truth-value and are cognitively relevant. However, one wonders whether, to be so relevant, a proposition ought to be verifiable by *every* competent investigator, or whether verifiability by a *single* investigator would do. Each alternative raises considerable difficulties.

Suppose the principle be construed as stipulating that scientifically relevant propositions must be verifiable by every investigator. Yet statements concerning a future when I shall not be alive any longer seem to be unverifiable by me. The same holds true of statements concerning a past which I did not witness and of which no traces discoverable by me remain. Both kinds of statements would be banned if verifiability by every investigator were required. By so restricting the meaning of verifiability, we fail to do justice to the cognitive achievements of science.

One may point out other reasons which explain why it would make no sense to confine the range of scientific information to statements which are verifiable by every investigator. Science is a social enterprise, and the scientific community includes in principle the whole of mankind. It does not matter, from a scientific point of view, which member of the community has been designated to verify a required item of information; all that matters is that the verification be carried out. Instead of dispatching a given individual to a given region in order to use subsequently his report in checking on a theory, the region may also be explored by automatic recording devices which will fully make up for the traveller's report. From this point of view, the human observer is just a particular recording instrument, capable of keeping track of certain phenomena in the space-time region which he occupies. Science has no universal recording device capable of providing data for checking on any and every bit of information. The human detector is no exception in this respect.

Should we then conclude that "verifiable" means in our context "capable of being checked by at least one human observer," so that only information which is unverifiable by every human observer would be placed beyond the pale of science? This view is too liberal, since it disregards another aspect of the social function of science, namely, the objectivity of scientific information—information verifiable by a single observer, though vitally interesting from several points of view, is illegitimate in science. A mystic may have the same degree of certitude as to the presence of a supernatural agency as he has of the presence of a material object which he happens to perceive. Still, his information concerning this agency is of no use within science because it cannot be verified by other investigators.

We thus see that the difficulty in properly delimiting the scope of verifiability with a view to determining the range of science is due to the necessity of meeting two ostensibly incompatible requirements: (1) to be objective, scientifically relevant information must be verifiable by every investigator; (2) scientific information may refer to facts which are not accessible to every investigator. Since history, palaeontology, and cosmology provide scientifically valid information that does not seem to be verifiable by every investigator whereas information verifiable only by the mystic lacks scientific status, we cannot consider either the verifiability of a piece of information by every investigator, or the possibility of its being verified by at least one investigator as essential for its scientific relevance.

Now it is apparent that the source of the difficulty is the emphasis unduly laid upon direct verifiability. We shall say, for example, that the investigator who has relied entirely on non-inferential, say observational, evidence available to him in solving some problem has verified *directly* the proposition or propositions reporting on his observations and providing a solution to the problem. In more complicated cases which involve both observational evidence and logical processing on the part of the investigator, we may say that by establishing the correctness or incorrectness of a proposition answering his problem he has *indirectly* verified this proposition, on the understanding that to obtain such indirect verification he had first to verify directly the statements reporting his observational evidence and then to submit them to an appropriate logical processing. Accordingly, the fact that the boundary between evidence and logical processing depends upon the choice of the investigator and his spatio-temporal location may be expressed by saying that the same proposition can be verified directly by one observer and indirectly by another observer, depending upon their individual characteristics and their situation.

In view of this dependence of the verificatory operations upon the observer, the question arises as to how the roles of direct and indirect verification should be apportioned in formulating the Principle of

Verifiability. It is obvious that this principle must not confine scientific relevance to statements which can be directly verified by every investigator, even if we artificially stretch the range of direct verifiability so as to include, for example, astronomical hypotheses. Propositions concerning the conditions which prevail within the planetary system might be admitted to be directly verifiable by man since the perennial dream of interplanetary travel may come true some day. But the verifiability of statements about remote periods of the past, especially those preceding the emergence of man, can hardly be based on an impossible travel through time. It is much simpler to acknowledge that indirect verifiability is as good as direct, in so far as the cognitive relevance of verifiable statements is concerned.

Thus a statement which is directly verifiable by a given group of observers, may be indirectly verifiable by another group, provided the statement be connected by appropriate logical ties with a set of propositions which are directly verifiable by the second group. It may also happen that a statement is directly verifiable by no individual or group; it would nevertheless be scientifically relevant if it were indirectly verifiable by every competent individual or group. We thus obtain three kinds of propositions:

(1) Those which are directly verifiable by at least a single individual and indirectly by other individuals. We shall call such propositions *directly verifiable*. A special subclass of directly verifiable propositions consists of those which are directly verifiable by exactly one individual and indirectly by other ones. These are the introspective statements, the legitimacy of which seems therefore unquestionable from our point of view.

(2) Propositions which are not directly verifiable by any investigator, but accessible in principle to indirect verification by every competent individual. We shall call them *indirectly verifiable* propositions.

(3) Finally, propositions which do not belong to any of the two classes just referred to, we shall call *unverifiable*. They include those which are unverifiable (either directly or indirectly) by any individual, as well as statements which, though verifiable by some individual investigators, are not verifiable in principle by all investigators.

The upshot of this discussion is that scientifically relevant information consists of statements which could be checked by every competent investigator since they are either directly verifiable by him or connected by appropriate logical ties with other propositions which are directly verifiable by him. This provisional formulation is admittedly unsatisfactory since it includes such vague terms as "being connected by appropriate logical ties with a set of propositions" and "being directly verifiable by a particular investigator." What are these appropriate logical ties? What are the directly verifiable propositions? This vague preliminary formulation is chosen in order to indicate that the finding of a suitable definition of verifiability can be split into two separate tasks, namely, a delimitation of directly verifiable propositions on the one hand and a definition of the logical relationships responsible for indirect verifiability on the other hand. We shall discuss the concepts of direct and indirect verifiability in the next chapter. An attempt to prove the Principle of Verifiability (construed as requiring that a cognitively relevant proposition must be either directly or indirectly verifiable by every investigator) and to establish thereby the universality of science will be made in the last chapter.

Versions of the Principle of Verifiability

In thus embarking upon a determination of the range and limits of scientific knowledge in terms of empirical verifiability, we are entering a field of contemporary philosophy of science and epistemology which has been a centre of interest for a considerable time and has witnessed innumerable discussions. Empirical verifiability was promoted to the rank of a fundamental concept in the philosophy of science and general epistemology by neopositivist, pragmatist, operationalist, and related schools of thought; it also played a prominent role in the evolution of science by influencing the methodological and epistemological outlook of the leading scientists. In view of the special nature of this study, I should like to stress from the outset the particular point of view from which we shall have to investigate the problem of empirical verifiability. Both a specialization of this general problem and several departures from conventional discussions of it will prove inevitable.

Since we are interested in determining the inherent, not the actual, range of science, we shall not have to dwell upon the historical role of the Principle of Verifiability. This means that we can disregard both its important place in the present scientific outlook and the numerous violations of it in the history of science until about the middle of the last century, when it came to exert an ever increasing influence, owing partly to the work of A. Comte and partly to the spontaneous evolution of the scientific outlook. More particularly, the fact that contemporary science exhibits an undeniable tendency to guide its research according to the Principle of Verifiability may be important in connection with our problem, since we are anxious to reach a result applicable to actual science, and not to deal with any fictitious substitute for science.

However, the problem hinges upon the logical implications of the Principle of Verifiability and is not decisively affected by the historical aspects of the principle. In the logical analysis of science, a misuse of historical considerations is as dangerous as the misuse of psychological data. In this field the fallacy of *historicism* has to be avoided just as the fallacy of psychologism has been avoided since Husserl's[1] classical discussion.

In other words, we shall not be interested in the *de facto* question as to whether empirically unverifiable statements do occur in scientific papers and monographs of the past or of the present: this is a historical rather than an epistemological problem. We shall rather investigate the *de jure* aspect of the requirement of verifiability and try to find out whether empirically unverifiable hypotheses do play some cognitively essential part in scientific theory, in the sense, for example, that, by abandoning such hypotheses, science would actually be curtailed, since it would by the same token have to give up some of its empirically verifiable laws and theories as well. This is the position we have already indicated and which will have to be substantiated in detailed analysis. It constitutes one of the basic differences between the role ascribed to the Principle of Verifiability in several trends of the philosophy of science and in the present study.

This difference is connected with the fact that, traditionally, the Principle of Verifiability has been construed as asserting that empirically unverifiable statements and theories lack any literal meaning in spite of their apparent significance. Of course, with such a view the presence of unverifiable statements in science cannot be tolerated, and an "anti-metaphysical" verificationalist purge becomes imperative. However, once the Principle of Verifiability is dissociated from the verifiability theory of meaning and re-formulated to assert that verifiability is a criterion of definiteness of truth-value, there is no more reason for banning all unverifiable statements. In other words, the re-formulated Principle of Verifiability is more comprehensive and "liberal" since it does not prevent the scientist from using unverifiable hypotheses and theories, but still restricts the scientifically knowable to what is describable by empirically verifiable statements and thus retains the gist of the traditional formulations.

We can also be more "liberal" in the very choice of the concept of verifiability, by defining it in such a way that it is applicable to statements the complete verification of which would require an infinite number of observations, and even to those which are at most capable of being

[1] E. Husserl, *Logische Untersuchungen* (1900).

rendered more or less probable by a finite or infinite set of observational data. The need for a "liberal" definition of the concept of verifiability is suggested by the main phases of its historical evolution.

Needless to say, the choice among various definitions of verifiability will have to be made on epistemological grounds. We shall not be moved by non-cognitive considerations as to the desirability of either stretching as much as possible or narrowing down the bounds of verifiability and the limits of science: our problem is one of epistemology, and not of ethics or of scientific policy. It goes without saying also that by framing a definition of verifiability which is not met by all the statements actually occurring in science we shall have to account for the presence of non-conformist statements without urging their removal, and, rather, try to show that their role is an auxiliary, though indispensable, one and perfectly consistent with ascribing to science the task of providing reliable and verifiable information.

The epistemological clues in selecting the appropriate concept of verifiability are clearly derivable from our previous discussions. We shall have to define verifiability in such a way that both the range of scientific methods in actual use and the universality of science will be accounted for: the definition of verifiability has to secure that information obtainable by all conventional scientific methods is demonstrably verifiable, and that demonstrably no unverifiable statement is either true or false.

The multiplicity of possible definitions of verifiability may be illustrated by what could be termed unilateral *versus* bilateral possibility of verification. The empirical verifiability of a statement consists in the possibility of checking upon whether it is true or false by making suitable observations and by drawing suitable conclusions from them. At first sight, it seems irrelevant whether the observational check would result in showing the statement under consideration to be either true or false. In other words, only the possibility of an observational check matters, not its result, which may turn out to be either favourable or unfavourable to the statement. Thus false statements may be verifiable as well as the true ones. It seems to be a general rule that whenever a statement is empirically verifiable, the same holds of its negation, the only difference being that the observations which would be favourable to the statement would be fatal to the negation, and *vice versa*.

Yet although this certainly obtains in all the usual cases, it is easy to construct a statement which is, as it were, only half verifiable, since observations are conceivable which would prove it to be right, but none which could refute it. It is also easy to construct a statement which could be refuted but not proved by observational data. For example, if somebody tried to explain the evolution of the animal kingdom by assuming the simultaneous operation of natural selection and a supernatural agency, his hypothesis would be only unilaterally (negatively) verifiable, since some observations might disprove the operation of natural selection, but no conceivable observations could actually prove that natural selection in conjunction with a supernatural agency is responsible for evolution. Conversely, a hypothesis claiming that biological evolution can be explained either by natural selection or by supernatural intervention (to the exclusion of any other agency) would be positively verifiable by observational data favourable to natural selection, without being negatively verifiable: it would be impossible, indeed, to refute it on observational grounds, since, even if one succeeded in eliminating natural selection, there would still remain the observationally irrefutable possibility of a supernatural intervention.

Thus a statement may be provable on observational grounds without being disprovable on these grounds; we may then say that it is positively, but not negatively, verifiable. Similarly, a proposition which is only negatively verifiable could possibly be refuted but not proved on observational grounds. We may term unilaterally verifiable all those statements which are either positively or negatively verifiable, but not both, in contradistinction to bilaterally verifiable statements for which some observational data would be favourable and other data fatal. Generally speaking, any conjunction of a bilaterally verifiable and a completely unverifiable statement is only negatively verifiable, whereas a disjunction of a bilaterally verifiable and a completely unverifiable statement is only positively verifiable.

The question naturally arises as to the kind of verifiability which scientifically relevant statements should possess: is bilateral verifiability required, or would unilateral verifiability do, or is perhaps only positive (or negative) verifiability essential? To answer this question and similar queries about the right choice of the concept of verifiability, we shall have to ascertain in detail whether the resulting version of the Principle of Verifiability can be shown to account for the actual range of scientific methods and for the inherent universality of science. These two aspects of the problem of verifiability, referring to the range and the limits of science respectively, will be dealt with in the following two chapters. In this chapter, it will be necessary to comment on the concept of truth involved in the reformulation of the Principle of Verifiability which we have suggested and thus clarify the general logical setting of this principle.

29   TRUTH AND VAGUENESS

The Principle of Verifiability maintains that unverifiable statements are neither true nor false. The existence of statements having no definite truth-value, which may be called *indeterminate statements*, seems to be incompatible with the basic laws of logic. We shall therefore investigate the logical status of indeterminate statements, examining first the logical mechanism mainly responsible for the existence of such statements in ordinary language, and then ascertaining whether their presence has any bearing on the laws of logic valid in this language.

It seems obvious that the presence of statements which, ostensibly at least, are neither true nor false, is often due to the vagueness of their terms. That "life can be beautiful" cannot be shown to be true or false, because the concepts involved in this saying are too vague. Shall we nonetheless say that since, according to the Law of Excluded Middle, every statement is either true or false, this statement must also fall under one of these categories, although we never shall be able to make out under which? This would amount to assuming an occult quality of truth or falsehood residing within this statement and concealed forever from the human mind. It seems more in keeping with a sober scientific outlook to suppose that it will never be discovered whether this sentence is true or false, because it is neither. Within such an outlook, truth and false-hood are to be construed as empirically ascertainable properties of only those statements to which they are justifiably ascribable, rather than as occult qualities one of which is attributable to any and every statement.

One might argue, of course, that this saying is neither true nor false because it is not a genuine statement at all and that only genuine state-ments have a definite truth-value and are referred to by the Law of Excluded Middle. Such a view would be taken by the follower of the verifiability theory of meaning; since the saying is obviously unverifiable, it would have to be regarded as meaningless. Yet some verifiable state-ments have no definite truth-value either, unless truth and falsehood are construed as occult qualities. Consider, for example, the prediction "The first man I shall meet today will be bald." This statement is verifiable according to all the current criteria, since we know under what possible observable circumstances we would have to admit it as true, and under what other equally possible circumstances we would have to reject it as false. If the first man I came across were like the Count of Monte Cristo just before he escaped from prison, the statement would be found to be false; if he closely resembled Picasso, the statement would have to be accepted as true. The statement is therefore verifiable in principle.

Nevertheless, should the man I meet have his hair spread thinly enough to make the vagueness of the adjective "bald" apparent, we would be unable to make out whether the statement is true or false. Under such circumstances, ascribing to the statement a definite though unknowable truth-value would make an occult quality out of truth.

It goes without saying that the vagueness of the adjective "bald" would be responsible for the statement's misbehaviour. However, the responsibility would not rest entirely with the vagueness of this term, since other statements containing the same term may turn out to have a definite truth-value. Picasso is certainly bald and Monte Cristo certainly was not. Moreover, since every empirical term is vague to some extent, we would be left with no determinate statements at all, if all those which contain vague terms had to be rejected as indeterminate. In other words, the mere presence of a vague term in a statement does not automatically lead to its indeterminacy. Some additional condition must be fulfilled in those cases where the vagueness of a term in a sentence does affect its truth-value. The logical mechanism of this phenomenon depends obviously upon the nature of linguistic vagueness.

A vague term may be characterized tentatively as one the correct use of which is compatible with several distinct interpretations. The term "Toronto" is vague because there are several methods of tracing the geographical limits of the city designated by this name, all of them compatible with the way the name is used. It may be interpreted, for instance, either as including some particular tree on the outskirts of the city or as not including it. The two areas differing from each other with respect to the spot where this tree is growing are two distinct individual objects; the word "Toronto" may be interpreted as denoting either of these two objects and is for that reason vague. Of course the vagueness of this name is much greater than is suggested by the two areas just referred to, since there are a great number of admissible interpretations.[2]

A statement including vague terms may nevertheless be either true or false if its truth-value is not affected by the multiplicity of their admissible interpretations. Such a statement is true (or false, as the case may be) if it remains true (or false) under every admissible

[2]Vagueness must not be confused with ambiguity. A proper name is vague if it can be interpreted as applying to any individual of a number of overlapping individuals. A proper name is ambiguous if its correct interpretation depends upon its context. The name "Toronto" is vague. It is also ambiguous because there are several non-overlapping localities denoted by it. The expression "Toronto, Ontario" is no longer ambiguous but it is still vague.

To remove the ambiguity of a word, it suffices to place it in a suitable context. A vague term, however, cannot cease to be vague unless it acquires another meaning and thus becomes another term.

interpretation of the vague terms it contains. Thus, although both "Toronto" and "Canada" are vague terms, it is nevertheless true that Toronto is in Canada, because this statement remains true under any admissible interpretation of the two geographical terms it contains. Similarly the sentence "Toronto is in Europe" is false, because its falsehood is not altered by the choice of any admissible interpretation. The statement "The number of trees in Toronto is even" becomes true under some of the admissible interpretations of its subject, and false under the remaining interpretations; it is therefore neither true nor false. We may say that in this last statement the vagueness of the word "Toronto" becomes relevant and affects the truth-value of the statement; and this is why the statement is indeterminate. Accordingly, the presence of vague terms in a language always entails the indeterminacy of some of its statements, namely, those whose truth-value happens to be affected by the vagueness of their terms.

Let us now discuss the apparent incompatibility between the existence of indeterminate statements and the Law of Excluded Middle. Since we are prepared to acknowledge the validity of this law for ordinary language, we must square it somehow with the indubitable vagueness which prevails in the vocabulary of this language. It is important to distinguish carefully the *metalogical* Law of Excluded Middle asserting that every statement is either true or false from the *logical* Law of Excluded Middle which is exemplified in such sentences as "Either the earth is round or the earth is not round," "Either John is bald, or John is not bald." The difference between these laws is obvious. The metalogical law refers to statements, it involves the concept of truth, and it is expressible in ordinary language. The logical law involves neither the concept of truth nor a reference to sentences; moreover, although every instance of this law is expressible in ordinary language, the law itself is not so expressible, since this language lacks propositional variables. One may be tempted to minimize the difficulty connected with indeterminate statements by pointing out that their existence is incompatible with the metalogical law, but not with the logical law, so that the admission of such statements entails only the minor calamity of abandoning the metalogical law. The validity of the logical Law of Excluded Middle would not be affected by the presence of indeterminate statements.

This advice is, however, of little comfort since the two laws are by no means independent of each other. Although the logical law is not expressible in ordinary language, all its instances are logically true statements of this language. According to ordinary rules, the truth of any disjunction implies the truth of at least one of its members. Since, moreover, the disjunction of any sentence and its denial is expressible in ordinary language, we must conclude that one member of any disjunction of this kind is true, hence that every statement of ordinary language is either true or false. The fact that it is logically true, for example, that "Either John is bald, or John is not bald" would compel us to admit that it is either true or false that John is bald, even if he manages to strike a troublesome balance between Picasso and Monte Cristo. There seems then to be no means of abandoning the metalogical Law of Excluded Middle without giving up, by the same token, the logical law as well. We are thus once more driven to the apparent incompatibility between the requirement of not making occult qualities out of truth and falsehood in reference to statements which are affected by the vagueness of their terms and the more imperative requirement of remaining on friendly terms with the laws of logic.

The answer seems to be that the ordinary connection between the truth-value of a disjunction and the truth-values of its members does not apply to statements with vague terms. Since a statement with vague terms is true provided it remains true under every admissible interpretation of its terms, a disjunction with vague terms may well happen to be true, even if its members are indeterminate. This will obviously be the case whenever neither member of the disjunction is true under every admissible interpretation, whereas the disjunction itself always remains true, owing to a suitable interrelatedness of its members. For instance, although neither the statement "The number of trees in Toronto is even," nor its denial remains true under every admissible interpretation of "Toronto," the disjunction itself does remain true under all such interpretations and is therefore true. Thus it seems that the indeterminacy of statements with vague terms is compatible with the validity of ordinary logic in ordinary discourse (viz., with the fact that all instances of the formulae of the propositional calculus expressible in ordinary discourse are logically true sentences), provided that the usual connection between the truth-values of compound sentences and of their components be readjusted in cases where the presence of vague terms becomes relevant.

Let us conclude that we need get into no trouble with the ordinary laws of logic if we attempt to justify the Principle of Verifiability by showing that the truth-value of unverifiable statements is always affected by the vagueness of their terms, whereas the truth-value of verifiable statements is not always so affected. From this preliminary discussion of the connection between the vagueness of terms and the indeterminacy of statements we may draw two further conclusions concerning the Principle of Verifiability:

(1) The verifiability of a statement is only a necessary condition of its possessing a truth-value, not a sufficient one; this is evident from the examples of verifiable and indeterminate statements quoted above. This fact must not be confused with the usual formulation of the Principle of Verifiability which asserts that empirical verifiability is a necessary but not a sufficient condition of meaningfulness, because this condition does not apply to analytic statements; even synthetic verifiable statements may turn out to be indeterminate if their truth-value is affected by the vagueness of their terms. In other words, the Principle of Verifiability, if valid, would only entitle us to classify as indeterminate those statements which are unverifiable. It does not guarantee the determinacy of verifiable statements.

(2) Accordingly, we cannot *explain* the impossibility of discovering the truth-value of an unverifiable statement by resorting to the Principle of Verifiability, or justify this principle by promoting it to the rank of an explanatory hypothesis. To be sure, the impossibility of finding anything which fits a given description may often be accounted for by assuming that there is nothing to find. One may argue, for example, that we shall never be able to find out whether it is true or false that time had a beginning, because the assumption of a finite time is neither true nor false. In view of conclusion (1), however, such an explanation would be misleading, since verifiability is compatible with indeterminacy; hence it may be possible to find out whether a given statement is true or false even though it happens to be neither. In other words, a statement is verifiable if it is possible to find out whether it is true or false; to ensure this, it suffices that the statement be possibly true or false, without being actually either.[3]

30 A DEFINITION OF TRUTH

Truth of Statements about Individuals

In the preceding section, we have pointed out that the truth-value of any statement is connected with the vagueness of its terms. No explicit definition of truth has been used to account for this connection. Yet it is obvious that the truth-value of a statement depends upon "what it is about," upon the admissible interpretations of its terms. For instance, the statement "There are more than two billion people alive at present"

[3]The concept of possibility involved in the principle will be discussed in § 36.

is true if "billion" is interpreted in the American way, as applying to any collection of one thousand million individuals; the same statement is false by British usage, which construes "billion" as equivalent to one million million. It is not surprising, therefore, that the truth-value of a statement may be affected by the vagueness of its terms, since the vagueness of a term consists in the multiplicity of its admissible interpretations.

However, to establish that the verifiability of a statement is connected with its having a truth-value, this general insight into the relationship between the truth-value of a statement and the interpretation of its terms is not sufficient. What we need here is a definition of a true statement. It goes without saying that the present context does not allow opportunity for a detailed discussion of the problems connected with defining truth, or for furnishing a thorough substantiation of the unavoidably oversimplified definition to be outlined in the following pages.

Let us first be clear as to what sort of definition of truth is called for in connection with the Principle of Verifiability. Since our aim is to justify the application of this principle to statements of ordinary language, we need a definition of truth which comes as close as possible to its actual use as a property ascribed in ordinary language to statements couched in ordinary language.[4] We must therefore look for a defining expression interchangeable *salva veritate* with "true" in all the relevant contexts, by virtue of their actual meanings in ordinary discourse. Such an expression should enable anyone who is not familiar with "true" or who prefers to dispense with this word, to replace every context of "true" with a context which contains the defining expression and would be recognized as equivalent by everyone who understands both expressions correctly. We shall not claim that "true" is just an abbreviation of the defining expression (or would be felt to be such by anybody familiar with it), or that it is exactly synonymous with the defining expression (or would be felt to have exactly the same meaning as the defining expression by anybody familiar with both). We shall not hold either that familiarity with the defining expression is required in order to understand "true" correctly (except in contexts where both occurr), or that the defining expression is the only one to meet the

[4]"True" is, of course, ambiguous in ordinary language. The cognitive use of the term, exemplified in such contexts as "X made a true statement," "Some consequences of this hypothesis turned out to be true," etc., is the only use of interest to us. In this sense, truth is a non-relational property ascribed to statements or to beliefs expressed by statements. Contexts in which truth is applied to statements in a relational way (e.g., "This statement is true of X," "This statement is true for X"), or to entities which are neither statements nor beliefs (e.g., "a true friend," "it is true that . . .") are irrelevant in this discussion.

aforementioned requirements. All that is necessary is that the defining expression, say "having the property P," be coextensive with "true" by virtue of their respective meanings alone; in other words that the equivalence "Any statement S is true if, and only if, it has the property P" be analytic, in the sense that disbelieving it or rejecting it would be a sign of an incorrect use of the two terms.[5] Any analytic equivalence of this type may be called an analytic definition of truth in terms of the property P.

The proposition just quoted that "the truth of any statement depends upon what it is about" is an analytic context of this sort for "true." The dictum "what follows from a true statement is true" is another context of this kind. These two analytic contexts of "true," connecting it respectively with the concepts of denotation and consequence, do not add up to an equivalence. They therefore fail to constitute a definition of truth in terms of denotation and of consequence. We may use them, however, as a clue for constructing a defining expression of truth in terms of these two concepts.

It should be noticed that in trying to use such concepts as denotation, consequence, analyticity, proof, etc., studied so far mainly in the modern logical theory of artificial symbolic languages, we must take special precautions to ensure their applicability to ordinary language. The discussion of the validity of the Law of Excluded Middle in § 29 illustrates this necessity sufficiently.

For similar reasons we shall not be able to apply the powerful "semantic method" used by Tarski[6] in his investigations into the concept of truth in formalized languages and underlying the alternative approaches to the concept of truth discovered and explored by Carnap.[7]

[5]The term "analytic" will be used here in its usual, wider sense, as connoting a property ascribable not only to statements whose rejection would amount to an incorrect use of their respective terms, but also to statements deducible from such statements. "Friday is the day after Thursday" is analytic in the narrower sense. "Friday comes 113 days after Thursday" is analytic in the wider sense.

In construing the analyticity of statements of a given language in terms of the admittedly fluctuating rules of correct usage that are valid in this language, we do not draw an absolutely sharp boundary-line between the analytic and the synthetic. Consequently, the weighty arguments mustered by W. V. Quine and other investigators against the possibility or significance of such a sharp boundary do not affect our use of the terms "analytic" and "synthetic." As a matter of fact, the gist of our argument would not be altered if only full-fledged and conditional definitions were granted the status of analyticity.

[6]Cf. A. Tarski, *Pojecie prawdy w jezykach nauk dedukcyjnych* ("The Concept of Truth in the Languages of Deductive Sciences"), 1933. The English version of an enlarged German translation of this work is contained in A. Tarski, *Logic, Semantics, Metamathematics* (1956).

[7]Cf. R. Carnap, *Introduction to Semantics* (1946).

So far, the most illuminating result obtained by applying this method has been an exact definition of the elusive concept of truth (in formalized languages) in terms of unproblematic logical and linguistic (morphological or syntactical) concepts expressed in appropriate meta-languages. However, these writers assume explicitly that no vague terms occur in the languages for which truth is to be defined and accordingly admit the validity of the metalogical Law of Excluded Middle, whereas the justification of the Principle of Verifiability requires both the presence of such terms and the abandonment of this law.

Strictly speaking, Tarski's semantic method could be used to define the concept of truth in a given object-language L in such a way as to avoid the validity of the metalogical Law of Excluded Middle in the meta-language L_m. To achieve this, it would suffice to include the logical Law of Excluded Middle neither in L nor in L_m by assuming that these two languages have an intuitionist logic built into them.[8] But the rejection of the metalogical Law of Excluded Middle for L would then have no connection with the presence of vague terms in L.

How are we then to define a true statement? This question is readily answerable in respect to the simplest statements about individuals of the subject-predicate type. We may notice, to begin with, that the statement "Toronto is a city" is true because its subject applies to a particular individual, while its predicate applies to this individual as well as to the others. We can therefore say that this statement is true because it is decomposable into two expressions both applicable to the same individual. The same would be held of any statement so decomposable.[9] We thus reach a partial definition of truth; *a statement is true if its subject and predicate apply to the same individual.*[10] We shall call *referentially true* all statements to which this definition applies.

It seems natural to extend this partial definition to relational statements about individuals, for example, to statements consisting of a two-place predicate accompanied by two proper names. That "Toronto

[8]Cf. S. C. Kleene, *Introduction to Metamathematics* (1952), pp. 499 ff.

[9]Instead of asserting that a linguistic expression applies to an individual, I shall also say that this expression denotes (or refers to) this individual. The individuals, if any, to which an expression applies will be called its referents; the totality of such individuals will be termed the denotation of this expression.

The concept of reference or denotation, as used here, applies therefore to linguistic expressions, regardless of their syntactical roles in subject-predicate statements about individuals. The proper name "Toronto," the common name "city," and the corresponding predicate "is a city," denote (apply to, refer to) the same geographical individual. Conversely, this individual is a referent of the proper name, of the common name, and of the corresponding predicate occurring in this statement.

[10]This is in substance Hobbes's definition of a true statement.

is larger than Hamilton" may be construed as consisting of the subject "Toronto" and the predicate "is larger than Hamilton," obtained by lumping together the relational expression and the second individual name. The definition to the effect that a statement is referentially true if it is decomposable into two expressions both applicable to the same individual would thus apply to any *simple* sentence about individuals (involving neither sentential composition nor quantification), of either the attributive or relational type.

Compound unquantified statements about individuals are built from simple sentences by means of connectives such as "and," "or," "if," etc. A statement of this kind can be defined as being *inferentially true*, if it follows[11] from a set of referentially true sentences, and as being *inferentially false* if its denial so follows. "Toronto is a city and is larger than Hamilton" is inferentially true, because it follows from the two referentially true sentences of which it is a conjunction. Similarly, "Toronto is neither a city nor a village" is inferentially false because its denial follows from its first referentially true component. More precisely, we shall call inferentially true all sentences which follow from referentially true sentences, but which are not themselves referentially true. *A statement will be said to be true if it is either referentially or inferentially true.* Accordingly, compound unquantified sentences about individuals, if true at all, are inferentially true.

From simple and compound unquantified sentences considered so far, quantified sentences can be built by applying such words as "always," "sometimes," "everywhere," "somewhere," etc. to some of the relevant terms: for example, "Every city is larger than any village," "Some cities are larger than Toronto." The simplest quantified sentences can be obtained from the formulae "Everything is P" and "Something is P" by substituting suitable values for P. We thereby obtain universal and existential sentences respectively. Now a universal sentence cannot possibly be false if all its instances are true. A universal sentence may therefore be said to *follow* from the set of its instances; its inferential truth may still be defined as consisting in the fact that it follows from a set of referentially true sentences, namely the set of its instances. Similarly, an existential sentence follows from any of its instances, since it cannot possibly be false if any of its instances is true. We thus see that

---

[11]To explain provisionally the meaning of this word, it may suffice to point out that the statement S is *a consequence* of (*follows from*) the statements C, if S could not possibly be false, on the assumption that all the statements C are true. The apparent circularity of this explanation which is preparatory to our definition of truth will be removed in § 31.

the definition to the effect that a statement is inferentially true if it follows from a set of referentially true statements applies to these two kinds of quantified statements as well.

Not all quantified sentences are of these two simplest kinds. The applicability of the aforesaid definition of inferential truth may therefore be doubtful in more complicated cases. Even if there is only one quantifying word in a sentence, its instances may be compound instead of simple, so that its truth may not be identified with its following from a set of referentially true instances. This holds even of a universal categorical sentence of the type "Every A is B" if it is construed in the usual way as meaning that "if something is A, then it is B as well"; its instances are then compound conditionals and hence, at most, inferentially true.

However, on closer inspection, this complication is quite compatible with the above definition of inferential truth since the consequence-relation is transitive in the sense that whenever a statement S follows from a class of statements C, each of which follows in turn from sentences included in a new class of sentences C', the statement S follows also from the class C'.[12] Thus if S is a universal categorical sentence whose instances are unquantified compound conditionals, and if each instance is inferentially true because it follows from some referentially true sentences included in the class of statements C, then the statement S will also follow from the referentially true sentences comprised in C and will therefore be inferentially true.[13] The definition of inferential truth will thus apply to every universal categorical sentence all of whose instances are inferentially true. Similar considerations hold in regard to existential categorical statements of traditional logic.[14]

[12]This follows from our provisional explanation of consequence (cf. the previous footnote). If the truth of S is guaranteed by the truth of all the sentences in C, each of which, in turn, could not possibly be false on the assumption that all the sentences in C' are true, then this assumption guarantees also the truth of S.

[13]For example, the class of referentially true statements corresponding to the inferentially true statement "Every metal is a conductor" consists of all referentially true instances of the formula "X is a metal" together with the corresponding instances of the formula "X is a conductor."

[14]On the other hand, a technically more serious complication arises if the statement whose truth-value has to be defined contains several quantifying words some of them possibly universal, some of them existential, and not all of them necessarily placed at the beginning of the statement. It is not immediately clear, for example, what we should consider as the unquantified instances of the statement "For every gas there exists a critical temperature such that no pressure exerted on the gas will liquefy it, unless its temperature is lower than this critical value." The uncertainty in determining such unquantified instances may cast some doubt on the possibility of our attempt to reduce the inferential truth of quantified

Extending the Definition of Truth to Statements
about Non-Individual Objects

In sum, by defining a true statement as one which is either referentially true or inferentially true, we may account for the concept of truth in respect to all sentences about individuals, regardless of the logical complexity of their structure. Simple, compound, and quantified statements about individuals do not, however, represent all the statements considered true either in daily life or in science. "The colour of this table is brown" is, ostensibly at least, a true subject-predicate statement about a colour, not about some individual coloured object. If we applied the above definition of referential truth to this statement and declared that its truth consists in its subject and predicate both applying to the same entity (namely, to the same colour, not the same coloured individual) we would commit ourselves to the existence of such universals as colours. This liability, although it might not lead to bankruptcy, would very much weaken our argument, for it would make the justification of the Principle of Verifiability dependent upon a Platonist solution to the nominalist-realist controversy. Since this principle is the common denominator of various brands of empiricism (phenomenalism and realism, positivism and pragmatism, operationalism and instrumentalism, and so on) its justification should be as non-partisan as possible.

Of course, the example just quoted is trivial and may be construed as an awkward paraphrase of the simpler subject-predicate statement about individuals, asserting that this table is brown. This easy way out is, however, not always available. When a referentially true statement specifies the number of some ostensibly non-individual entities (in asserting, for example, that there are seven colours in the rainbow) one

statements to the truth of their unquantified instances and eventually to the referential truth of the statements from which these instances follow.

However, it is shown in text-books of logic that every quantified statement can be transformed into an equivalent statement consisting of all its quantifiers followed by an expression whose instances are unquantified compound or simple sentences. These sentences may be called the instances of the original statement. It is easily seen that in the limiting case, where all the quantifying words are universal (e.g., "Every piece of sugar is soluble in every portion of water"), the statement is true if all its instances are true. When all the quantifying words are existential (e.g., "Some piece of sugar is soluble in some portion of water"), the truth of the statement follows from the truth of any of its instances. Finally, in the "mixed" case, where some quantifying words are existential and the remainder are universal (e.g., "Some piece of sugar is soluble in every portion of water") it can be shown that the original statement still follows from certain sets of its instances. In view of the kind of transitivity pertaining to the consequence-relation, we may therefore say, once more, that a quantified statement is inferentially true if it follows from a class of referentially true statements, regardless of the complication of quantificational procedures involved in this statement.

can hardly escape the conclusion that the entities counted must exist.[15] If colours did not exist at all, there would not be a single colour, let alone seven distinct ones. Similarly, the biologist estimates the number of presently existing biological *species* (not of biological individuals) at about one million, the chemist determines the number of chemical species (chemical elements and compounds), the psychologist counts the humanly discernible colours.

Even the most trivial applications of mathematics seem to involve us in such Platonic commitments. If there are, for example, four chairs in a room then one can easily prove that the number of couples of chairs in the room is six, the number of trios of chairs is four, and so on. Since these sets of chairs are countable, they must exist. But this starts us on a vertiginous journey, because the sets of chairs once brought into existence are countable anew, sets of such sets are formed and the helpful mathematician indicates readily the number of any kind of sets of sets of chairs in the room: couples of couples of chairs, couples of trios of chairs, etc. This procedure can be continued indefinitely, so that starting with our four chairs in the room, we get an infinite array of entities, all of them existing, since their numbers are specifiable in true propositions. It is but small comfort to an empiricist, not inclined to over-populate the universe, that the total number of entities so originating out of the four chairs is only denumerably infinite. Yet one has to admit that all the statements referring to the respective numbers of the various sets of chairs, sets of sets of chairs, etc., are verifiably true, because they follow from the verifiably true statement which specifies the number of individual chairs.

Our attempt at defining truth may thus involve us in the whole controversy over the existence of abstract entities, such as classes of individuals, or classes of classes of individuals, whereas the aim of this definition is a justification of the Principle of Verifiability which will be

[15]Strictly speaking, a statement about the number of entities which meet a specifiable requirement is not of the subject-predicate form and hence, if true at all, only inferentially true. Such a statement is usually interpreted as a quantified sentence with instances whose subjects and predicates refer to the entities counted. When speaking about referentially true statements concerning the numbers of certain entities, I had in mind the referentially true instances involved in such statements.

For example, the statement "there are two chairs in the room" can be interpreted as meaning that there are objects x and y such that x is not y and every chair in the room is either x or y. The unquantified sentences about the entities counted are, in this case, instances of the formula "x is a chair in the room." Accordingly, when a statement specifies a number of colours, or of sets of chairs etc., the terms of the unquantified instances involved in it denote colours or sets of chairs respectively.

compatible with both nominalist and Platonist brands of empircism.[16] I think that a reasonable way out of this predicament consists in confining the above definition of referential truth to subject-predicate statements about observable individuals and in considering any other statement, even if it be of subject-predicate form, as true only if it is inferentially true. Thus the empirically verifiable statements "about" numbers of discernible colours, of chemical species, of sets of chairs, etc., would be true only because they follow from referentially true statements about observable individuals. There is nothing to compel us to grant anything more than inferential truth to those verifiably true statements which, if literally interpreted, would refer to non-individual entities. In other words, we have to grant the truth of the statement that there are six different couples of chairs in the room without asserting thereby that couples of chairs are entities distinct from individual chairs and somehow referred to by the terms of the statement. We can take this position if we assume that the truth of this statement about the number of couples of chairs consists in its deducibility from the true statement concerning the number of individual chairs. At least, as far as this analysis of such true statements is concerned, we can argue that in order to account for their being true, we need acknowledge only the referential truth of the statement about the individual chairs and the deducibility of the former from the latter. Admitting the inferential truth of a statement ostensibly concerning abstract entities does not force us to countenance the existence of such entities.[17]

The applicability of the above definition of truth to statements ostensibly dealing with abstract entities ultimately based upon individuals (i.e., classes of individuals, relations between individuals, classes of classes of individuals, and so forth) is thus ensured to the extent to which such statements follow from statements about individuals. The same device may be used to extend the concept of inferential truth to statements which are neither about observable individuals nor about abstract entities based upon individuals, but refer to such elusive entities

[16]Platonic or conceptual realism is implicit in most contemporary views of the subject-matter of mathematical knowledge. Significant advances towards establishing the possibility of a nominalistic interpretation of mathematics are mainly due to W. V. Quine, *From a Logical Point of View* (1953). A systematic attempt to implement the nominalistic programme of eliminating references to non-individual entities from all cognitive contexts has been made by T. Kotarbinski in his doctrine of "reism." Cf. his *Elementy logiki, teorii poznania i metodologii nauk* ("Elements of Logic, Epistemology and Scientific Method"), 1928.

[17]An alternative approach to the problem of "abstract" and "classical" entities is explored by T. A. Goudge in "Peirce's Theory of Abstraction," *Studies in the Philosophy of C. S. Peirce*, ed. P. P. Wiener and F. H. Young (1952).

as sense-data, introspective data, and mental dispositions, which are not locatable on any level of abstractions with regard to perceivable individuals. "I have now an elliptical sense-datum of a round penny" is inferentially true since it follows from the referentially true statement about individuals, "This round penny now looks elliptical to me." "The mind of X is logical" is inferentially true, since it follows from the referentially true statement that X thinks logically. In other words, statements ostensibly about elusive entities, if true at all, are only inferentially true. Admitting their truth does not commit us to any position concerning the ontological status of such entities.

Another device for extending the concept of inferential truth to abstract and elusive entities without offending nominalist or positivist susceptibilities is to treat symbols referring to such entities as syncategorematic and to lump them together in order to form expressions denoting observable individuals. "There are slow-moving electrons in this box" may be construed as meaning "This box contains slow-moving electrons"; the latter can be treated as a referentially true statement about the observable individual box, the predicate being "containing slow-moving electrons" (cf. § 18). The truth of statements which involve questionable scientific "constructs" may thus be admitted without interpreting such constructs as categorematic.

Loosely speaking, entities which are not observable individuals may be said to reflect themselves in the world of such individuals and thereby become accessible to knowledge. This figurative way of speaking actually amounts to saying that if expressions called symbols of non-individual entities occur in propositions deducible from true statements about individuals, then such propositional contexts of these expressions are inferentially true and may contain genuine knowledge. Even though we assume, for example, that electrons are not observable by man, still if some statements asserting that a particular box contains slow-moving electrons are verifiably true, then electrons may be said to become accessible to human knowledge through their being reflected in this observable individual box. I see no reason why statements which are not translatable into statements about individuals, yet which are verifiably true through being deducible from them, may not be said to contain knowledge of unobservable entities, and may not play an important, or even indispensable part in scientific knowledge. A similar consideration may be used to show that our argument in favour of the Principle of Verifiability avoids taking sides in the controversy over the proper "universe of discourse" of the scientific language, which underlies the issue of phenomenalism *versus* realism.

To sum up:

(1) The definition of truth in terms of denotation and consequence obviously provides for indeterminate statements and does not imply the validity of the metalogical Law of Excluded Middle. For instance, if the statement S is not referentially true, and neither S nor its denial follow from referentially true statements, then S is indeterminate. Thus the statement "This book has incommensurable edges" will prove to be neither true nor false, because it has neither a referential nor an inferential truth-value.

(2) As explained above, the definition entails no commitment in the nominalist-realist controversy and avoids any crippling restrictions on the range of true statements. Statements which, if literally interpreted, would refer to abstract or other non-individual entities, may nonetheless be considered true without being interpreted in this way. They may be viewed as inferentially true instead.

A similar non-partisanship in the phenomenalist-realist controversy could be shown to be inherent in the definition.

(3) The definition automatically ensures the truth of all analytic statements and substantiates the view that such statements are true "come what may," and hence not subject to the Principle of Verifiability. Since, by definition, a true statement is one following from referentially true statements, it may as well be defined as one which *follows logically*[18] from referentially true and analytic statements. An analytic statement is therefore always true, because the requirement of being a logical consequence of referentially true statements supplemented by analytic statements is vacuously fulfilled in this case.

On the other hand, the intra-linguistic nature of analytic truth may be seen from the following consideration. According to the above definition of a true statement, referentially true statements owe their truth to the fact that they are directly related to extra-linguistic objects denoted by their terms. An inferentially true statement may acquire such a reference only indirectly, through the referentially true statements from which it follows. Hence a statement which is inferentially true regardless of any particular referentially true premises, as are analytic statements, can be said to be true come what may and not to assert anything about the factual world.

This may also explain the failure of attempting to connect the analyticity of statements with their subject-matter (e.g., by defining an

[18]The difference between "consequence" and "logical consequence" is explained in the next section.

analytic statement as tautological, or valid in every possible world). Analytic statements are true come what may not because they say so little, but rather because the meaning of truth requires it.

31   TRUTH AND LANGUAGE

Truth and Consequence

The concepts of consequences and of denotation, crucial in the above definition of truth, have been successfully investigated in the modern logical theory of artificial symbolic languages. We have therefore to ascertain to what extent the findings of this theory can be utilized for explaining the concepts of consequence and denotation in ordinary language.

This could be achieved by reinterpreting both concepts in terms of rules of correct usage which are valid in ordinary language. Roughly speaking, a statement S is a *consequence* of the statements C if it would be *correct* to infer S from C; the relevant rules centre around the use of logical words such as "hence" and "so."[19] An expression *denotes* an individual, if it would be *correct* to make statements which include this expression in order to talk about this individual. We shall outline such an explanation of the concept of denoting in terms of rules of correct usage in the next subsection. A similar explanation of "consequence" would be too lengthy. We have to confine ourselves to some essential distinctions.

We have utilized the concept of consequence to extend the definition of truth beyond the range of the simplest statements about individuals. Tentatively, a statement S has been said to be a consequence of, or to follow from, the class of statements C, if S could not possibly be false on the assumption that all the statements in C are true. This explanation may be useful to give an approximate idea of the nature of the consequence-relation, but it is unsatisfactory in this context for two reasons. In the first place, it involves the vague concept of possibility. This would prevent the explanation from being effective in the most typical cases. Secondly, it resorts to the very concept of truth which we are anxious to define in terms of denotation and consequence and which therefore must not be used in explaining them. Moreover, on closer inspection, we realize that in order to extend the concept of truth to statements about abstract and elusive entities, we actually had to broaden the concept of consequence which was used in defining the inferential truth of state-

[19]Cf. G. Ryle, "'If,' 'So' and 'Because,'" *Philosophical Analysis* (1950), pp. 323–40.

ments about individuals. A few remarks about these various concepts of consequence and the possibility of explaining them independently of the concept of truth are therefore indispensable.

(1) A compound sentence, such as a disjunction, is said to be a consequence of some of its components. In this special case, "consequence" means "logical deducibility." It is well known that logical deducibility can be defined explicitly in terms of the rules of inference governing the correct use of those very logical connectives which serve to form the compound sentences out of their components. It suffices to list these rules of inference and to declare that a statement S is logically deducible from the statements included in the class C if it can be obtained from these statements by applying to them any number of times in any succession the rules occurring in the list. Thereby the concept of consequence in the sense of logical deducibility, which is involved in defining the inferential truth of unquantified statements about individuals, can be explained independently of the concept of truth.

(2) A universal sentence has been said to be a consequence of the class of its instances. This once more fits the preliminary explanation, since a universal statement could not possibly be false on the assumption that all its instances are true. By adding to the rules of inference implicit in logical deducibility this new rule which extends the concept of consequence to cases where the class C consists of all the instances of the statement S, we obtain a broader concept. This concept may be termed logical consequence, in contradistinction to logical deducibility. The concept of logical consequence which is involved in defining the inferential truth of quantified statements about individuals can still be explained independently of truth. The only additional relation required to obtain it from deducibility is that of instantiation of a universal sentence by a set of unquantified sentences.

It is true that the application of the term "logical consequence" to the relation obtaining between a universal sentence and the class of its instances may arouse some misgivings, since this class of instances may be an infinite one. We could of course dispense with calling this relation "logical consequence." Our only reason for not so doing is the desire to obtain a concise definition of truth. An equivalent, though more awkward, formulation of this definition can be obtained without extending the concept of consequence, namely, by explaining differently the inferential truth of compound and of quantified sentences, the first in terms of logical deducibility, the second in terms of instantiation. As a matter of fact, extending the concept of consequence so as to consider a universal sentence as a logical consequence of its instances is not only natural (because the truth of the sentence is guaranteed by its instances) but has proved a useful tool since its introduction by Tarski[20] in purely logical research.

(3) A further broadening of the concept of consequence is involved when, for instance, a sense-datum statement is said to follow from a statement about a material object, or the statement "Tomorrow is Friday" is said to follow from "Today is Thursday." The former statement is not obtainable from the latter by applying to it the logical rules of inference governing the correct use of connectives and quantifiers. Actually "Tomorrow is Friday" follows from "Today is Thursday," because it follows logically from the latter statement in conjunction with the analytic statement "If today is Thursday, then tomorrow will be Friday." In other words, this new concept of consequence, which may be called "entailment," is obtained from the concepts of logical consequence and analyticity. A statement S is entailed by the statements C if it follows logically from the statements C, possibly supplemented by some additional analytic statements. Thus entailment is still definable independently of truth.

It goes without saying that these remarks are put forward in order to prevent the most likely misunderstandings. The concept of consequence itself, or rather the family of related concepts classed under this heading, plays a major part in the whole of logical theory and raises many problems which have not yet been solved. We have selected the three simplest concepts of consequence—logical deducibility, logical consequence, and entailment—because their applicability to ordinary language and to the concept of truth in this language is comparatively safe.

Truth and Denotation

How do we ascertain what people are talking about when they are talking sense? What do we mean by saying that at a certain juncture they are talking about A, not about B? Defining "denotation" or "reference" amounts to answering these two questions. They may also be formulated as follows: How can we ascertain what objects, if any, are denoted by the terms of a statement which is made by someone who uses these terms correctly? What do we mean when we assert that an expression of some language, if used correctly, applies to some particular individual or individuals?

[20]Cf. A. Tarski, "Betrachtungen über die Begriffe der Widerspruchsfreiheit und der Vollständigkeit," *Monatshefte für Mathematik und Physik* (1933).

There is no doubt that what expressions denote depends upon the rules for their correct usage. As used in the United States, the expression "A continent with more than one billion inhabitants" denotes Asia. By British usage this expression does not denote Asia. To ascertain what people who speak a certain language are talking about, one has to find out the rules governing the correct use of relevant linguistic expressions, the patterns of correct linguistic behaviour in relevant situations.

By correct linguistic behaviour in a given situation, or, by correct use in this situation of the language spoken by a given group of people, I mean the behaviour that would be considered correct by the members of the group or is in accordance with the norms of linguistic behaviour valid within the group. These norms are not necessarily formulated in text-books, dictionaries, or in teaching the language. They may be implicitly inculcated by examples. Departure from them is frowned upon as is departure from other group-norms. Such departures may also prevent the delinquent from obtaining from other members of the group results which he would have secured by complying with the rules. Language is, basically, a facility for formulating, recording, and conveying information; when justifiably believed, such information constitutes knowledge; when acted on, such beliefs help to ensure the success of action. Belief or disbelief with regard to linguistically formulated information is thus crucial in both the cognitive and the practical rôle of this information. That is why the correct use of a linguistic expression consists essentially in the user's readiness to believe or disbelieve, under appropriate circumstances, a specifiable set of sentences in which this expression occurs, and to act or refrain from acting on such beliefs in appropriate situations. Hence, to ascertain what people are talking about when they are using certain expressions of their language, we have to find out under what circumstances it is correct to believe certain sentences containing these expressions.

For illustrative purposes, we may assume in what follows that the language for whose expressions we are anxious to define the concept of denotation, has the same vocabulary, phonetics, spelling, and grammatical syntax as ordinary English. In other words, the members of the group under consideration use correctly all the expressions of English as far as phonetics, spelling, vocabulary, and syntax are concerned. The words they utter sound like those uttered by English-speaking people. A sequence of words, is, by definition, a declarative sentence, if it is correct to believe or to disbelieve it. Accordingly, we assume that the strings of words correctly believable or disbelievable by members of the group are exact replicas of declarative sentences in ordinary English. By determining the patterns of such strings of words, we can find out the rules of grammatical syntax which are valid for the language spoken by the group, as well as the classification of all the words into parts of speech and its modern counterpart, the theory of logical types. All this does not, however, guarantee that the members of the group speak English correctly in other respects, for they may still, in spite of all the aforementioned coincidences, use the expressions of their vocabulary to speak about different things from those the British or the Americans speak of. Consequently the denotation of any expression used by them may differ considerably from the denotation of the same expression in ordinary English. How can one make sure whether this is actually the case?

We may start by examining their *ostensive terms*—those terms, that is, the correct use of which by any member of the group depends essentially upon his perceptual situation. We can check, for example, whether members of the group use the word "window" in conformity with English-speaking people by ascertaining whether they look at the window, and not at some other object, when trying to make up their mind about the statement "The window is open." We may also check on their use of this word by finding out whether, when told to "open the window," or to "point at the window," they take the right course of action instead of trying, for instance, to open the door or to point at it.

More precisely, we can say that the proper name "A" as understood by the group *denotes*, or *refers ostensively* to, the object X if there is some statement, "A is B," which no member of this group can fail to make if he *considers* this statement (viz., tries to answer whether A is B), perceives the object X, and correctly uses "A." For instance, "John Smith" denotes ostensively some given person, because the statement "John Smith is tall" must be made by everybody who understands this name correctly, considers the statement, and perceives the person.

Similarly, a common name, say "B" (and the corresponding predicate "is B"), denotes ostensively every object of the class C, if there is a statement "A is B" such that the correct use of "B" compels everybody to make this statement, whenever he considers it and perceives any object X belonging to the class C and denoted ostensively by the proper name "A." For instance, the adjective "tall" refers ostensively to every person of six feet or more, because whenever an ostensive name "A" denotes such an individual, the statement "A is tall" will be made by everybody who uses "tall" correctly, sees the individual denoted by "A," and considers this statement.

By interpreting a proper name and a predicate as denoting the same individual if their correct use compels everybody to make the statement consisting of these two expressions whenever he considers this statement and perceives this individual, we ensure the referential truth of this statement, since it is then decomposable into two expressions both applicable to the same individual. Conversely, we may say that this interpretation of the two ostensive terms is admissible since it ensures the truth of the ostensive statement consisting of these two terms, provided the statement is made in the appropriate perceptual situation.

The denotation of ostensive terms used by the group, and its conformity with ordinary usage, can thus be ascertained empirically by watching the correct linguistic behaviour of members of this group in perceptual situations. It can also be defined in terms of the rules which govern such behaviour.

Denotation is not, however, an exclusive property of ostensive terms. Non-ostensive names and predicates also apply to definite objects. For example, the adjectives "electrically charged," "soluble in water," "built three centuries ago" are to be interpreted as applying to certain individuals and not to others. By analogy with ostensive terms, we may be tempted to define as admissible those interpretations of non-ostensive terms which guarantee the truth of the *analytic* statements in which they occur. We have seen, however, that analytic statements are always true, so that there is no point in trying to ensure their truth by selecting a suitable interpretation. The proper analogue for the admissibility of non-ostensive terms is their being compatible with, rather than providing a guarantee of, the truth of their analytic contexts. In other words, we must not interpret non-ostensive terms in such a way that their analytic contexts, which are in any case inferentially true, would become referentially false by virtue of the denotation assigned to these terms. An example will illustrate what I have in mind.

We have seen that the non-ostensive term "being one yard long" is conditionally definable by an ostensive test (the superposition on a yardstick) and an ostensive response (coincidence with the yardstick). Owing to this definition the statement "A is one yard long" is true if it is entailed by the true ostensive statements asserting that a yardstick was superposed on A and that coincidence followed. Similarly it would be true that the negative of this non-ostensive term applies to object B, if the statement ascribing the negative term to B were entailed by the true ostensive statement asserting that the yardstick did not coincide with B when it was superposed upon it. If a yardstick is never superposed on object C, and if no other analytic context is available for the term apart from the two contexts just quoted, we would have to say that the term could be interpreted either as applying or as not applying to C. Both interpretations are admissible in this case, because neither is ruled out by the way the term is used.

We propose to consider the statement just made about the denotation of being one yard long as a general definition of this relation.[21] *The term P denotes the object whose ostensive name is "A" if the statement "A is P" is entailed by some true ostensive statements.* In the special case in which P is itself an ostensive predicate, this condition is fulfilled automatically. If we assume that P applies to A, the ostensive statement "A is P" would be true and would of course entail itself.

It should be noticed that although truth has been defined in terms of denotation, and the above definition of denotation resorts in turn to the concept of truth, no vicious circle is actually involved, because the proper order of the definitions is as follows: (1) The concepts of deducibility, analyticity, and entailment have been explained without resorting either to the concept of truth or to that of denotation. (2) The ostensive denotation of linguistic expressions is then defined in terms of correct linguistic behaviour in perceptual situations, independently of the concept of truth. (3) A true ostensive sentence may then be defined as one whose subject and predicate both denote ostensively the same individual. (4) The denotation of non-ostensive terms is then defined in terms of ostensive truth, ostensive denotation, and entailment. (5) Finally, truth is defined in terms of denotation and entailment.

Vagueness and Referential Truth Redefined

In the present subsection I shall add certain comments on the concept of vagueness which are indispensable for the main argument. We have seen that a term is vague if it can be understood in several ways without being misunderstood. In other words it is vague if several distinct interpretations are compatible with its correct use. We may also say that linguistic vagueness arises whenever a question concerning the cognitive

---

[21]An analysis of the way a single non-ostensive term comes to have a denotation may seem too slender a basis for defining the denotation of all non-ostensive terms. However, a more comprehensive discussion would be too lengthy. Let us point out at least that our only clues towards making out what the users of a given language are talking about, are provided by the rules of correct linguistic behaviour which they follow in various situations. For instance, the rules of linguistic behaviour in perceptual situations determine the reference of ostensive terms. Similarly, an examination of the rules of linguistic behaviour in other, non-perceptual situations, would determine the reference of non-ostensive terms and substantiate the assumption which underlies the above definition, viz., that the analytic contexts of a non-ostensive term completely determine its referential function.

use of language by a human group proves unanswerable, because the rules of correct linguistic usage adopted by the group provide no answer to it. Thus it is impossible to make out whether, on the whole, the word "or," as ordinarily understood, has the exclusive or the non-exclusive meaning. The word is vague to this extent; either interpretation is admissible, since neither is ruled out by the way this word is actually used. Similarly, discussions about whether the universal categorical sentence "Every A is B," as ordinarily understood, entails the existence of at least one A, can go on forever, since the actual use of this type of sentence rules out neither view. The word "every" is, to this extent, vague.

The kind of vagueness exemplified in these two cases may be called intra-linguistic, because it consists in a shortage of suitable analytic contexts for the vague terms. The removal of the vagueness would result in new entailment relations between the relevant statements of the language. Thus, if the sentence "If every A is B, then some objects are A" were as self-evident as that two plus two make four, the vague feature of "every" would disappear. Similarly, if either the statement "If P is true, then P or Q is true," or alternatively, the statement "If P is true and P or Q is true, then Q is false" were self-evident, then the non-exclusive or exclusive meaning of the connective would prevail respectively and the connective would cease to present intra-linguistic vagueness.

In connection with the Principle of Verifiability, however, we are mainly interested in what may be called the extra-linguistic or denotational vagueness of linguistic expressions. An expression is denotationally vague if the rules of correct linguistic behaviour do not suffice to determine the individual or individuals denoted by it. The expression can therefore be interpreted as either applicable or inapplicable to certain individuals without any violation of the relevant rules of linguistic behaviour. Owing to our non-committal position in the nominalist-realist controversy, only proper and common names and one-place predicates corresponding to the latter can be denotationally vague in this sense.

For example, the name of a city is denotationally vague, if it can be interpreted correctly as applying to any of a number of overlapping areas situated within a specifiable boundary. Strictly speaking, such a term is only partly vague, because some of its grammatically possible interpretations are admissible (namely those referring to the areas just mentioned), while others are not. It is out of the question to interpret "Toronto, Ontario" as applying to any place outside Canada. In connection with the Principle of Verifiability, completely vague terms are of central interest. Such terms, which admit any and every one of their grammatically possible interpretations, are responsible both for the unverifiability and the indeterminacy of the statements in which they occur. We shall therefore have to examine, in § 37, the conditions under which a linguistic expression comes to be so vague that any conceivable interpretation of it becomes compatible with its correct use.

But let us try first to define the general concept of vagueness (both complete and partial) in terms of the denotational relation discussed in § 31. This will enable us to improve our definition of referential truth in one important respect, apart from preparing for the discussion of the conditions of complete vagueness.

A proper name is vague if it denotes several distinct individuals overlapping in space and time. This applies not only for names of geographical places, but also for names of any and every observable individual, owing to the impossibility of fixing in a unique way the spatio-temporal boundaries of any individual.

A common name (as well as the predicate corresponding to it) is vague if it denotes certain individuals, its negative denotes other individuals, and there exist other individuals denoted neither by the predicate nor by its negative. Thus "bald" applies to some human individuals, "not bald" applies to some other individuals, but there exist also individuals with hair spread thinly enough so that neither predicate applies. A vague predicate thus divides the class of all individuals into three mutually exclusive subclasses which may be called the positive precision-area, the negative precision-area, and the vagueness-area of the term. A term is vague if its vagueness-area is not empty.

A difference will be noted between the ways we have defined the vagueness of proper names and the vagueness of predicates. This difference corresponds to the difference between the concept of denotation of proper names and the concept of denotation of predicates (in § 31). It is a purely terminological matter since the reference of proper names and predicates could be re-defined in a uniform way. It is convenient, however, for the ensuing discussions and will therefore not be modified.

To sum up: A proper name is vague if it has several overlapping referents; a predicate (or the corresponding common name) is vague if its referents together with those of its negative do not exhaust the class of all individuals.

We can now determine how the truth-value of a subject-predicate statement can be affected by the vagueness of its terms. That "Toronto has an even number of trees" was found to be indeterminate, because the statement does not keep its truth-value under all the admissible interpretations of "Toronto." For reasons of simplicity, we may assume that the vagueness of the word "tree" does not affect the statement, in other words, that there is no plant in the area which could not be unambiguously classified as either being or not being a tree. This might well occur, for example, if the haziness of the boundary between "tree" and "bush" became relevant. Thus only the vagueness of the subject is relevant in bringing about the indeterminacy of the whole statement. The statement is indeterminate because the predicate applies to some referents of the subject-term, but not to all of them. Conversely, "Toronto, Ontario, is in Canada" is true because the predicate "being in Canada" applies to every referent of the subject. We may therefore say that a subject-predicate statement is referentially true if its predicate applies to every referent of its subject, and referentially false if its predicate does not apply to any referent of its subject, as, for example, in the statement "Toronto is in Europe." This shows immediately why a subject-predicate statement about individuals may happen to be neither true nor false. This will be so whenever the predicate applies to some, but not to all of the referents of the subject.

A subject-predicate statement about individuals will also be indeterminate if some or all referents of its subject fall within the vagueness-area of the predicate. This is the case with the statement "John is bald" discussed in § 29. Such examples of indeterminacy due to the vagueness of the predicate and not of the subject are more important for our discussion, because those completely vague terms which are responsible for the interdependence of indeterminacy and unverifiability are predicates, not subjects.

CHAPTER TWO

The Range of Science and the Principle of Verifiability

---

32 DIRECT VERIFIABILITY

THE VERIFIABILITY of a statement has been tentatively explained as consisting in the possibility of finding out its truth-value by carrying out appropriate observations and submitting the results to an appropriate logical processing. We shall now have to construct this concept of verifiability in a more precise way with a view to determining both the range and the limits of science. In other words, we shall try to show that verifiability can be construed in such a way as to meet the following three requirements:

(1) Verifiable statements and theories provide the range of all actual scientific methods; that is, the solution to any problem obtainable by any well-established scientific method is either a verifiable statement or a verifiable theory.

(2) Verifiable statements and theories have, in principle, a definite truth-value and, if well established, constitute genuine knowledge. Verifiability is, therefore, coextensive with the epistemological range of science.

(3) Unverifiable statements and theories have no definite truth-value and cannot constitute knowledge, whether scientific or non-scientific. Hence, the range of science as expressed in terms of verifiability could not possibly be extended. Science is, in this sense, universal.

Requirement (3) will be dealt with in the next chapter. In this chapter, the concept of verifiability will have to be constructed in such a way as to secure that requirements (1) and (2) are met. In accordance with conclusions previously reached, we shall start with some comments on *directly verifiable* statements.

Problems to which direct fact-finding methods can be applied are solvable by directly verifiable propositions. Thus, a proposition is directly verifiable if it is possible to find out its truth-value without resorting to inference, and, more particularly, without having first to determine the truth-value of any other proposition. Directly verifiable propositions affect the scope of science for several reasons. We have noticed, for instance, that the universe of discourse of the language of science consists of objects dealt with in such propositions, since the definitional techniques used by the scientist are unable to extend his universe of discourse beyond the realm of objects accessible to direct fact-finding methods. Furthermore, the body of directly verifiable propositions lies at the foundation of the whole of scientific knowledge in the sense that whatever is known by the scientist is either expressed in directly verifiable propositions or has been validated by being inferred from directly verifiable propositions. The very multiplicity of terminological proposals concerning the designation of directly verifiable propositions testifies to their importance: they are referred to by various writers as "basic propositions," "reports," "protocols," "observation-statements," "ostensive statements," "expressive statements," "constatations," etc.[1]

In what follows, we shall often use the phrase "observation-statement" to denote propositions capable of being validated by sensory, perceptual, or introspective observation, without recourse to inference. Perceptual observation-statements, which are concerned with common-sense things, will also be referred to as "ostensive statements." Observation-statements are not the sum-total of directly verifiable propositions, since these also include propositions the validation of which is memorative.

There are a few preliminary logical questions about directly verifiable statements which we shall discuss before examining the bearing of these propositions upon the universe of discourse of the language of science and the related phenomenalist-realist controversy over the subject-matter of scientific knowledge.

Direct Verifiability and Determinacy

A directly verifiable statement may come to be indeterminate if the vagueness of its terms happens to affect its truth-value. We have seen, for example, that the ostensive statement asserting that a given human individual is bald may have no truth-value if this individual strikes a

---

[1]Cf. A. J. Ayer, *The Foundations of Empirical Knowledge* (1940); R. Carnap, "Testability and Meaning," *Philosophy of Science*, vol. 3 (1936); C. I. Lewis, *An Analysis of Knowledge and Valuation* (1946); M. Schlick, *Gesammelte Aufsätze* (1938).

troublesome balance between Picasso and Monte Cristo. Such indeterminacy, however, is logically accidental; it is due to the contingent
fact that the referent of the subject of this proposition is located within the
vagueness-area of its predicate. Other statements which have the same
logical structure will have a definite truth-value.

To put it in another way: since the verifiability of a statement consists
in the *possibility* of finding out its truth-value, one may feel that if a
statement has no truth-value at all, it cannot be verifiable, because it is
impossible to find out a non-existent truth-value. We have already
pointed out, however, that the connection between verifiability and
determinacy is not exceptionless: for it to be *possible* to find out the
truth-value of a statement, it suffices that the statement possibly have a
definite truth-value, without, perhaps, actually having one. Yet in directly
verifiable statements, the lack of a definite truth-value is only accidental.
In principle, every directly verifiable statement is either true or false.
The determinacy of ostensive statements follows from our analysis of
truth and of ostensive denotation in the preceding chapter. Similar considerations can be shown to apply to other types of directly verifiable
propositions.

Logical Aspects of Directly Verifiable Propositions

As already pointed out, propositions which can be validated by direct
fact-finding methods must have the simplest logical structure and not
involve either logical connectives or quantifiers. Sentential composition
and quantification acquire meaning through inferential operations, and,
accordingly, inference is essential to the validation of compound and of
quantified statements. In other words, directly verifiable statements
always have an "atomic" structure and are either of the attributive or of
the relational type. The only debatable point is concerned with the role of
negation in direct verification. Negation is a logical connective, and the
negative of a directly verifiable proposition should therefore be viewed
as compound and susceptible of only indirect verification. This is in
keeping with the usual way of construing direct observation: one can see
that snow is white, but one cannot see that snow is not black; this negative proposition is thus not directly verifiable. To make sure that snow
is not black, an inference seems indispensable. It is not black because
it is white and whatever is white is not black. The negation of a directly
verifiable statement would thus follow logically from another directly
verifiable premiss supplemented by an analytic premiss and would be
indirectly verifiable.

Classifying negatives of directly verifiable propositions as indirectly verifiable has, however, certain drawbacks. Thus, the analyticity of the above premiss that "white is not black" is essential to the indirect verifiability of the negative statement that "snow is not black" but this analyticity has sometimes been doubted (cf. our definition of analyticity in § 30). Moreover, apart from such examples of analytic statements required to ensure the indirect verifiability of negations of directly verifiable statements, only the non-controversial analyticity of full-fledged and of conditional definitions will be relevant in most of the subsequent discussions: a statement will be considered to be entailed by a set of directly verifiable statements if it follows logically from this set, possibly supplemented by analytic premises of a definitional kind. We shall find it useful to resort to other types of analyticity in the definition of entailment and of indirect verifiability as little as possible. This is why it seems preferable to regard the negations of directly verifiable propositions as being directly verifiable as well. This amounts to assuming that one may be trained to answer questions as to whether or not a directly perceivable object has a directly perceivable property by merely relying on one's observation and without drawing inferences. This is certainly feasible and the assumption would hardly change the meaning of ostensive terms.

Another feature common to directly verifiable statements is the inclusion of words the complete meaning of which depends upon the spatio-temporal circumstances under which these words are used: words like "I," "now," "here" belong in this category of "egocentric particulars." Such particulars appear in every memoratively validated statement, which involves an egocentric tense, and also in all sense-data statements and introspective statements: "This looks so and so to me," "I feel so and so" are typical expressions of sensory and introspective observation. This applies in most cases to perceptual statements as well, since the natural way of expresing the belief that an object I perceive has a certain property consists in referring to this object by a demonstrative pronoun. However, one may wonder whether this linguistic procedure is inevitable in principle: there seems to be no difficulty in referring to oneself in sense-data and introspective statements by one's proper name, instead of a personal pronoun; some statements of the form "A is B" (say "Casa Loma is in Toronto") are perceptually verifiable, although A is a proper name of the object, and not an egocentric particular. Egocentric particulars in memorative statements could also be avoided by attaching individual labels to time-moments etc. There is no doubt that in order to avoid egocentric particulars in directly verifiable statements, one would have to speak in a way more cumbersome and less suitable for conversational exchange of information. However, the possibility of getting rid, in principle, of such terms in formulating the basic types of scientific information is important, because the very objectivity of such information seems to depend upon the possibility of conveying it without referring to any particular individual apart from the objective facts relevant to this information.

Phenomenalism and Realism in Science

According to the phenomenalists, scientific knowledge is only of the appearances which things present to the human observer and not of the things themselves. On the other hand, the realists claim that science provides not only reliable information about how things look or could possibly look to a human observer but also, and primarily, about how things are. The controversy between these two views on the scope of scientific knowledge begins in two different assessments of the role to be attributed to directly verifiable propositions of the *sensory* type. The phenomenalists maintain scientific knowledge is of appearances only because they assume that all the verifiable statements of the language of science are either literally concerned with sensory appearances or can be faithfully translated into statements about sensory appearances: to say how a thing is, is tantamount to saying how this thing does or possibly would look to a human observer. Phenomenalism claims such trans-latability into statements about sensory appearances not only for directly verifiable propositions about pre-scientific common-sense objects, but also for indirectly verifiable statements about pre-atomic and atomic scientific objects. The latter would be translatable, in the first place, into statements about pre-scientific common-sense objects, and, eventually, into statements about appearances. Thus a proposition asserting that a common-sense thing X is brown would be tantamount to claiming that X would look brown to an appropriate observer under appropriate circumstances. Similarly, the flow of an electric current in a wire would be synonymous with the potential deflection of a directly perceivable magnetic needle of a galvanometer which could be connected with the wire. Similarly, ascribing a Euclidean (or non-Euclidean) nature to space would amount to specific predictions concerning the potential behaviour of yardsticks and clocks under specifiable circumstances.

In this context we shall deal only with the alleged faithful translatability of directly verifiable statements about common-sense objects into statements about sensory appearances; similar considerations are easily seen to apply, however, to the alleged translatability of other phases of the scientific world-picture into statements about sensory appearances.[2]

[2]At the atomic and sub-atomic level, the phenomenalist translatability thesis has proved to be incompatible with the unobjectifiability of quantum-theoretical concepts (cf. § 20, pp. 177–80).

In evaluating the phenomenalist claim that directly verifiable statements about common-sense things can be faithfully translated into statements exclusively concerned with sensory appearances, let us first state precisely what is the subject-matter of directly verifiable statements of either kind. It is obvious that a proposition about a directly unobservable object could not be directly verifiable. Moreover, if a proposition which ascribes a particular property to a particular object or asserts that a particular relation obtains among a few particular objects, is to be directly verifiable, it is not sufficient that these objects be directly observable; the properties and relations involved must be directly observable as well. The two propositions which assert respectively that this book is brown and that this book is beside that other book are both directly verifiable because the proper names "this book" and "that book" denote directly observable objects, "brown" denotes a directly observable property and "beside" a directly observable relation.

To ensure the direct verifiability of a statement, the direct observability of objects, properties, and relations this statement refers to, seems necessary. Neither a proposition ascribing a directly unobservable property to a directly observable object, nor one which attributes a directly observable property to an object which is not directly observable would be directly verifiable. The colour of this book is directly verifiable but its chemical composition is not. Similarly, the closeness of two books is directly verifiable in contrast to that of two electrons.

Thus we might conclude that the direct verifiability of a proposition of the form $f(a)$, constituted by a one-place predicate f and the proper name a, requires that f denote a directly observable property and a a directly observable object. Similarly, the proposition $R(a,b,c, \ldots)$, consisting of the predicate R and the proper names a, b, c, etc. would be directly verifiable only if R denoted a directly observable relation among the directly observable objects a, b, c, etc. It is clear, however, that propositions of this type can account, at most, for all observation-statements. The subject-matter of directly verifiable propositions which are memoratively validated does not consist of directly observable objects, properties, and relations.

It is also worth pointing out that the concept of an observation-statement may raise serious difficulties if the term "directly observable object" is not construed widely enough. Thus, one may wonder which directly observable object is referred to in an introspective statement or in a statement where the verb is impersonal.[3] Thus, when noticing that

[3]Propositions of this grammatical type are important since they constitute the prototype of "field-laws" which describe the spatio-temporal distribution of measurable quantities, without involving corpuscular concepts.

it is warm here, I certainly make a directly verifiable statement. But what directly observable object is this statement about? I could give it the symbolic form $f(a)$ by construing it as attributing an observable thermal property, denoted by the predicate f, to an observable place denoted by the individual constant a. But am I warranted in considering a spatial region as such to be a directly observable object, since it differs so much from objects usually referred to as directly observable?

It is obvious that this difficulty is but terminological. To say that a place is a totally different entity from a thing is to say that a local adverb is not a noun and that they belong to two distinct logical types. But there is no reason for restricting direct observability to entities which are not represented by expressions of only a single logical type in a given linguistic system.

Introspective observation-statements raise similar difficulties. Thus, I can directly verify whether I have a headache right now, and other observers can check indirectly on this statement by interrogating me, watching my overt behaviour, and so on. The statement is therefore directly verifiable; there is little doubt that it has often been directly verified by human beings. But is the being which the statement refers to, the ego or the self, directly observable? Irrespective of the particular philosophical theory of the ego which we may adopt, by construing it, for example, as a "bundle of ideas" or as a "polyp made up of memories," or by erecting it into a "transcendental subject," we may wonder whether it is directly observable. Yet, this difficulty, too, is purely verbal: the ego and the ordinary, directly observable material objects are admittedly separated from each other by a logical gap since the personal pronouns and the nouns which form, respectively, their linguistic counterparts, are not meaningfully interchangeable in meaningful statements and pertain, accordingly, to different logical types. This, however, need not prevent us from granting direct observability to the ego.[4]

[4]Classifying sensory, perceptual, and introspective verification under the heading of observational verification is neither unobjectionable nor essential to our argument. It is natural to say that we observe a physical thing when we try to find out carefully how it looks to us. It is certainly correct to say that we observe a physical object when we look at it in order to ascertain its actual size and shape (in contrast to its apparent size and shape). One may wonder, however, whether the introspective attitude a person takes up in order to answer correctly and responsibly a question concerning his present feelings or mood is best described as a sort of observational attitude. A person who understands the term "pain'" correctly must be prepared to assert the statement "I am in pain" whenever he considers this statement (cf. § 31) and experiences this feeling. The readiness to make or to accept this statement under these circumstances is part and parcel of correctly understanding its crucial term. But there seems to be no reason for assuming that such readiness can be acquired or exercised only when the person who learns or makes the linguistic response not merely has pain but also "observes" in some

The misgivings to which these examples give rise suggest that it is preferable not to explain an observation-statement in terms of the direct observability of the objects, properties, and relations involved in such a statement, but to define it rather in terms of its ability to be validated by direct (sensory, perceptual, or introspective) observation. We may then, in turn, define directly observable objects, properties, and relations as referents of names and predicates which occur in observation-statements. To avoid nominalistic objections, direct observability may be ascribed to predicates occurring in observation-statements rather than to what such predicates refer to, so as to dodge the question of their referential function.

More important, however, than such doubts concerning the best way of delimiting the concept of observation-statement and the status of elusive entities referred to in some important types of observation-statements, is the uncertainty as to the entity which is "really" observed on any particular occasion. What am I "really" observing, for example, when I am looking at the sun? There seem to be at least three distinct groups of admissible answers, which may be distinguished as those involving the sensory, the perceptual, and the physical object of the observation:

---

sense his having this feeling. The mere presence of pain seems sufficient for either acquiring or exercising the corresponding linguistic habit.

In other words, in order to learn the correct use of the word "pain" we have to become conditioned to answer "I am in pain" whenever we experience pain and are anxious to give responsible information about our feelings. Since the psychological stimulus and the linguistic response occur in the same person, there is no need to explain their conditional association by assuming an intermediary activity called "introspective observation." The situation is different in perceptual verification. To learn to call a dog a dog, and thus to master this word, we have to see dogs and to associate the word with the sight. The presence of a dog in the vicinity of a human organism and the appropriate linguistic response of the human organism cannot come to be correlated with each other without an intermediary perceptual or observational process. Accordingly, to verify directly whether a given individual is a dog, the human observer must see this dog in order to activate his linguistic response.

In sum, the basic similarity between introspective and perceptual verification is in their non-inferential nature. Both are reliable verificatory procedures and neither could be dispensed with without curtailing science. But the link between the two facts described by an ostensive and by an introspective statement and the corresponding linguistic responses varies in these situations. In the second case, the mere presence of a fact in an observer enables him to verify directly the statement describing this fact. In the first case, the perceptual or observational activity of the observer is also necessary.

I am indebted for much of what is contained in this account of the nature of introspective statements to K. Ajudukiewicz's pioneer investigations; cf. *O znaczeniu wyrazen* (The Meaning of Expressions), 1931, and several subsequent publications of this author.

I. While watching the sun, I am really observing a yellow disc against a blue background; the disc is localizable vaguely in my visual field and has an apparent size, but is neither separated by a definite distance from me, nor possessed of an inside, nor of an unseen part of its surface. It is easy to produce two discs instead of a single one, by exerting a pressure on my eye and to remove the disc by placing an opaque object before my eyes.

IIa. The directly observed object is the common-sense thing called sun, which, when watched from the earth, presents to the human on-looker the appearance of a yellow disc. The sun is round, yellow and shining, exactly as the disc is, but in contrast to the disc, it has an inside and an unseen part of its surface; it can neither be duplicated by squint-ing nor removed by insertion of an opaque object.

IIIa. The directly observed object is really a gaseous incandescent globe which was located 8 minutes before the observation took place at a distance of about 150 million kilometres from the observer.

IIIb. The observed object is the part of the surface of the gaseous globe determined by its intersection with the luminous cone which has reached the observer's eye when the observation was made.

IIIc. The observed object is the physical event which occurred in the aforementioned part of the surface of the sun and consisted in the emission of radiations which eventually reached the observer's eye. These radiations have a maximum of energy in the neighbourhood of the wave-length the usual physiological effect of which is a photochemical process in the retina which leads eventually to the emergence of a sensa-tion of yellow in the observer.

This list is by no means complete. It could be extended by adjunction of new groups: by declaring, for example, that the images on the retina (IV) or the hypothetical process in the cortex of the observer which underlies his having a sensation of yellow (V) are the "really" observed objects on this occasion. Groups I-III might also be extended by adding to the aforementioned unique version I, which corresponds to IIIb, a variant Ia which identifies the observed object with the "sensory thing" discussed by phenomenologists ("Sehding"). A more important sup-plementation would ensue by adding (to groups I and II) the variants Ic and IIc which correspond to IIIc. The variant Ic, for example, would claim that the really observed object is not the sensory thing as such, nor any part of its surface, but the sensory event consisting in this sensory thing having definite sensory properties or standing in definite sensory relations to other sensory objects.

We notice that in all the groups 1–3 the variants (c) come closest to what is observed; while the variants (a) involve the whole sensory, perceptual, or physical object in the observation, and the variants (b) refer to the spatio-temporal fragment of such object relevant to the observation, the variants (c) retain, from this fragment of the observed object, only the group of characteristics and relations which are directly given to the observer. Thus, IIIc, for example, is neither the sun in its entire spatio-temporal extension, nor the spatio-temporal fragment of this extension which is involved in the observation, but merely the event of emission which occurred in this fragment. We may disregard variants IV and V since we confine ourselves to the subject-matter of sensory and perceptual knowledge.

We are thus faced with about a dozen answers to the question as to what we really observe when looking at the sun and we may wonder how this multiplicity affects the phenomenalist view that scientific knowledge is of (observed or observable) appearances only. Does this mean that only Ic, which comes closest to the content of a sensory observation, is legitimate, and that statements about groups IIa–IIIc can express knowledge only in so far as they are translatable into statements about Ic? Or does it mean that only the appearance Ia which the sun presents to me is real, whereas the common-sense things IIa called "sun" and the scientific object termed "sun" (IIIa) are but indirect ways of referring to Ia? I think that either claim is incompatible with the actual meaning of the relevant terms: the sun which looks like a yellow disc to me, the sun which is a common-sense thing whose shape is not disc-like at all, and the sun dealt with by the astronomer and the astrophysicist are one and the same sun. Hence, if the phenomenalist claims that scientific knowledge is of appearances only, he unjustifiably disregards the fact that our knowledge of the sensory appearances of the sun, of its common-sense (perceptual) aspects, and of its scientific aspects is knowledge of one and the same entity. If, on the other hand, the phenomenalist maintains that whatever is knowable about the sun is translatable into statements exclusively concerned with Ic, then his claim can be shown to be at variance with the nature of perceptual or scientific concepts involved in IIa–IIIc.

To realize this, consider the directly verifiable sensory and perceptual statements: "This looks like a yellow disc" and "This is yellow." There is hardly any doubt that the subject "this" is intended to refer to the same entity in both statements. The reference is even more conspicuous in those cases where a sense-illusion is unmasked: when a person says "This stick looks bent to me, but is actually straight," he certainly ascribes the sensory and perceptual predicates to the same object. But what about the logical interrelation between the predicates "looking like a yellow disc to me" and "being yellow"? If asked "How do you know that this thing which admittedly looks yellow to you actually is yellow?" I would probably answer that I know reliably that this object is yellow since it looks yellow to me. I would not preclude the possibility of an illusion or even a hallucination on my part. Were I in a position to rule out my being subject to such an illusion or hallucination, then I would answer that not only do I know reliably that this is yellow because it looks yellow to me, but that I could not possibly be mistaken in attributing this colour to this object in view of its presenting to me such an appearance. The fact that I might be subject to an illusion prevents me from claiming infallible knowledge of the colour of the object under consideration. But it does not prevent me from claiming reliable knowledge thereof—my main reason for claiming such reliable, though not infallible, knowledge is the fact that this object appears to me this way.

Logically speaking, the situation is that the perceptual concept ("being yellow") is conditionally definable in terms of the corresponding sensory concept ("looking yellow"); the fact that the conditional definition must also be "probabilistic" in order to take into account the existence of sense-illusions and hallucinations is of secondary importance (cf. § 27). In other words, the actual meaning of "being yellow" is such that, by definition, whatever looks yellow to me probably is yellow.

The epistemological interrelation between "reality" and "appearance" is a definitional one. The meaning of perceptual terms referring to "reality" is such that statements about "reality" follow from corresponding statements describing appearances. Doubting that what looks yellow to me is yellow, is on a par with doubting that the reading of a thermometer applied to a physical object corresponds to the temperature of that object.

It is important to realize that regardless of whether the conditional definitions explaining perceptual terms by means of sensory ones be probabilistic or not, such definitions do not provide a faithful translation of the statement about how an object is into synonymous statements about how the object looks to me. For, even if there were no illusions or hallucinations, and the perceptual term P (e.g., "being yellow") were strictly definable by means of the sensory test T ("being looked at") and the sensory response R ("looking yellow"), the perceptual statement ascribing the property P to the "real" object A would follow, by definition, from the two sensory statements asserting that A was submitted to the test T and yielded the response R, whereas the perceptual statement denying the property P to the object A would be inferable from the two sensory statements to the effect that A was submitted to test T and failed to respond in the way R. Thus, $P(A)$ is provable by the sensory statement $T(A) \cdot R(A)$ and disprovable by the sensory statement "$T(A) \cdot -R(A)$," which is not the negation of the statement $T(A) \cdot R(A)$. Thus $P(A)$ is not equivalent to, let alone synonymous with, this sensory statement $T(A) \cdot R(A)$ in spite of being definitionally connected with it.

Similar considerations can be shown to apply to the remaining groups of directly verifiable statements: for example, on closer inspection introspective predicates prove to be probabilistically, but not fully, definable in terms of perceptual tests and responses. Hence, introspective statements are not translatable into perceptual ones. The four groups of directly verifiable statements are therefore all of them essential, since dispensing with any one of them would curtail science.

The phenomenalist claim that perceptual statements can be translated into statements about sensory appearances is not born out by an investigation of the logical connection of perceptual and sensory terms. We may add that if they could be so translated, this would not substantiate the phenomenalist view that scientific knowledge is only of appearances, for, if statements about common-sense things can be faithfully translated into statements about appearances, the former must be analytically equivalent to the latter. Hence if the latter contain reliable information about their subject-matter, so do the former. The translatability would only establish a close relation between our knowledge of appearances and our knowledge of reality, without destroying the latter.

Of course, it is essential that perceptual statements, while untranslatable into statements about appearances, are nevertheless verifiable, for only on this assumption can they express knowledge. This, too, follows from the conditional definability of perceptual concepts in sensory terms. If the perceptual concept P has a sensory test T and a sensory response R, then the perceptual statement $P(A)$ follows from the conjunction of two sensory statements $T(A) \cdot R(A)$; similarly, the negation of $P(A)$ follows from another conjunction of sensory statements, namely, $T(A) \cdot -R(A)$. Hence, if sensory statements are verifiable, perceptual statements must be verifiable as well.

The upshot of this discussion is that the phenomenalist claim that the whole of scientific knowledge is ultimately concerned with sensory appearances is untenable. The successive stages of the scientific world-picture dealing with sensory appearances, with common-sense objects, with pre-atomic and pre-relativistic scientific objects, and with scientific objects pertaining to the atomic and relativistic level are irreducible to each other in this sense, that at each level there are statements which are demonstrably untranslatable into statements of other levels and which convey genuinely new reliable information.

33 FINITE VERIFIABILITY

The concept of direct verifiability accounts for the range of direct fact-finding methods. But it fails at once in the simplest cases of measurement and indirect observation. Even if a direct measurement is performed on a directly observed object with a view to ascertaining the value of a quantity defined by a single definitional criterion couched in ostensive terms, the quantitative statement expressing the result of such a measurement is not directly verifiable in the sense explained in the preceding section. For instance, to verify a statement ascribing a particular length to a particular object, one has to *deduce* either this statement or its negation from a set of directly verifiable statements about the mensural operations, supplemented by the definition of length. For such a deduction to be possible, the quantitative statement must *follow* from the aforementioned set of observation-statements enlarged by an analytic statement expressing the conditional definition of length. The number of observation-statements in this inference is obviously rather small, and, in any event, finite. Accordingly, we shall stipulate that a statement S is finitely verifiable if S follows logically from a finite consistent set of directly verifiable statements possibly supplemented by some analytic premises. Unless otherwise stated, these additional analytic premises will be supposed to be either full-fledged or conditional definitions. Regardless of whether any use of analytic premises is made in deducing a finitely verifiable statement, its finite verifiability can also be defined as consisting in its being *entailed* by a finite, consistent set of directly verifiable statements.[5]

The concept of finite verifiability, as just defined, indicates of course, not only the range of direct mensural methods, but also that of any other mensural methods and of indirect observation as well. Thus, when the

[5]If the quantitative concept Q is probabilistically defined in terms of a test T and a response R, the quantitative statement $Q(A,a)$ asserting that the quantity Q takes on the value a for the individual A, is not strictly entailed by the ostensive statements $T(A,B)$ and $R(B,a)$ which indicate that the instrument B has been attached to A and that, thereupon, the reading of B was a. Hence, strictly speaking, such a quantitative statement is not finitely verifiable. Yet, we may stipulate that if, by definition, almost every object which responded in the way R to the test T has the property P and almost no object which failed to so respond to this test has this property, then the statement attributing the property P to an object A may be said to be *probably entailed* by the ostensive statements asserting that the object A reacted in the way R to the test T. Accordingly, finite verifiability can be attributed to any statement which is strictly or probably entailed by a consistent finite set of observation-statements. On this extended definition of finite verifiability, quantitative statements are to be considered to be finitely verifiable, regardless of whether or not the conditional definition of the relevant quantity is probabilistic.

quantity involved is defined in terms of other directly measurable quantities, the quantitative statement which specifies the value this quantity takes on for a given object will still follow from a finite set of directly verifiable statements supplemented by some definitional premisses, the only difference being an increase in the number of both the directly verifiable and the definitional premisses. In the case of a computational measurement, which involves the utilization of a theory with a view to determining the value of the measured quantity, the number of observational premisses is still finite, since the only contribution which a theory makes towards the computation of a quantity is that it provides new definitional criteria for this quantity. The only observational premisses required in such a case describe the behaviour of measuring instruments and their final readings, exactly as in fundamental and indirect measurement.

The concept of finite verifiability which we have introduced in order to account for the range of measurement and of indirect observation, turns out to cover a much wider field, since statements which are neither quantitative nor concerned with indirectly observable objects prove to be finitely verifiable. This wider scope is, of course, not unexpected, because our main reason for resorting to the concept of verifiability is the desire to account for the joint range of well-established scientific methods and of all discoverable scientific methods. The simplest type of finitely verifiable statements that have nothing in common with either measurement or indirect observation is seen in compound statements consisting of directly verifiable components (or, for that matter, of any finitely verifiable components). For instance, the compound statement "If this book is not on the desk, then it is in the drawer," is finitely, though not directly, verifiable. As a rule, every compound statement consisting of finitely verifiable components is also finitely verifiable; for such a compound proposition can always be expressed in a "normal form,"[6] for example, as a disjunction of conjunctions of statements each of which is either a component of the compound proposition or the negation of such a component. Since a disjunction follows from any of its members, it also follows from the class of all directly verifiable statements which occur among the premisses of any statement included in such a conjunction.

It is worth noticing, however, that, in contradistinction to direct verifiability, finite verifiability, in the sense just defined, does not entail essential determinacy. In other words, a statement may follow from a finite set of directly verifiable propositions and may yet inherently lack

[6]Cf. D. Hilbert and W. Ackermann, *Principles of Mathematical Logic* (1950), pp. 17ff.

any definite truth-value. Thus, a statement S ascribing a definite position to a definite elementary particle is finitely verifiable according to the Quantum Theory since, as already mentioned, the only relevant contribution of this infinitist theory consists in providing new definitional criteria for quantitative statements about such particles. Similarly, a statement S' which ascribes a definite velocity to the same particle is also known to be verifiable, because according to quantum-theoretical assumptions instruments for measuring such velocities can be constructed in principle. However, a statement which specifies the position and the velocity of this particle is no longer verifiable; in other words, the conjunction of the two finitely verifiable statements about position and velocity of a given particle is not verifiable in turn, because the two sets of observation-statements O and O' which are required to ensure the verifiability of S and of S', respectively, add up to an inconsistent set $O + O'$ which is the only set of observation-statements available for validating the conjunction. This classical quantum-theoretical example shows that the requirement of deducibility from a finite consistent set of observation-statements is not sufficient to ensure the inherent determinacy of the finitely verifiable statement.

To put it in other words: Since the conjunction of two finitely verifiable statements may happen to be unverifiable, we can distinguish *joint* and *separate* verifiability of statements. Two finitely verifiable statements will be said to be jointly (or separately) verifiable if their conjunction is also (or is not) verifiable. Now the point is that unverifiable statements have no definite truth-value. This restriction applies, in particular, to the unverifiable conjunction of two separately verifiable statements. Since, according to elementary rules of logic, the truth-value of a conjunction could easily be determined if each component of the conjunction had a definite truth-value, we must conclude that one at least of its verifiable components has no definite truth-value. Thus in cases of separate verifiability, the finite verifiability of a statement may fail to ensure in principle the determinacy of this statement.

Other examples of finitely unverifiable statements are afforded by conjunctions or disjunctions consisting of one finitely verifiable and one unverifiable component (cf. § 28). A disjunction of this type is finitely verifiable, but its negation is not; a conjunction of this type is finitely unverifiable while its negation is capable of being finitely verified. These examples suggest that, in order to ensure, in principle, the finite verifiability of compound statements, we may stipulate that all compound and simple statements be bilaterally verifiable (cf. § 28). Since any bilaterally verifiable statement S follows from a consistent set of observation-statements, whereas the negation of S follows from another set of observation-statements, S can be neither a conjunction nor a disjunction involving unverifiable components. As a matter of fact, the most typical fact-like statements within science are bilaterally verifiable.[7] Thus, if the quantitative statement $Q(A,a)$ involves a quantity Q defined in terms of an ostensive test T and an ostensive response R, then $Q(A,a)$ (asserting that the value of Q for A is a) follows from the ostensive statements $T(A,B)$ (the instrument B for measuring Q has been properly applied to A) and $R(B,a)$ (the reading of B is a), whereas the negation of $Q(A,a)$ follows from $T(A,B)$ and the negation of $R(B,a)$. However, the bilateral verifiability of all the components is easily seen to be neither sufficient nor necessary for the finite verifiability of a compound statement.

First, consider a non-ostensive concept P whose only analytic context is a conditional definition involving the ostensive test T and response R. The statement asserting that a given individual was not submitted to the test T and yet possesses the property P would then be unverifiable although its components are bilaterally verifiable. This shows that the bilateral verifiability of components does not guarantee the verifiability of the corresponding compound statements. On the other hand, in order to realize that the bilateral verifiability of all the components of S is not necessary for the bilateral verifiability of S, it is sufficient to notice that the disjunction of one bilaterally and one merely negatively verifiable statement is bilaterally verifiable and that a merely negatively verifiable statement can always be constructed by forming the conjunction of any negatively verifiable statement with a completely unverifiable one; this provides the possibility of smuggling completely unverifiable components into bilaterally verifiable statements. For example, the statement "Either this man will do his job shortly, or else he is lazy and the absolute is lazy" is bilaterally verifiable in spite of the presence of a completely unverifiable indeterminate statement in its second component.[8]

[7] It is important to realize, however, that statements about *continuous* quantities are only unilaterally (negatively) verifiable. We have seen in § 17 that in view of the fact that measuring instruments for such quantities always involve a least count, a measurement can only substantiate the hypothesis that the value of a continuous quantity Q for a given object X is located somewhere between N and $N+1$ smallest units of the measuring instrument (N being a positive integer). Accordingly, the hypothesis that the above value is exactly equal to a units (where a is a real number) can be refuted by measurement whenever the number a is outside the interval $(N,N+1)$ which the measurement has yielded. Thus the statement to the effect that the value of the continuous quantity Q for a given object X is expressible by a given real number a is in principle observationally refutable without being so provable.

[8] The fact that bilaterally verifiable statements may include completely unverifiable components seems hardly compatible with the verifiability theory of mean-

The upshot of this discussion is that, in general, the finite verifiability of compound statements neither presupposes nor follows from the bilateral verifiability of all their components. Separate verifiability (e.g., in Quantum Theory) or paucity of definitional criteria for relevant concepts may interfere with the finite verifiability of statements the components of which are all bilaterally verifiable. The only way of securing the finite verifiability of compound statements is to investigate whether any pre-assigned class of these statements raises such difficulties as occur, for example, in Quantum Theory or when terms are used which are linked up with the ostensive level by but a few conditional definitions.

We may also conclude that, in spite of frequent suggestions, it is impossible to realize the Principle of Verifiability within science and to eliminate unverifiable scientific assumptions by adopting a language whose terms are either ostensive or conditionally definable in ostensive terms, since statements couched in such terms may turn out to be unverifiable. For the same reason, it is impossible to define the verifiability of a statement by requiring that it be translatable into a language whose extra-logical vocabulary consists of terms which are either ostensive or conditionally definable in ostensive terms. As a matter of fact, the elimination of unverifiable scientific assumptions would be disastrous in view of the major part they play in all basic scientific theories. The implications of the Principle of Verifiability within science are nevertheless decisive, since this principle separates the cognitively relevant from the subsidiary in the information science provides. But this separation cannot be obtained in a wholesale way or by squeezing scientific information into an artificial linguistic medium; the basic scientific theories must be examined on their respective merits, in so far as their verifiability is concerned.

34 INDUCTIVE VERIFIABILITY

Inductively Verifiable Laws

Finite verifiability does not account for the inductive validation of scientific laws and theories. Law-like statements are, in general, finitely unverifiable. This applies even to the simplest example of a qualitative law asserting that the directly observable quality P is always and every-

---

ing since, on this theory, such compound statements would have to be considered as meaningful in spite of the meaninglessness of some of their components. No such difficulty arises if verifiability is construed as a criterion of definiteness of truth-value.

where associated with the directly observable quality Q. Of course, if the evidence produced in support of this law consists of a finite number of observations concerning individuals in whom the property P was found to be associated with the property Q, then a finitely verifiable proposition (namely, a finite conjunction of statements such as "If the individual X has the property P, then it has the property Q as well") would suffice to summarize such evidence. However, the situation changes entirely when we do not confine ourselves to a summary of actual observations concerning the association of the two qualities in a finite number of individuals, but proceed to the assertion that this association holds true of every individual object, even if the number of such objects should turn out to be infinite. This assertion is no longer expressible as a logical conjunction of a finite number of finitely verifiable propositions. In order to take into account such finitely unverifiable items of scientific information we have to extend the concept of verifiability once more so as to make it applicable to universal statements of the kind exemplified by this law. We can stipulate, for example, that a finitely unverifiable statement is nevertheless *inductively verifiable* if every instance of this statement is directly or finitely verifiable. This definition of inductive verifiability, suggested by laws which are universal in regard to quantification, can be extended to laws involving a different (existential or mixed) quantificational pattern, provided the concept of instantiation be appropriately extended. To make this extension, it is sufficient to replace any law whose quantificational pattern is not universal with its law-like variant obtainable by substituting a purely universal pattern for the actual one and to declare that, by definition, every literal instance of this law-like variant is also an "instance" (in the new, generalized sense) of the law itself (cf. § 22). In this terminology, which is not without artificiality but which proves very convenient for our purposes, we may say that a general law is inductively verifiable if all its "instances" are finitely verifiable.

This is not yet the most general definition of inductively verifiable laws; it may happen, for example, that all the instances of a law are inductively verifiable in the sense just defined; then the law itself would not be so, according to the above definition. Yet such a situation arises whenever a concept involved in a law has been introduced by a quantified definitional criterion (for example, when a *perfect* liquid is said to keep its volume unchanged under *any* pressure, by definition). I see no reason why such laws should be ruled out. To include them within the range of the inductively verifiable statements, we have to stratify the concept of verifiability and stipulate, for example, that a statement is inductively verifiable in degree 1 if all its instances are finitely verifiable, but that its inductive verifiablity is of degree 2 if the verifiability of all its instances is of degree 1, and so on. Generally speaking, a law will be said to be inductively verifiable if all its instances are verifiable to any finite degree. This comprehensive concept of verifiability would include as special cases all those problems dealt with in contemporary science which are susceptible either of a finite verification or of an inductive verification of the 1st degree.

It will be noticed that finite verifiability has been defined in terms of entailment by observational premisses, whereas the definition of inductive verifiability has involved instantiation (direct or indirect) by observation-statements. In § 31 we have seen, however, that these two types of definitions can be reduced to a common denominator since any statement all the instances of which are ostensive can be construed as a consequence of at least one class of such instances. We can therefore stipulate that a statement is verifiable if it follows from a consistent set of observation-statements, and then subdivide all statements that are verifiable in this sense into the two classes of finitely and inductively verifiable statements depending upon whether the corresponding set of observational premisses is finite or infinite. In other words, a statement is finitely verifiable if it follows from at least one finite and consistent set of observation-statements. A statement S is inductively verifiable if every consistent set of observation-statement which entails S is infinite.

The most general definition of empirical verifiability to be used in most of our subsequent discussions can be phrased as follows: *the statement S is empirically verifiable if either S or its negation follows (strictly or with probability) from a consistent (finite or infinite) set of directly verifiable statements.* Positive, negative, unilateral, and bilateral verifiability can be defined in a similar way. For instance, the statement S will be said to be bilaterally verifiable if S and the negation of S follow respectively (strictly or with probability) from two consistent (finite or infinite) sets of directly verifiable statements.

Inductively Verifiable Theories

We have seen that the validation of theories can be thought of as being inductive and as involving the same rules of probable inference which are relevant to the validation of laws, provided we make up for the greater logical complexity of theories by translating each of them into a universal law of a meta-language associated with the object-language of the theory. Every theory T would thus have to be replaced with the law-like statement "Every verifiable consequence of T is true" in order that T should be amenable to inductive treatment. This suggests that to account for the range of scientific methods of theory-formation, we may simply identify any verifiable theory T with the sum-total of verifiable consequences which can be derived from the basic assumptions of the theory. In other words, we have to define a verifiable theory as being the sum-total of all the verifiable consequences which follow from a finite and consistent set of synthetic assumptions. So defined, a theory can be inductively validated since it is then equivalent to the inductively justifiable law couched in the meta-language and asserting that all the verifiable consequences of the basic assumptions of the theory are true.

This definition of verifiable theories calls for some comments. We have stipulated that the number of assumptions from which the theory is derivable is finite simply because all well-established scientific theories are axiomatizable and, hence, deducible from a finite set of assumptions. The consistency of the set is required since, from an inconsistent set of assumptions, any and every consequence can be derived; thus a theory derivable only from inconsistent sets of assumptions would be neither consistent nor empirical. Finally, we have stipulated that the assumptions are synthetic because we are interested only in *empirical* theories. Theories derivable from analytic assumptions are analytic and non-empirical by the same token.

It is also important to realize that when a verifiable theory is defined as the sum-total of the verifiable consequences of a given consistent set of assumptions, then such a theory cannot be identified with an axiomatized deductive system in the usual sense of the word. (We call here "deductive" any set of statements which includes all the consequences of any finite subset of these statements; an axiomatic system associated with a given deductive system of statements originates whenever a finite set of these statements has been selected in such a way as to ensure that the remaining statements can be derived therefrom.) I am not alluding here to the fact that there are demonstrably non-axiomatizable deductive systems, that is, deductive sets of statements which cannot be deduced from any finite subset of the relevant set. On a previous occasion we assumed tentatively that empirical theories are axiomatizable since they are deducible from a finite set of assumptions; for example, the whole of electro-dynamics follows from Maxwell's equations. We have now to readjust this assumption in accordance with the new definition of an empirical (or verifiable) theory. Let us call an "axiomatic basis" of a class of propositions any consistent and finite set of synthetic assumptions from which all the propositions of the class can be derived by deductive operations. We may then say that a verifiable theory admits of several axiomatic bases without being an axiomatic system. The main differences lie in the parts played by axiomatic bases in ordinary axiomatic systems and in verifiable theories.

(1) In contrast to axiomatic systems, an empirical theory does not contain all the logical consequences of its axiomatic bases, but only those consequences which are themselves verifiable. Thus the composition of the axiomatic basis of a theory does not uniquely determine the composition of the theory itself. A separation of all the consequences into those which are and those which are not verifiable is also needed, and this transcends the purely logical structure of the theory.

(2) More particularly, an empirical theory need not contain the specific propositions included in its axiomatic basis. This is certainly not the case when the axiomatic basis of the theory contains unverifiable propositions. In other words, the propositions included in an axiomatic basis of an empirical theory may serve only to delimit the sum-total of the statements which make up the theory, by requiring that these statements be both verifiable and deducible from the axiomatic basis; in general the axioms themselves need not belong in the theory.

It goes without saying that when a verifiable theory happens to be deducible from a consistent set of verifiable assumptions, the latter do belong, all of them, in the theory. Such verifiable theories will be said to have an "internal basis." Yet, important as this type of theory may be, in so far as the Principle of Verifiability is concerned, there is nothing to prevent us from granting verifiability and observational justifiability to verifiable theories which are not deducible from a consistent set of verifiable assumptions and may therefore be called empirical theories with an external axiomatic basis.[9]

[9]It is worth while to notice that the class V of all the verifiable consequences of any theory T always constitutes a deductive system since the consequences of verifiable premises are also verifiable. If T is transcendent or has only external bases then the associated deductive system V will not be axiomatizable.

Another significant and more restricted deductive system, say O, can be associated with every empirical theory T, according to a theorem of W. Craig (cf. "On Axiomatizability within a System," *Journal for Symbolic Logic*, vol. 18, (1953) and "Replacement of Auxiliary Expressions," *Philosophical Review*, vol. 55 (1956). This theorem implies that all the observational consequences of any theory T (that is, consequences which involve only terms possessed of a direct empirical meaning) form a deductive system O. It goes without saying that this new deductive system does not account for all the verifiable consequences of the theory under consideration since a verifiable consequence need not be formulated in terms possessing a direct empirical meaning; terms endowed either with explicit definitional criteria or with criteria implicit in relevant semi-definitional laws and theories, may also occur in the verifiable consequences of a theory. The deductive

The two kinds of empirical theories defined in terms of the verifiability of their axiomatic bases correspond roughly to the historically important distinction between "phenomenological" and "transcendent" theories. However, this distinction has been mostly derived from considerations of the subject-matter of the relevant theories; a theory was taken to be phenomenological or transcendent according to whether it dealt with observable or with unobservable entities. Yet we have seen that, on closer inspection, the concept of "observability" proves unsatisfactory and elusive, and that it seems preferable to define the distinction between the two types of theories in terms of the accessibility of their basic assumptions to the basic fact-finding and law-finding methods of science. If this step is taken, the distinction between phenomenological and transcendent theories and that between empirical theories with internal or with external axiomatic bases tend to coincide with each other. The only difference which still remains is the fact that the content of a theory with an external basis is identified with the verifiable consequences of its assumptions, whereas, in a transcendent theory, assumptions and consequences are defined in terms of their inferential functions within the theory, but otherwise dealt with on a footing of equality, especially in so far as the claim of the theory to make true statements is concerned.

We may say that a verifiable theory with an external basis consists of all the verifiable consequences of the assumptions of a transcendent theory. Furthermore, the entities dealt with in a theory with an external basis need not be unobservable in any straightforward sense of the word.

---

system O associated with any pre-assigned theory T possesses nevertheless some remarkable properties.

For example, although the system O is not axiomatizable in our sense (that is, deducible from some of its finite subsets) it possesses an effectively definable infinite basis. In other words, O includes an infinite effectively definable subset A such that any statement in O can be derived from a finite part of A. If we widen the concept of axiomatic system by considering as axiomatic any deductive system D provided D has a (finite or infinite) effectively definable subset d such that every statement in D is deducible from a finite part of d, then the system O may also be said to be axiomatizable (although not finitely axiomatizable, that is, axiomatizable in our sense). The fact that O is axiomatizable, even in a generalized sense, may suggest that it would be possible to handle all the observational consequences of T within the framework of O and without resorting to T. However, on closer inspection, the generalized axiomatizability of O does not make T superfluous even for handling its purely observational consequences, let alone the class V of all its verifiable consequences. For example, any valuable theory T is capable of predicting some of its observational consequences C from assumptions established independently of C (and hence, of indirectly validating these consequences; cf. § 26). This possibility of indirect validation would vanish should the theory T be replaced with the associated observational system O.

The decisive circumstance is that the theory be derivable from some unverifiable assumptions about the entities it deals with without being derivable from any finite set of verifiable statements concerning this subject-matter. Thus all theories whose assumptions are differential or partial differential equations (the independent variables being spatio-temporal) have external bases only, although some of them would be classified as phenomenological on a literal interpretation of the classical distinction between the two types of theories. In view of the afore-mentioned drawbacks of the concept of observability, it will be more helpful to base our subsequent discussions on the distinction between theories with internal and those with external bases.

Requirement of Joint Verifiability of Empirical Theories

We have suggested that a verifiable theory be identified with the sum-total of the empirically verifiable consequences which follow from the basic assumptions of the theory, regardless of whether these assumptions are themselves verifiable. If they are not, they have to be excluded from the core of the verifiable theory and viewed as playing a merely auxiliary, though possibly indispensable, part of an "external basis" of the theory. We shall now have to investigate an additional condition to be met by the statements which make up the core of an empirical theory. It turns out that the mere verifiability of each statement included in the theory does not actually suffice to remove certain paradoxical consequences and to ensure the validity of the Principle of Verifiability. The circum-stances we are going to discuss have been brought into focus by the un-easy conceptual situation in Quantum Theory, although, basically, they are implicit in the application of the Principle of Verifiability to any scientific theory.

The verifiability of a theory, that is, of the class of verifiable conse-quences of a given set of assumptions, might be open to doubt should it so happen that the impossibility of observations (i.e., the falsehood of observation-statements) required to ensure the verifiability of some proposition included in the theory were entailed by other propositions of the same theory. If this were the case, the truth-value of the whole body of verifiable statements which constitute the theory under con-sideration could not be determined by carrying out suitable observations, in spite of the fact that each single statement in this body is verifiable.

A simple example may illustrate the logical mechanism operating in such a case. The hypothesis asserting the uniform expansion of the whole universe or the uniform acceleration of all the cosmic processes is held to be unverifiable in principle, since the spatial and temporal changes involved would also affect the instruments of spatio-temporal measurement, that is, clocks and yardsticks. For instance, if the dimensions of all the bodies inclusive of all yardsticks increase n times, then the ratio of the dimensions of any object to those of a yardstick would remain unchanged and the whole alleged cosmical disturbance could not be detected (cf. § 14).

It goes without saying that an interpretation of such unverifiable cosmological hypotheses depends upon the definitions of the spatio-temporal quantities involved. Should length, for example, be defined in terms of coincidence with the yardstick, with no supplementary definitional criterion available, then any hypothesis concerning the comparative length of two non-simultaneous objects would be unverifiable, and this would hold true, in particular, of the question as to whether any given object changes its spatial dimensions in the course of time. On the other hand, if the initial definitional criterion is supplemented by an analytic statement to the effect that the yardstick does not change its length as time goes on, then a comparison of the length of non-simultaneous objects would become verifiable, but the assumption of a universal expansion would involve a contradiction, since a yardstick would now become invariable, by definition. This contradictory assumption would therefore not be a genuine synthetic hypothesis.

These two interpretations, however, are hardly satisfactory. What one tries to say while asserting the unverifiability of the hypothesis of a uniform universal expansion is that the *human observer* would be unable to notice it. For this interpretation to make sense we have either to replace the aforementioned definitional criteria of length, or to supplement them by an additional definitional criterion to the effect that two objects which look conspicuously unequal to a normal human observer under normal circumstances are actually unequal. Such a definitional convention, though less precise than the preceding ones, is nevertheless acceptable from a purely logical point of view and can be substantially improved by specifying the accuracy of human perception of length. If we regard the new criterion as contributing to a determination of the meaning of length, then the verifiability of statements concerning either simultaneous or non-simultaneous material objects would be established (in view of the human observer's ability to compare the length of two non-simultaneous objects by using his memory). However, the hypothesis of universal expansion would now become verifiable, since nothing seems to prevent a normal human observer or a group of such observers from noticing that, from a specifiable time moment on, all material objects, including human organisms, have kept increasing their dimensions in the same ratio, so that their measures expressed in˙yardsticks remain constant.

Those who still consider the hypothesis unverifiable, probably use another tacit hypothesis concerning the interconnection between the physiological processes in the organism of any human observer and the psychological processes ascribable to him. If all the objects change their dimensions, then, the same applies to the retinas of the observer and the images projected on the retinas by any object whatsoever. Now it has always been observed, they would argue according to this additional hypothesis, that the apparent size of an object as perceived by a given observer is completely determined by the relative size of the corresponding retinal images: accordingly, the apparent size of all material objects, which is decisive in assessing their actual size given the interpretation under consideration, would still remain unchanged, and a universal expansion would still be undetectable.

Hence the hypothesis of a uniform universal expansion is verifiable in principle according to the new interpretation, since we know in advance what kind of observations would have to be regarded as supporting it, and what other types of observational results would have to be viewed as contrary to it. Yet this intrinsic verifiability of the hypothesis is completely inefficient in view of another hypothesis which is also accepted, namely, the hypothesis of a partial or complete psycho-physiological parallelism. This second hypothesis implies that those observations which are either favourable or unfavourable to the first hypothesis will never be carried out. We may express this situation by stating that on the interpretation of length now considered, the hypothesis of universal expansion, though inherently verifiable, is yet not verifiable jointly with other consequences of the theory which includes also assumptions about psycho-physiological parallelism.[10]

We are thus led to put an additional restriction on the verifiability of theories: a theory is verifiable if it consists of all the verifiable consequences of a finite synthetic set of assumptions, provided that all these consequences be jointly verifiable within the theory, i.e., provided that the assumptions of the theory do not preclude the joint verification of any two of its intrinsically verifiable consequences. This requirement of joint verifiability within a theory must not be confused with the concept of joint verifiability of statements discussed on a previous occasion. Two verifiable statements are jointly verifiable if their conjunction is also

[10]Cf. H. Mehlberg, "Sur quelques aspects nouveaux du parallélisme psycho-physiologique," *Proceedings of the International Philosophical Congress*, Paris (1937).

verifiable. Two verifiable propositions are jointly verifiable within a theory T if there exists a single set of observation-statements which entails the conjunction of these two propositions and is also compatible with the assumptions of the theory T.

Our main reason for inserting this additional condition in the definition of verifiable theories is that only by so doing can we secure that every verifiable theory is capable of being validated, that is, shown to be probably true, by producing relevant observational evidence. For if all the verifiable consequences of a theory are jointly verifiable within it, we can lump together into a single class all the sets of observation-statements which are required, respectively, to secure the verification of each constituent of the theory. This class of observation-statements to which all these sets add up will then both be consistent and entail all the verifiable consequences of the theory. Such a theory can therefore be said to follow from a consistent set of observation-statements, exactly as any single verifiable statement does. This does not imply, of course, that this comprehensive class of observation-statements can be condensed into a single law-like assumption; this is impossible whenever the theory under consideration has no internal basis. But the truth of the theory, that is, the truth of all the verifiable statements it consists of, will be secured in the same way as the truth of every single verifiable statement is secured, namely, by the fact that the whole theory follows from a consistent class of observation-statements.

The requirement of joint verifiability within a theory is implicit in several methodological applications of the Principle of Verifiability to fundamental scientific theories. To illustrate this, we shall comment briefly on the concepts of absolute simultaneity and location. The theory of absolute space and time, usually condemned in view of its alleged unverifiability, can easily be interpreted as consisting of verifiable statements; the point is, however, that these statements which go to make up the theory are not jointly verifiable within the theory.

The concept of absolute, that is, invariant simultaneity can readily be defined in such a way as to imply the intrinsic verifiability of the relevant statements.[11] It suffices, for example, to stipulate that two events are simultaneous with each other, if it is impossible to connect them by a signal travelling with a finite speed. On the assumption that no signal is faster than light, one finds that according to this definition simultaneity is not absolute, since two events found to be simultaneous with each other by an observer attached to a given frame of reference may nevertheless prove to occur one after another when the observer switches to a

[11] Cf. H. Mehlberg, *Essai sur la théorie causale du temps* (1935).

new frame of reference. Yet, on the contrary assumption (which is itself intrinsically verifiable) that, although all physical signals travel with finite velocities, there is no upper limit to these velocities, one can easily show that the concept of simultaneity defined in the aforementioned way becomes invariant. Thus, when considered in isolation, the proposition asserting that two events are absolutely simultaneous with each other is verifiable. But such a proposition is not verifiable jointly with other propositions within a theory that holds that there is an upper limit to the speed of all physical signals.

The relativity of physical space, often asserted on epistemological grounds, gives rise to similar remarks. It is often contended that the question as to whether two non-simultaneous events have occurred at the same place has no empirical meaning, unless the frame of reference with regard to which the spatial coincidence of the two events is supposed to take place is specified. I can verify whether I budged from my place with regard to the walls of my room or with regard to the sun; but a proposition asserting merely that I did not budge, without any frame of reference such as the walls of my room being mentioned, seems unverifiable in principle.

Yet this is once more a case of joint unverifiability of an intrinsically verifiable proposition within a theory. To realize the intrinsic verifiability of absolute location, we can assume provisionally that space is spherical and that light travels along geodesics in such a space. To make sure that an object did not budge "absolutely," that is, independently of any frame of reference, it would then suffice to notice that a light beam emitted by this object has rejoined it after the determinate lapse of time which light requires in order to "go around a spherical world." Thus absolute spatial coincidence is intrinsically verifiable and comes to be unverifiable only within a theory which rejects the above assumptions about the nature of space and the propagation of light.

35   VERIFIABLE THEORIES BASED ON UNVERIFIABLE POSTULATES

We shall now discuss the question concerning the legitimacy of basing verifiable theories on unverifiable assumptions. Such a procedure is unacceptable from the point of view of the verifiability theory of meaning. We have already pointed out that the fact that unverifiable assumptions play an important, or even an indispensable, part in several fundamental scientific theories constitutes the strongest argument against this theory of meaning. However, in this section, we shall not be concerned with the identification of verifiability and meaningfulness which makes the role of unverifiable assumptions in science utterly unintelligible. We shall rather examine the misgivings that the presence of such assumptions in fundamental theories seems to have aroused in some important philosophies of science which do not share the verificationalist view on meaning, such as Poincaré's conventionalism and the Cambridge neo-realism (Whitehead, Russell, Broad, Nicod, etc.).

Let us start with a historical remark concerning a striking change in the epistemological attitude towards fundamental assumptions in the natural sciences which took place in the last decades. In the nineteenth century, the hypotheses concerning the molecular and atomic structure of matter bore the brunt of positivistic and related attacks on the "metaphysical elements of science," that is, on scientific assumptions transcending every possible observational check. On the other hand infinitist concepts, for example, those implicit in the use of differential equations for describing the observable behaviour of observable entities, were not considered objectionable. Systems of differential equations interrelating changes of measurable quantities have even acquired the status of a standard "positive" theory. Thus the analytic theory of heat phenomena developed by Fourier without recourse to hypotheses about any molecular or atomic structures underlying such phenomena has profoundly impressed Comte and has been regarded ever since as a model of an empirical theory, whereas atomic and molecular assumptions were frowned upon as being unverifiable.

At present, science has gone beyond both the molecular and the atomic levels and seems to concentrate rather on the sub-atomic level of elementary particles and nuclear transformations. The problem concerning the "physical reality" of molecules, which is tantamount to the question of the empirical verifiability of hypotheses concerning molecules, is no longer the subject-matter of sceptical discussions. After having succeeded in measuring the dimensions of a molecule, in determining the spatial arrangement of atoms within the molecule, in effectively, though indirectly, observing the molecule by means of ultra-microscopes and electronic microscopes, the scientist can hardly be expected to make any difference between the "physical reality" of the microscopic and the molecular level.

It is true that at the atomic and the sub-atomic level, the situation is more complicated in view of the inevitable and appreciable interaction between the observational set-up and the observed object, which limits to a certain extent the objectifiability of certain quantum-theoretical concepts. Yet the joint verifiability of quite a few fundamental hypotheses within the Quantum Theory of atomic and sub-atomic phenomena is not affected by these complications. In principle, an atom is as indirectly observable by means of a gamma-ray microscope as a microbe is observable through an optical microscope, a table with the naked eye, and a star through a telescope. The problem of "atomic reality," so impassioned at the turn of the century, would certainly have come to a head at the start of this atomic era if the empirical verifiability of basic hypotheses concerning atoms had not been definitively established by improved observational facilities.

On the other hand, the epistemological difficulties implicit in the use of infinitist concepts and assumptions within empirical science seem more conspicuous today than ever, partly, perhaps, because of the notable progress in the logical foundations of infinitist methods which has disclosed their complexity and paradoxical nature. At bottom, these infinitist difficulties do not differ from those which have always been associated with the applicability of Euclidean geometry to observable objects. Since we never observe geometrical points, straight lines, planes, or any other of the idealized entities which constitute the only subject-matter of this science, how is it possible that geometry can be so effectively applicable to observable objects?[12] The situation is essentially similar in any other "exact" science. In kinematics, which forms, at bottom, a fragment of space-time geometry, the concepts of a mobile point, of uniform motion, of acceleration, of frame of reference, etc., give rise to the same questions. In physics, the examples are legion: material points, rigid bodies, incompressible liquids, ideal gases, isolated systems, reversible processes—all such concepts serve to form propositions the verifiability of which seems highly doubtful.

This is the case whenever the mathematical language of infinitesimal analysis is resorted to. Yet this is the language used in the majority of the basic theories of science: mechanics, thermodynamics, electromagnetism, relativity, and Quantum Theory are based, each of them, on a small number of differential or partial differential equations which, when interpreted literally, transcend empirical verification.

[12]In terms of verifiability as defined in the preceding sections, the situation may be described as follows: since points, straight lines, planes, and similar entities are not directly observable objects, the geometrical propositions concerned with such entities are not observation-statements. If verifiable at all, geometrical propositions must be deducible from specifiable sets of observation-statements about directly observable objects. The difficulty consists in specifying the set of observation-statements from which any pre-assigned geometrical proposition can be deduced. The only serious effort in this direction is due to A. N. Whitehead and his followers. Cf. *infra*.

There are several ways of accounting for the presence of such unverifiable assumptions in empirical science without either rejecting them
as meaningless, according to some versions of the verificationalist
dogma, or relapsing into an obsolete rationalism which confers the rank
of *a priori* propositions upon the unverifiable assumptions of science:

(1) The verifiability of the fundamental assumptions of science
might be secured by suitably reinterpreting their crucial concepts. This
solution has been advocated mainly by the neo-realist philosophy of
science represented by Broad, Russell, and Whitehead, and centring
around Whitehead's Method of Extensive Abstraction.

(2) These assumptions might be held to be empirically unverifiable,
and yet unobjectionable in empirical science, since they actually play
the role of expedient *conventions*. This is the solution originally put
forward by Poincaré with regard to geometry and mechanics, and then
variously extended by his followers to several branches of science—
sometimes to all law-finding sciences (Le Roy, Ajdukiewicz); sometimes, strangely enough, to particular scientific facts (Popper);
sometimes to metrical (quantitative) theories in contradistinction to
topological (ordinal) ones which were granted the status of genuine,
not merely conventional, knowledge (Reichenbach).

(3) These assumptions are empirically unverifiable and hence do
not constitute items of scientific knowledge in their own right: they are
neither true nor false. Science is not committed either to asserting them
or to denying them and only accepts the sum-total of their verifiable consequences. This is the solution we shall adopt later in trying to account
for the indispensable role of such assumptions in fundamental scientific
theories.

The basic tendency in the formation of scientific concepts is to attain
the maximum simplicity of scientific laws regardless of the complexity
of their constituent concepts. The principles underlying classical mechanics, Einstein's theory of gravitation, and quantum mechanics can each
be squeezed into a single line. The unsurpassable conciseness of these
principles is offset, in a sense, by the lengthy and complicated inferential
operations required to connect them with direct observational data. The
question is whether these operations can be performed in such a way as
to ensure the empirical verifiability of the principles themselves.

The British authors just mentioned have concentrated their efforts on
an analysis of the spatio-temporal concepts used in science. The answer
they provide consists in a system of definitions which attempt to connect the refined scientific concepts related to space and time with the

relevant observational data. Thus, instead of literally interpreted geometrical propositions about an unobservable extensionless point, A. N. Whitehead considers an infinite convergent sequence of finite solids, all of which are observable in principle and bear to each other the spatial relation of part and whole; this relation may still be considered observable, when it obtains among observable parts and wholes. The unobservable geometrical point could thus be replaced by an infinite sequence of observable solids standing to each other in an observable spatial relation, viz., the very sequence which, in ordinary parlance, would be said to have only this point in common. We need not enter into the technical details of this method, which actually requires the replacement of a point by a class of infinite sequences of finite and indefinitely decreasing solids rather than by a single sequence of this kind and construes spatial concepts as limiting cases of spatio-temporal ones. The prohibitively complicated definitions of the simplest geometrical concepts provided by the Method of Extensive Abstraction are neither surprising nor fatal. The method replaces statements about points and similar unobservable entities with equivalent statements about convergent sequences of observable solids because it aims at replacing the literally unverifiable statements about points by inductively verifiable statements about certain complicated sequences of solids. The effect of substituting such complex entities for single geometrical points is that geometrical statements, though finitely unverifiable, acquire inductive verifiability. The objective of securing the verifiability of the basic assumptions of mathematical theories involving spatio-temporal "idealizations" would, nevertheless, be obtained, if this infinitist difficulty were the only one involved.

The trouble is that the Method of Extensive Abstraction ensures only the intrinsic verifiability of geometrical propositions. But we know that this is not enough. In addition to intrinsic verifiability, joint verifiability within the cluster of basic theories of science is required. Yet it is clear that these basic theories condemn the observable indefinite divisibility of space and time presupposed by the Method of Extensive Abstraction. Thus an infinite convergent sequence of indefinitely decreasing solids could be observed in principle by utilizing a microscope whose magnifying power can be indefinitely increased. Not only is such a magnifying power lacking in any microscope which has so far been manufactured, but a microscope of this kind will never be set up because its structure is incompatible with the basic assumptions of well-established scientific theories. Thus the geometrical propositions reinterpreted according to the Method of Extensive Abstraction, though intrinsically verifiable, fail to meet the requirement of joint verifiability within the context of well-established theories.[13]

We see that the real difficulty does not reside in the infinitist nature of geometrical and physical concepts the definition of which, in terms of direct observational data, would require some limiting process. Such infinitude need not prevent the intrinsic verifiability of the relevant scientific propositions, as the neo-realist construction actually shows. The essential obstacle is that well-established theories prohibit the limiting process required to link up the reinterpreted propositions with the observational data.

More particularly, the differential form of the fundamental laws of contemporary science does not automatically preclude their inductive verifiability. The free fall of bodies in a gravitational field may be described either by the differential equation $d^2s/dt^2 = g$, or by its general integral $s = \frac{1}{2} gt^2 + c_1 t + c_2$. The two descriptions are exactly equivalent to each other and hence they are either both verifiable or both unverifiable. The differential form of the first description does not matter. The trouble is not connected with the differential form of the first equation, but rather with the existence of laws of nature precluding the absolutely precise spatio-temporal measurements which would be required in order to verify the differential or integral laws of free fall.

The Method of Extensive Abstraction tries to secure the positive verifiability of basic scientific assumptions and fails to attain this objective because joint verifiability cannot be attained in this way. It is important to realize that the negative verification of several fundamental scientific assumptions is not obtainable either. This fact has been discovered and stressed by the conventionalist analysis of science.[14] The

---

[13]A simpler example is afforded by the concept of electrical field-strength whose usual definition is exactly one which might be suggested by Whitehead's method. It is usually said that the field-strength at a given point is equal to the limit to which the force exerted on a tiny electrical charge placed at this point would tend, on the assumption that the value of this charge decreases indefinitely. On such a definition, the local value of an electric field becomes intrinsically verifiable. But it fails to meet the requirement of joint verifiability, since, according to contemporary physical theory, an electric charge can never decrase below a finite value, for example, the charge of an electron. The proposition ascribing a definite field-strength to a point of an electric field and similar assumptions of the theory of electro-magnetic phenomena are therefore unverifiable in the context of contemporary atomic theories.

[14]H. Poincaré has also discussed important scientific hypotheses which are intrinsically incapable of a positive verification, though their infinitist nature is not responsible for the trouble. Thus, according to him, the verification of the laws of motion would require that a particular clock keep time correctly (say

relevant results of foundational research will be summarized in order to illustrate the difference between the conventionalist interpretation of the unverifiable assumptions of science and the view which assigns to such assumptions the role of constituents of external axiomatic bases of verifiable scientific theories.

Thus, in the case of Euclidean geometry, Poincaré's classical discussions tend to show that its basic assumptions could not possibly be refuted by any observational data, since in order to check on these assumptions, one has to measure lengths, angles, areas, volumes, and similar geometrical quantities. These measurements require mechanical and optical instruments the use of which is based on the theories of mechanics and optics, and these, in turn, assume the validity of geometry. Thus, we shall have submitted to an observational check a combination $G + P$ of geometrical assumptions G supplemented by physical (mechanical and optical) assumptions P, instead of verifying the isolated geometrical set G. Only the sum-total of the assumptions G and P has verifiable consequences and is capable of being refuted by observational data, should some of these consequences turn out to be at variance with the results of actual observations.

Other examples of hypotheses which are intrinsically unverifiable when considered in isolation, but do lead to definite observational predictions in conjunction with other appropriate hypotheses have been analysed by Duhem. His discussions are of considerable historical interest since they seem mainly responsible for the widely held view that no scientific assumption is separately verifiable and that only the whole system of science is susceptible of being confronted with observational data. I think that such an extreme view of the relationship between scientific theory and observation is untenable and comes from confusing the language and the meta-language of science (cf. § 15). However, logically speaking, the fact that there are hypotheses which are intrinsic-

---

the earth, as a uniformly rotating body). This hypothesis concerning the correctness of a clock could not be checked, according to Poincaré, without recourse to the very laws of mechanics which the clock is required to verify. In other words, the positive verification of the hypothesis H_1 related to the laws of motion presupposes the use of a hypothesis H_2 (concerning the behaviour of a clock) in addition to some observation-statements O_1, whereas the hypothesis H_2 could not, in turn, be substantiated without adding to some other set of observational data O_2 the hypothesis H_1. It goes without saying that should this analysis of Poincaré's be correct then neither H_1 nor H_2 would be verifiable, since neither of them follows from a set of observation-statements (in each case, the set of observation-statements would have to be supplemented by an additional non-analytic premiss). This, however, would not prevent us from including both H_1 and H_2 in the external basis of mechanics.

ally unverifiable but lead, in conjunction with other hypotheses, to verifiable consequences, is rather trivial: If H_1 is unverifiable and H_2 is
verifiable, then H_1 alone may have no observational consequences,
whereas the combination "H_1 implies H_2" and "H_1," obviously has the
verifiable consequence H_2. This does not detract from the import of
Duhem's discussions which tend to show that a similar logical situation
prevails in fundamental scientific theories.

The presence of empirically irrefutable hypotheses in empirical science
is thus undeniable. The only question is whether the conventionalist
interpretation of this fact is correct. Since, according to the above
analysis, a combination G plus P of geometrical and physical hypotheses
may have verifiable consequences, whereas the two components G and
P are each of them intrinsically unverifiable, science could always afford,
according to Poincaré, to retain one of these components and modify
the other one, if their combination is at variance with observational
data. Only reasons of expediency and convenience induce the scientist
to keep one of these components unchanged. As Poincaré saw it, the
scientist would not only retain the geometrical component but would
even make it safe and impregnable to any observational attack by promoting the geometrical assumptions to the rank of "principles," that is,
of analytic conventions.

Clearly, the motives of expediency or usefulness which may prompt
the scientist to take one course or another in adjusting his geometrical
and physical assumptions to observational data are beyond a purely epistemological analysis. It is quite possible that, the theory G plus P being
refuted on observational grounds, the scientist will prefer for extra-
logical reasons to construct a new theory G plus P', in which only the
physical component will have been replaced with a new one. Any other
theory, either of the form G' plus P, or even G' plus P' would also be
acceptable, provided it fit the observational data. As a matter of fact,
the last course is the one taken by Einstein in his General Relativity,
where both components are modified in such a way as to bring about
a better agreement with observation.

Yet conventionalism is not just an historical hypothesis about the
motives which might have induced the scientist to cope with recalcitrant
observational data by modifying the old theory in one way or another,
especially if the assumptions of the theory are intrinsically unverifiable.
It may also be viewed in this context as an epistemological doctrine
concerning the "admissibility" of scientific theories. The basic postulate
of this doctrine would then consist in requiring that whenever a scientific theory involves unverifiable assumptions, these assumptions are to

be construed as conventions, or analytic propositions. On this interpretation, conventionalism would amount to asserting that the positivistic programme of freeing science from unverifiable assumptions can be carried out by a simple change of terminology, for that is what the promotion of synthetic unverifiable assumptions to analytic rank actually amounts to.

Before taking up any attitude with regard to this position, we shall ask, of course, whether the elimination of unverifiable assumptions should be accomplished by granting analytic rank to every unverifiable assumption, or whether such rank must be reserved for only as many assumptions as will prove necessary to ensure the verifiability of the remaining assumptions. The example of physical geometry shows that the first alternative is unacceptable: if all the separately unverifiable assumptions of both G and P were to be given analytic rank, we would be left with no synthetic assumptions at all, and the whole theory would cease to be an empirical one. On the other hand, if only some unverifiable assumptions were turned into analytic conventions so that the remaining assumptions become thereby verifiable, we would have to condemn altogether theories with external bases, and physical geometry in particular. Indeed, if we render analytic the component G, we may secure the negative verifiability of P. Similarly, by following a suggestion of Reichenbach's, we might attribute analyticity to P and thus make G negatively verifiable. However, in neither case would we rid ourselves of separately unverifiable assumptions: for example, in Reichenbach's case, we would insure the global refutability of G, but single geometrical axioms would still remain separately unverifiable. Now since we are prepared to put up with the presence of separately unverifiable hypotheses within G, why should we not tolerate the unverifiability of the whole component G (or P, as the case may be), within the theory constituted by the verifiable consequences of $G + P$?

Let us conclude that the conventionalist postulate of so replacing some separately unverifiable assumptions of any verifiable theory with analytic quasi-definitional statements or conventions that the remaining assumptions of the theory thereby become verifiable, is as unsatisfactory as the alternative suggestion to replace all separately unverifiable assumptions by conventions. In other words, we cannot get rid of unverifiable assumptions in empirical science simply by changing its terminology.

CHAPTER THREE

The Limits of Science and the Principle of Verifiability

---

36 A SURVEY OF CONCEPTS OF VERIFIABILITY

IN THE PRECEDING CHAPTER, the verifiability of statements and of theories has been defined in terms of observation-statements and of entailment. According to these definitions, a statement S is verifiable if either S or its negation is entailed (with certainty or probability) by a consistent set of observation-statements. A theory is verifiable if it is made up of all the verifiable statements which are entailed by a finite consistent set of assumptions. So defined, the verifiability of statements and theories has proved to indicate the range of all the basic well-established methods of science: all problems to which such methods can be applied are solvable by verifiable statements or theories. As a matter of fact, the extension of the concept of verifiability just referred to exceeds by far the range of presently known scientific methods. It might thus be possible to utilize this concept in order to delimit all problems which science could possibly solve, over and above those it can now solve by applying presently known methods. However, the requirement of accounting for the range of presently well-established scientific methods does not uniquely determine the concept of verifiability; accordingly, the inherent range of science might be delimited in various ways, depending upon the choice of the concept of verifiability. This is why an additional criterion is needed for the making of this choice. Such a criterion will be derived from the need to prove the Principle of Verifiability; in other words, the concept of verifiability will have to be so defined as to ensure both that problems accessible to well-established scientific methods are solvable by verifiable statements or theories, and that demonstrably unverifiable statements are neither true nor false. Needless to say, the task of suitably defining verifiability is different from that of defining truth; verifiability is a technical term used in epistemological investigations, and there is no point in trying to reproduce its actual meaning in ordinary discourse. The real task is that of selecting a definition capable of accounting for the present range of scientific methods and of showing clearly the connection between the verifiability of statements and the definiteness of their truth-value.

As a matter of fact, several definitions involving distinct concepts of verifiability have been put forward by different writers. We shall have to ascertain which of these concepts is best suited for our purpose. Should the verifiability of a statement be conceived, for example, as consisting in the logical possibility of confirming it observationally, or as involving, in addition, the empirical possibility of so doing? Should the Principle of Verifiability require that a statement be "strongly" (conclusively) verifiable, or is "weak" (inconclusive, confirmatory) verifiability all that is required? Is indirect inferential verification admissible along with the direct confrontation of a statement with the relevant observational data? Should any verification involve at most a finite number of relevant observational data, or is inductive as opposed to finite, verifiability also admissible? Is public verifiability mandatory, or is private verifiability acceptable? And so forth.

We shall have to characterize the main differences between the various *concepts* of verifiability, as well as comment briefly on the questions connected with formally correct *definitions* of these concepts. The multiplicity of concepts of verifiability raises such broad epistemological issues as those of the existence of indubitable knowledge (compare conclusive and confirmatory verifiability), of communicable knowledge (private *versus* public verifiability), of the epistemological status of the laws of nature (finite *versus* inductive verifiability), etc. On the other hand, the task of framing formally correct definitions of verifiability has so far proved rather technical and refractory.[1] The most elaborate attempts raise serious difficulties, because the definitions turn out to be either too comprehensive, or too narrow, or both. This applies in particular to the definitions of Ayer[2] and Carnap.[3]

We may start a survey of the concepts of verifiability by clarifying the distinction between the "logical" and the "empirical" possibility of verification. To verify a sentence is to come to believe it or to disbelieve

[1]Cf. C. G. Hempel, "Problems and Changes in the Empiricist Criterion of Meaning," *Revue internationale de philosophie*, 1950.

[2]Cf. A. Church, review of "Language, Truth and Logic," *Journal of Symbolic Logic*, 1949, pp. 52–3, and D. J. O'Connor, "Some Consequences of Professor A. J. Ayer's Verification Principle," *Analysis*, 1949/50, pp. 67–76.

[3]Cf. H. Mehlberg, *Science et positivisme* (1948), pp. 31–2.

it under circumstances which guarantee or render reasonably probable the correctness of the decision reached. The attitude of belief or disbelief which results from the verification is correct if a true statement comes to be believed, or a false one to be disbelieved, and is incorrect, when either of the remaining two possibilities materializes. The verificatory operations performed in order to bring about circumstances which would guarantee or render probable the correctness of the decision to be reached as to the truth-value of the verified statement, may consist, for example, in looking at the object or objects the sentence is about, in trying to remember these objects, in applying some measuring instruments to them, in trying to deduce the sentence from other sentences, and so on. Thus, in order to verify the statement "This book weighs 1 lb." we may put the book on the pan of a balance and ascertain whether it remains in equilibrium when a 1-lb. weight is put on the other scale. The verificatory operations include in this case the bringing about of the physical situation which consists in the respective locations of the book and the weight, the observation of the final outcome of the arrangement, and the drawing of the relevant conclusions from the ostensive statements describing this final outcome.

Had these verificatory operations not been carried out, the statement would not have been verified. But it would still remain verifiable, because the verificatory operations, even if unperformed, are at least performable. In other words, a statement is verifiable if it is possible to carry out operations which would guarantee or render probable the correctness of the decision concerning its truth-value, reached as a result of these operations.

The concept of possibility intervenes twice in this vital passage from actual to possible verification. An actual verification of the statement consists in bringing about a physical situation describable in a set of ostensive statements from which the verified statement or its negation is actually inferred. Consequently, a possible verification requires (1) that it be possible to bring about the relevant physical situation and (2) that it be possible to infer from the statements describing this situation either the statement to be verified or its negation. In other words, a statement is verifiable if it is possible to infer it or its negation from ostensive statements describing a possible physical situation. A statement will therefore be unverifiable if one at least of these possibilities is not available.

For example, the statement to the effect that this book has incommensurable edges is unverifiable, because in order to verify it, we would have to ascertain that the edges are not multiples of any length however small, and this would involve the use of a microscope whose magnifying power could be indefinitely increased or some other similar device providing for measurements of arbitrarily small bodies. It is well known that the construction of such instruments (and consequently, physical situations which include their presence) is impossible because it is precluded by well-established laws of nature. Similarly, the statement "This book is a manifestation of the absolute" is unverifiable, because it is impossible to infer this statement or its negation from any set of ostensive statements, describing any conceivable physical situation.

There is an important difference between the respective reasons which prevent these two statements from being verified. The statement about the revelatory nature of the book is unverifiable because it could not possibly be inferred from any set of ostensive statements, for the simple reason that it does not follow from any such set of statements. On the other hand, the statement concerning the incommensurability of the edges of the book is unverifiable because every set of ostensive statements from which it follows describes an impossible physical situation. The impossibility of verifying the first statement may be termed a logical one since it is grounded in the purely logical fact of a lack of entailment between the unverifiable statement and any set of ostensive statements. The impossibility of verifying the statement concerning the incommensurability of the edges may be termed an empirical one, because it comes down to the logically contingent fact that the relevant sets of ostensive statements are incompatible with the conceptually universal laws of nature, that is, with true synthetic statements involving neither proper names nor local spatio-temporal constants (cf. § 19). The existence of an upper limit to the magnifying power of any microscope, responsible for the unverifiability of the latter statement, is such a logically contingent law of nature.

Statements unverifiable on empirical grounds may therefore be verifiable on logical grounds, and the requirement of a logical possibility of verification, that is, of being a consequence of some consistent set of ostensive statements, is less stringent than the requirement of an empirical possibility of verification, that is, of being a consequence of a set of ostensive statements compatible with the univeral laws of nature. In philosophical discussions concerning the legitimacy of metaphysical speculation, the less stringent requirement of verifiability on logical grounds is ordinarily involved. In scientific thought, the more stringent requirement of an empirical possibility of verification has played a decisive role and has come to be embedded in the methodological outlook of present-day science.[4]

Another distinction concerning verifiability is based on the kinds of verificatory operations considered essential. A statement is said to be directly verified if the operations leading to its verification do not include the previous verification of any statement. A verification is indirect or inferential if the set of operations it consists of includes the verification of at least one other sentence. For instance, in order to verify that the letter "o" occurs twice in the word "book," it suffices to look at this word. The only operation involved is that of looking at the object the statement is about; no verification of any other statement is required.[5] The situation is different when the statement that the letter "o" occurs an even number of times in the word "book" is being verified: one would have to verify first that "o" occurs twice in this word. Thus the statement asserting that "o" occurs an even number of times in "book" is indirectly or inferentially verifiable.

A statement is said to be conclusively or strongly verified if the decision as to its truth-value has been reached as a result of a set of operations which, if repeated by the same person or by different persons on the same statement or on a different statement, would always lead to a correct result. A statement is said to be inconclusively or weakly verified, or confirmed to a degree D, if the set of operations performed in connection with its verification would lead, if repeated a large number of times, in a fraction D of all the cases to correct results.

The two classifications of verificatory operations just referred to are distinct and overlapping.[6] Introspective and sense-data statements are ordinarily considered as directly and conclusively verifiable, whereas statements about perceived material objects and about remembered sense-data are held to be directly and inconclusively verifiable. Indirect verification involving inductive inference is always confirmatory. A deductive inference from introspectively verified premisses would be

[4]Some writers, however, require only that it should be logically possible to verify scientific hypotheses (cf. Braithwaite's discussion of the verifiability of statistical laws; *Scientific Explanation*, pp. 160 ff.).

[5]I take it that the expression "The letter 'o' occurs twice in the word x" forms a single ostensive predicate whose meaning can be conveyed by pointing at appropriate words. The number 2 involved in this predicate must not be analysed separately, after the fashion illustrated in § 30, note 15.

[6]The failure to realize the difference between these two classifications of verificatory operations seems to be mainly responsible for the controversy over the nature and existence of statements called "basic," "incorrigible," "expressive," "protocols," "reports," etc.

considered by most writers as an indirect, conclusive verificatory operation.

The question as to whether there are conclusive verificatory operations at all and, particularly, whether the verification of introspective and of sense-data statements falls under this category is historically important in connection with the "quest for certainty" and very much debated at present. Yet, as far as the justification of the Principle of Verifiability is concerned, the whole controversy is irrelevant. The principle seems to be justifiable regardless of whether some direct verificatory procedures are considered as conclusive, or all are viewed as confirmatory.

Among inferentially verifiable statements a distinction can be drawn which depends upon the number of directly verifiable statements from which the inferential process starts. For example, in the sentence just referred to concerning the evenness of the number of occurrences of a letter in a word, only one directly verifiable statement has been involved. This, of course, is but a limiting case. The important distinction is between sentences which follow from a finite number of directly verifiable statements and those whose complete verification would theoretically require the previous verification of an infinite number of statements.

Whether this piece of sugar would dissolve in this glass of water can be ascertained by a finite number of operations, but not whether every piece of sugar would do so, nor whether every piece of sugar would dissolve in some portion of water. The first quantified statement is finitely disprovable, but not finitely provable; in order to prove it, one would have to ascertain the behaviour of sugar in every portion of water, and there might be an infinite number of such portions. In the last example both proof and refutation would require an infinity of instances. To prove that each piece of sugar would dissolve in some portion of water, one has to examine all the pieces of sugar, which might be infinitely numerous. To prove the contrary, that some piece of sugar is not soluble in any portion of water, one has to examine all the portions of water.

There is thus no doubt as to the presence of finitely unverifiable statements even in the most elementary scientific theories, the essential reason being that quantified statements occurring in such theories often admit an infinite number of instances.[7] Nor is there any difficulty in defining

[7]The fact that universal statements are incapable of finite verification has been considered scandalous since Ramsay, because the fundamental laws of nature fall under this category, and would therefore have to be condemned as meaningless if finite verifiability were accepted as a criterion of meaning. Few writers, however, have been prepared to share this heroic condemnatory conclusion. Some

verifiability in such a way that it will be ascribable to statements which
follow from ostensive sentences, regardless of whether the number of
relevant ostensive sentences be finite or infinite. We may assume, for
example, that in contradistinction to finitely verifiable statements, a
statement will be said to be *inductively verifiable*, if all its instances are
finitely verifiable. This would ensure the inductive verifiability of the
three examples just referred to, but might still fail to endow with verifiability other scientific hypotheses which involve more complicated
quantificational procedures. A more general concept of inductive verifiability can be obtained by considering every statement as inductively
verifiable provided all its instances be either finitely or inductively verifiable. This definition is apparently circular, but it can be easily transformed into a correct one (cf. § 34).

The question whether verifiability should be defined so as to admit
only a finite number of ostensive premises, or, more liberally, so as to
include inductive verifiability as well, does not make much sense by
itself. The concepts of finite and inductive verifiability are distinct and
both legitimate. Our problem will consist in making out whether finite
unverifiability of itself entails indeterminacy, or whether only inductive
unverifiability can do so.

We shall interpret the other alternatives similarly. The concepts of
unilateral and bilateral verifiability (cf. § 28), of conclusive and confirmatory verifiability, of direct *versus* inferential verifiability, of verifiability involving a logical possibility *versus* one involving empirical
possibility, are all distinct and legitimate. The real problem is not one of
choosing the right definition of verifiability, but rather of ascertaining
what kind of verifiability of statements is a necessary condition for the
definiteness of their truth-value.

37   UNVERIFIABILITY AND VAGUENESS

While discussing the concept of truth, we have noticed that the presence of a vague term in a statement may prevent it from having any
definite truth-value, if it so happens that the hypothetical truth-value of

---

have stressed that universal statements are finitely refutable and have proclaimed
finite refutability as a criterion of meaning. The trouble with this solution is that
existential and mixed quantified statements would then have to be considered as
unverifiable (i.e., "meaningless" or "metaphysical" as the case may be), and this
would hardly be satisfactory. That this piece of sugar is soluble in some glass
of water is hardly more "metaphysical" than the solubility of this piece of sugar
in any glass of water. The fact is that the quantified empirical sentences are sometimes universal, sometimes existential, and sometimes mixed propositions. Giving
up any of these types of statements would unjustifiably curtail science.

this statement would not remain constant under all the admissible interpretations of this term. The mere presence of the vague term need not bring about the indeterminacy of the relevant statement, and it is rather a logically contingent fact that such indeterminacy does materialize on some particular occasion. The Principle of Verifiability asserts, however, that no synthetic unverifiable statement has a definite truth-value; the indeterminacy involved in this principle is therefore logically essential, not accidental. Since the indeterminacy of statements seems always to come from the vagueness of their constituent terms, the vagueness responsible for the logically essential indeterminacy of unverifiable statements must be of a particular kind. I think that this is actually the case and that the predicates occurring in unverifiable statements are not merely vague; they are completely vague. We shall therefore have to investigate the conditions under which predicates of fact-like or law-like statements come to be completely vague, and thereby establish a connection between the unverifiability of these statements and their logically essential indeterminacy.

A predicate is completely vague if its correct use is compatible with both its applicability and its inapplicability to any pre-assigned individual. In other words, the vagueness-area of a completely vague predicate is the class of all individuals. Both its precision-areas are empty. Thus, as will be shown later, the predicate "being the absolute" is governed by rules of linguistic behaviour which preclude neither its applicability nor its inapplicability to any observable individual. Accordingly, we shall have to consider this predicate as being completely vague.

It should be noted that by choosing this terminology, we are stretching the usual range of "vagueness." Completely vague terms in the sense just defined would not be called "vague" without further ado in ordinary discourse. We are induced to adopt this modified terminology because of the basic similarity between "vagueness" as defined by us and ordinary, incomplete, denotational vagueness. The latter is a result of the lack of linguistic rules precluding either the applicability or the inapplicability of an admittedly vague term to certain individual cases. Completely vague terms differ only with regard to the range of admissible interpretations, which is broader than for ordinarily vague terms, but still due to the same cause, namely the lack of restrictive linguistic rules.

Similarly, vagueness, as ordinarily understood, is not ascribed to quantitative scientific concepts, though both their applicability and their inapplicability to certain individuals may prove compatible with all the relevant rules. For example, the concept of temperature as defined by a specified thermometric substance is not completely vague. Its precision-areas include at least those material objects which are large enough not to prevent in principle the performance of the relevant measurements. When, however, it is applied to microscopic or sub-microscopic objects, it becomes increasingly vague. *Any* temperature may be ascribed to an electron, an atom, a molecule, without ever clashing with the rules governing the use of this term. Thus a quantitative concept may be as vague as a qualitative one, provided vagueness be defined as multiplicity of admissible interpretations.

It goes without saying that the whole argument would remain unchanged if instead of using "vague" in this generalized sense, we introduced another expression, say "at most partly definable." We would then acknowledge that we generalize the meaning of "vagueness" in dealing with partly definable quantitative concepts, which we also speak of as "incompletely vague," and in dealing with completely indefinable concepts, which we speak of as "completely vague." Only partial definability of qualitative concepts is coextensive with "vagueness" as ordinarily understood.

The fact that non-ostensive terms depend on *analytic* contexts in order to acquire any denotation whatsoever, in contradistinction to ostensive terms, which owe their denotational function to their occurrence in *ostensive* statements, has important consequences for the degree of vagueness of these two kinds of terms. Ostensive terms are always partly vague, whereas a non-ostensive term may happen to be completely vague if the analytic contexts available for it fail to put any restriction upon its admissible interpretations.

This is certainly the case when a non-ostensive term has no analytic context at all, for example, Kant's "thing in itself." *Any* interpretation of this term is admissible, since none is inconsistent with its non-existent analytic contexts. By asserting, for example, that this book is a thing in itself, I would not violate any of the rules of correct linguistic usage concerning this term. I would also be safe if I were to deny to the book the status of a thing in itself. The point is that the analytic contexts of a non-ostensive term are the only imaginable clue to what it refers to, that is, to what those who use it correctly are talking about. If no analytic context is available for a non-ostensive term, then no situation can ever arise where the correct use of the term would be prescribed by precise rules. Its use is, in this sense, always arbitrary.

For similar reasons, a non-ostensive term will still be completely vague, even if it does occur in an analytic context or in a consistent set of analytic contexts, provided that this context contain no other term with an independently secured denotation. An axiomatic system such as Spinoza's involving only non-ostensive terms—"substance," "attribute," "mode,"—cannot secure by itself any denotation for these terms. Consequently, they are all completely vague, in spite of their occurring in all the axioms and definitions.[8]

The last consideration can be supported by several logical results concerning the inescapable multiplicity of admissible interpretations of consistent axiomatic systems. Thus the Loewenheim-Skolem Theorem, already referred to on a previous occasion, implies that any consistent axiomatic system the formulation of which involves a single type of variable—as a matter of fact, all important axiomatic systems are known to be susceptible of such a formulation—can be interpreted in terms of natural numbers. The Lindenbaum-Tarski Theorem (cf. § 12) asserts that whenever a system can be interpreted in such a way as to hold true of a certain collection of individuals and of some abstract entities based

[8]We should perhaps add here that complete vagueness can be understood in different ways, when the vague expression occurs in several analytic contexts:

(*a*) A predicate P may be said to be *completely vague in the individual sense* if neither P nor its negative denote any individual.

(*b*) A predicate is *completely vague in extension* if it is admissibly interpretable as applying to any class of individuals. These two concepts of complete vagueness are not coextensive. There may be, for instance, no analytic statement preventing one from considering any pre-assigned individual as being the absolute. However, if some analytic statement specified that there is only one absolute, the rules of correct linguistic usage would thereby forbid one to confer the rank of the absolute on two distinct individuals; yet any single individual would remain eligible according to these rules. Thus a predicate which is completely vague in the individual sense may not be so in extension.

(*c*) A third concept of complete vagueness may be defined by considering as admissible only those interpretations of a term which are consistent with some over-all interpretation of the language which ascribes a denotation to each of its terms. Suppose, for example, that nothing is laid down axiomatically about the absolute and the thing in itself, or about the number of objects eligible to either rank, except that only one object can be both absolute and noumenal. If such an analytic context were available, the absolute would be completely vague both in the individual sense and in extension, but not in the over-all sense. For, although it would be admissible to promote two objects to noumenal rank, the status of absolute could not be granted to both of them, without conflicting with the linguistic rules.

For the purpose of this discussion, it will suffice to determine with some precision the conditions under which a term becomes completely vague in the individual sense. The other two kinds of complete vagueness are of interest mainly in connection with elaborate sets of axioms which, in the absence of any ostensive terms, do not prevent their concepts from being completely vague in the individual sense, although they put rather stringent limitations on their over-all interpretations. This is the meaning of Gergonne's often repeated view that an axiomatic system is an implicit definition of its undefined terms. This is quite compatible with such terms being completely vague in the individual sense.

on these individuals (classes of these individuals, relations between them, etc.), then any set of abstract entities isomorphic with the former will still be described by the same system of axioms. In other words, if we were to re-name all the individuals by re-distributing their former proper names among them in any way whatsoever, we would still obtain an over-all interpretation of the system as admissible as the former interpretation, provided all the predicates were re-interpreted in a suitable way. Thus the question as to whether a single predicate does apply to a single individual is never settled by the system, and under these circumstances each of its predicates is completely vague.

This also shows that, even though it illustrates the intrinsic denotational vagueness of axiomatic theories graphically, the Lindenbaum-Tarski Isomorphism Theorem supports the structuralist or relationalist view of scientific subject-matter only to a limited extent (cf. § 12). If we re-name the individuals of the universe of discourse of science and re-interpret the remaining scientific concepts (of classes and relations of any degree and level) based on these individuals in such a way as to secure the isomorphism of the new set of entities with the old set referred to in the conceptual apparatus of science prior to this process of re-defining the scientific terminology, then all that is claimed in the Lindenbaum-Tarski Theorem is the *validity* of the re-interpreted scientific theories with regard to the transformed set of entities. This is perfectly compatible with, and actually presupposes, the validity of the literally interpreted theories of the initial set of entities.

In other words, the Isomorphism Theorem assumes that the initial theory T is true on its initial interpretation (which we obviously have to think of as given by an effective list #5 of criteria for a suitable selection of the terms occurring in the theory since otherwise none of these terms would have any denotation whatsoever—cf. § 19) and claims that any other theory T' will also be true provided that T' can be obtained from T by retaining lists ##1,2,3,4 and replacing list #5 with a new list generated by any one-one relation in the universe of discourse of T. The knowledge supplied by the initial theory T is therefore a knowledge of the entire subject-matter of T with all the individuals and abstract entities that go to make up this subject-matter. The subject-matter of T (or, for that matter, of any of its isomorphic transforms) by no means reduces to a mere relational structure.

Let us, therefore, conclude that the Isomorphism Theorem points to the intrinsic denotational vagueness of axiomatic (or axiomatizable) theories presented without appropriate ostensive criteria rather than to a specific subject-matter of theories whose terms are linked up with the ostensive level by a list of effective interpretive rules.

Vagueness of Ostensively Definable Terms

We have seen thus far that in the absence of analytic connections with ostensive terms, non-ostensive terms are bound to be completely vague. Such a connection is therefore a necessary condition for a non-ostensive term to be at least partly precise. "Begriffe ohne Anschauung sind leer." However, it is by no means sufficient to prevent complete vagueness. Non-ostensive terms connected analytically with ostensive ones may still happen to be completely vague, if their analytic contexts fail to meet certain further requirements.

Let us first resume, for illustrative reasons, the example of unit of length. This non-ostensive term acquires a denotation from its analytic connection with the ostensive concepts of superposition and coincidence. By definition, every object on which a yardstick has been superposed with ensuing coincidence is one yard long. By definition equally, no object on which a yardstick has been superposed without ensuing coincidence is one yard long. The concept of unit length owes its partial precision to this analytic context which ensures its applicability to some individuals and prevents it from being applicable to some other individuals.

It is easily seen, however, that the restrictions put by this analytic context on the admissible interpretations of the term "unit length" would become ineffective if we chose for a yardstick a body which could not possibly be superposed on any other object, for example, a body so hot or so distant that any superposition would be ruled out. Although the analytic context connecting unit length with ostensive terms would still be available, it would not prevent this non-ostensive concept from being completely vague. Since all the restrictions on the admissible interpretations referred to objects on which the yardstick was superposed, the term could be either applied or not applied to any other object without conflict with the analytic context in question. Hence, this restriction would become completely ineffective if the yardstick chosen were never superposed on any object. This shows that a direct analytic connection with ostensive terms may fail to secure even partial precision for a non-ostensive term.

The general condition of complete vagueness of non-ostensive terms illustrated in this case can be easily described. It is of great importance in our whole argument, since it can be shown to comprise all cases of

complete vagueness of non-ostensive terms which are connected analytically with the ostensive level. We have called "conditional definition"
the type of analytic connection with ostensive terms which is exemplified
by the concept of length and which may fail to secure, on occasion, even
partial precision. It is obvious, however, that, as a rule, a conditional
definition of a non-ostensive concept P in terms of an ostensive test T
and an ostensive response R does restrict the range of admissible interpretations of P. Such a conditional definition, though incapable of delimiting the whole extension of P, establishes at least a lower and an
upper limit for this extension. By virtue of the conditional definition,
the term P certainly applies to all individuals which respond in the way
R to the test T and is inapplicable to any object which fails so to respond.
Consequently, the extension of P includes all objects which have yielded
the result R in response to the test T and is included in the class of
objects which were not submitted to this test at all or, in response to it,
yielded R.

This shows under what conditions a definitional criterion for the term
P may fail to prevent it from being completely vague. If the test T were
never applied with a positive result, the lower limit for the extension of
P would shrink to the null class, and this is a vacuous condition. Similarly, if the test T were never applied with a negative result, the upper
limit to the extension of P would coincide with the universal class; this
condition is vacuous as well. If the test T were never applied at all, the
presence of the definitional criterion would require only that the extension of P include the null class and be included in the universal
class. This condition is fulfilled by any extension and does not put any
restrictions on the admissible interpretations of P. We may call a definitional criterion, whose test has never been applied, *vacuous*, and, accordingly, we may state that a non-ostensive term whose only analytic context is a vacuous definitional criterion is completely vague.

It goes without saying that, as far as the concept of unit length is
concerned, the above definitional criterion based on superposition is by
no means the only possible one. There are other analytic contexts which
connect the concept of length with terms of the ostensive level and
which would prevent this concept from becoming completely vague even
if we were to choose an inaccessible yardstick. These contexts also
diminish the vagueness when the yardstick is chosen in a more reasonable way. A concept endowed with a non-vacuous definitional criterion
is of use only if further law-like contexts for it are discovered. We have
seen that apart from other services they render, such law-like contexts
provide additional conditional definitions for the concepts they involve;

they thereby secure the applicability of these concepts beyond their initial precision-areas and thus diminish their vagueness. We have also noticed that theories may provide conditional definitions of their own for the concepts they involve; this may happen even if such concepts have no independent conditional definitions and constitute "theoretical constructs." The precision-areas of a concept are jointly determined by all the conditional definitions which are available for it, either explicitly formulated, or implicit in the laws and theories involving this concept.

Conversely, the complete vagueness of a non-ostensive term does not follow necessarily from the vacuousness of its explicit definitional criteria; the definitional role of the laws and theories involving this term must also be taken into account. We may state, however, that a non-ostensive term is completely vague if, and only if, all its definitional criteria are either vacuous or non-existent.[9]

[9]In his paper on "The Methodological Character of Theoretical Concepts," *Minnesota Studies in the Philosophy of Science*, I (1956), R. Carnap has laid down a rule to the effect that a term M occurring in a theory T thereby acquires an empirical meaning relative to T if some purely observational consequence O is derivable from a set of premises which includes the postulates P of the theory T, the interpretive rules R of T as well as a statement $M(a)$ ascribing the property M to an object a. He then gives a second, more general rule stipulating that a term M_n of T acquires empirical meaning if there is in T a finite sequence of terms $M_1, M_2, \ldots, M_n$ such that the meaningfulness of all the terms preceding M_n in the sequence has already been secured by successive applications of the first rule and some observational consequence O happens to follow from a set of premises which includes, in addition to the postulates P and the interpretive rules R of T, a statement $M_n(a)$ ascribing the property M_n to an object a as well as a statement involving all the preceding terms M_1 through M_{n-1}.

Carnap proposes to replace his previous verificationist criterion of meaning (cf. "Testability and Meaning," *Philosophy of Science*, vol. 3, 1936) with the new criterion for the meaningfulness of terms constituted by the above two rules, on the understanding that a statement S in T is meaningful relative to T if all the terms occurring in S fulfil the new criterion of meaning relative to T.

In this investigation, we are trying to determine the range of science in terms of empirical verifiability. For reasons explained in §§ 5, 28, we have refrained from relating the meaning of scientific statements to their verifiability and have considered, instead, the verifiability of a statement as a criterion of the definiteness of its truth-value. It seems, nevertheless, that there is substantial agreement between the conditions under which a non-ostensive term may come to be completely vague and Carnap's new criterion of meaning. For if we tentatively (and not quite accurately) identify those terms in T which have meaning according to Carnap with terms that are not completely vague, then Carnap's criterion comes to make the incomplete vagueness of a term M conditional upon the availability of definitional criteria for M implicit in T.

To show this agreement, it will suffice to consider terms whose empirical meaningfulness is already secured by applying Carnap's first rule. The condition for the meaningfulness of M provided by this rule is obviously equivalent to the requirement that the postulates P and the interpretive rules R of T entail that $M(a)$ implies O. Since the constant a does not occur in $P+R$ this entailment holds not

38   THE INDETERMINACY OF UNVERIFIABLE STATEMENTS

We are now in a position to discuss the connection between the verifiability of a statement and the definiteness of its truth-value. The problem arises only in regard to synthetic statements; according to our definition of truth analytic statements have always a definite truth-value. There are various reasons why synthetic statements may be neither true nor false. It remains to be seen whether unverifiability is one of these reasons.

Let us consider first the bearing of the unverifiability of simple subject-predicate statements about individuals upon their indeterminacy. The verification of the statement "This book is a manifestation of the absolute" has been seen to be impossible on logical grounds, because neither the statement itself nor its denial follows from any consistent set of ostensive statements. Consequently, neither follows from any set of *true* ostensive statements. This implies, according to the definition of denotation in § 31, that the subject denotes individuals within the vagueness-area of the predicate. The statement is therefore neither referentially true nor referentially false according to the improved definition of referential truth in § 30.

Similar considerations apply to subject-predicate statements about individuals which are unverifiable on empirical grounds. Consider once more the statement "This book has incommensurable edges." It is empirically impossible to verify this statement because it is not entailed by any set of ostensive statements compatible with the conceptually universal laws of nature. Hence, it is not entailed by any set of *true* ostensive statements, because a set of true statements is compatible with any other set of true statements, particularly with the conceptually universal laws of nature (defined as true, synthetic statements with no local constants). The vagueness-area of the predicate "having incom-

---

only for a but also for any object x provided that appropriate substitutions be also made in O. Under reasonable assumptions, we can also expect that the term a will occur in O and, consequently, interpret the first rule as follows: M is empirically meaningful if the postulates P and the interpretive rules R of T supply an implicit definitional criterion for M ensuring the negative verifiability of any statement $M(x)$ ascribing the property M to an object x.

On this interpretation (admittedly approximate and involving some stretching) Carnap's new criterion of meaning seems to be in substantial agreement with our conditions of complete vagueness of non-ostensive terms. The only appreciable difference would consist in our additional requirement that in view of the fact that the mere availability of (implicit or explicit) definitional criteria for a term M does not suffice to prevent M from being completely vague, it must also be stipulated that at least one of these criteria is not vacuous.

mensurable edges" therefore includes the referent or referents of the subject, and the whole statement is neither referentially true nor referentially false.

We thus see that the unverifiability of subject-predicate statements about individuals always carries with it their referential indeterminacy, regardless of whether the impossibility of verification is due to logical or to empirical reasons.

Now, according to our definition of truth, only subject-predicate statements about individuals can be referentially true or false. Since, when such statements are unverifiable, they turn out to be neither true nor false, we must conclude that only verifiable statements can be referentially true or referentially false.

Statements of greater complexity than subject-predicate statements about individuals, if true or false at all, must be either inferentially true or inferentially false. In particular, any inferentially true statement must follow from referentially true statements, and hence, according to what has just been said, from a set of verifiable statements; it is therefore verifiable itself, according to the general definition of verifiability in § 34. Similarly, the denial of an inferentially false statement must follow from referentially true sentences, which entails the verifiability of the statement itself. Thus every inferentially true and every inferentially false statement must be verifiable. Since, moreover, every true and every false statement is either referentially or inferentially true or false, we must conclude that all true and false statements are verifiable, regardless of whether their truth-value be referential or inferential. This is the contention of the Principle of Verifiability as reformulated in § 5. This principle applies therefore to any kind of statement covered by our definition of truth: to simple, compound, and quantified statements about individuals, as well as to statements about abstract entities ultimately based upon individuals and to statements about elusive entities.

It is now clear how the vagueness of terms occurring in a statement may establish a link between its unverifiability and its indeterminacy. This is best illustrated by simple subject-predicate statements about individuals.

(a) If the statement "A is P" (say "This book is a manifestation of the absolute") is unverifiable on logical grounds, the predicate P can have no definitional criteria at all. For if there were such a criterion, involving test T and response R, the statement "A is P" would be entailed by the two ostensive statements "A was submitted to test T" and "A has yielded response R." Hence, if such a criterion were available, the statement would be verifiable on logical grounds. Consequently, the logical impossibility of verifying the statement "A is P" carries with it the lack of any definitional criterion for the predicate P. This, in turn, entails the complete vagueness of P according to the general conclusion reached in § 37. Now a subject-predicate statement with a completely vague predicate is necessarily indeterminate. The vagueness-area of its predicate includes all individuals and, in particular, all referents of the subject. The statement is therefore neither referentially true nor referentially false. A similar consideration would show that complete vagueness of the predicate is also responsible for the inferential indeterminacy of the statement under consideration.

(b) If the statement "A is P" (say "This book has incommensurable edges") is unverifiable on empirical grounds, all the definitional criteria of the predicate P must be vacuous. For if there were a non-vacuous criterion for P, with test T and response R, then at least one individual (say B) would have been submitted to test T, either with the result R or with the result non-R. In that case we would have two pairs of ostensive sentences: "B has been submitted to T" and "B has yielded R"; or "B has been submitted to T" and "B did not yield R." One of these pairs of statements would have to be true and, consequently, compatible with the universal laws of nature. Suppose that the first pair of ostensive statements is compatible with the universal laws of nature; then the pair of ostensive sentences which can be obtained from this first pair by replacing the individual name B with the name A will also be compatible with the universal laws of nature. (This follows from the fact that no set of statements in which some individual or local constants have been replaced with other constants can cease to be compatible with the universal laws of nature by virtue of this replacement, since the universal laws of nature contain, by definition, no such constants.) Hence, the pair of ostensive statements "A has been submitted to test T" and "A has yielded result R" would be compatible with the universal laws of nature and, consequently, the statement that "A is P" would be verifiable on empirical grounds, contrary to our hypothesis. We must conclude, therefore, that if the statement "A is P" is unverifiable on empirical grounds, all the definitional criteria of the predicate P are vacuous. This entails the complete vagueness of the predicate P (cf. § 37) and, eventually, the indeterminacy of the statement "A is P."

Versions of the Principle of Verifiability

The argument just outlined also throws some light on the question as to the right choice among the various versions of the Principle of Verifiability which have been put forward by different writers. The answer is to be found by asking another question: which concept of verifiability will make the Principle of Verifiability justifiable?

(1) Since inferentially true and false statements are determinate in spite of being only indirectly verifiable, it goes without saying that both direct and indirect verifiability are admissible, as far as the justification of the principle is concerned.

(2) The liberal version is still the right one in regard to the difference between finite and inductive verification. Even if a universal statement has an infinity of instances, it will still be determined, provided its instances are so (cf. § 36). Hence, inductive verifiability is on a par with finite verifiability as far as the justification of the Principle of Verifiability is concerned. There is no reason for banning universal laws of nature from the cognitive field, even if their complete proof would theoretically require an infinite number of observations. Once the determinacy of finitely unverifiable laws is secured, there remains the question how the truth of any particular law could be rendered reasonably probable by utilizing the finite observational evidence available in this particular case. This problem of inductive logic has been discussed in § 23.

(3) As to the difference between the logical and the empirical possibility of verification, our argument favours the less liberal version. Not only are all statements indeterminate which do not follow from any set of ostensive statements, but statements also lack definite truth-value if they are entailed by sets of ostensive statements incompatible with the universal laws of nature. The scientist is therefore as justified in looking suspiciously at statements which are unverifiable on empirical grounds, as the philosopher is justified in his distrust of speculative contentions which could not be decided upon by any conceivable observational evidence.

(4) The choice between conclusive and inconclusive verifiability is more complex. In this case, too, the liberal version is favoured by our argument. In view of the technicalities involved, it seems wise to confine ourselves to a brief comment on the admissibility of *direct* inconclusive verification.

The determinacy of directly verifiable statements (particularly, of ostensive statements) depends upon the denotation of their subjects and predicates. Previously we said that an ostensive predicate denotes a particular individual, if everybody who perceives this individual and who uses this predicate correctly would apply it to this individual. This definition of ostensive denotation presupposed the conclusive verifiability of ostensive statements. If such statements are only inconclusively verifiable, we shall have to define ostensive denotation in a slightly different way, by stating that the predicate P denotes the individual A if all who use P correctly and perceive A would in all probability apply this predicate to A. There are exceptions: when subject to an illusion or speaking negligently, they may perceive an individual denoted by a given predicate and yet refuse to apply this predicate to this individual.

If ostensive denotation is re-defined in this way, ostensive statements will be only inconclusively verifiable, yet referentially determinate in general. For, if a person who uses correctly the term P applies it to an individual A on perceiving him, the statement "A is P" is in all probability referentially true, since, in all probability, P applies to A. In other words, this statement is true in all probability, if it is made under the circumstances just mentioned, namely, by a person who perceives the object referred to by the terms of the statement and correctly uses these terms. These circumstances render very probable the truth of the statement and constitute therefore a reliable verificatory procedure for it.

For instance, if a person is prepared to apply the adjective "tall" to every dwarf he perceives, he could not possibly be said to use or understand this adjective correctly. Conversely, if he does understand the adjective "tall" correctly, those to whom he applies this term on perceiving them are in all probability tall. Thus the inconclusive verifiability of the statement "A is tall" does not prevent it from being referentially determinate.

To sum up: our argument favours, on the whole, the "liberal" versions of the Principle of Verifiability. A statement may have a definite truth-value regardless of whether the kind of verifiability it possesses is direct or inferential, conclusive or confirmatory, finite or inductive. Nor does it matter whether the verifiability of the statement is positive or negative, unilateral or bilateral, since our general definition of verifiability is non-committal in this respect. The only exception is the issue concerning the logical *versus* the empirical possibility of verification. Statements the verification of which is logically possible will nevertheless be indeterminate if it is empirically impossible to perform the operations required to check on their truth-value.

Status of the Principle of Verifiability

It has often been suggested that the Principle of Verifiability, while declaring synthetic unverifiable statements to be meaningless, may itself be synthetic and unverifiable and thus self-defeating. On the other hand, if it were synthetic and verifiable, very little would be gained, since the supporting evidence so far available is hardly impressive. Such statements, admittedly unverifiable and meaningless, as can be gleaned from the relevant writings (e.g., "the nothing nots" or "the absolute is lazy") are artificial and futile, in contradistinction to the impressive body of unverifiable statements present in empirical science (e.g., single geometrical statements empirically interpreted) the meaninglessness of which is highly questionable. Finally, if the Principle of Verifiability were considered an analytic proposition, beyond the scope of its own applicability, the situation would be hardly more favourable. For, according to the prevailing view of analyticity, the principle would then amount only to an arbitrary definition of meaning. It could not possibly have the importance ascribed to it in connection with basic epistemological problems, such as that of determining the limits of human knowledge, and in particular the reach of science.

The answer provided by the Principle of Verifiability to this last question has been the gist of a new empiricism, more radical than the empiricism of past philosophies. According to this principle, questions transcending possible experience (answerable by synthetic unverifiable propositions) are beyond man's reach, not because of his psychological limitations, as once was thought, but simply because they are unintelligible. The limits of human knowledge would thus be set by a logical analysis of relevant cognitive questions, rather than by a psychological investigation into man's cognitive apparatus. This shift has been felt to be an advance, or at least to provide a new approach to an old philosophical query.

There is little doubt that these serious misgivings are occasioned by interpreting the principle as a criterion of meaning. Is anything changed by our re-formulation? We declare synthetic unverifiable statements to be neither true nor false, but not to be meaningless. This would hardly improve the lot of the principle, if *it* turned out to be synthetic and unverifiable. On the other hand, should it be synthetic and verifiable, we can hardly claim to have adduced any observational evidence in support of the contention that unverifiability entails indeterminacy.

Nevertheless, since the argument involves both a re-formulation and a substantiation of the principle, I think that the situation is actually changed in several respects.

(1) The principle has been derived from definitions of truth, denotation, etc. These definitions are analytic in the sense explained in § 30, and hence, are neither tautological nor arbitrary. Consequently, the principle itself is analytic in this sense. Its substantiation, in the preceding argument, asserts that one who understands the words "true," "denoting," "consequence," and so on, in the way they are actually under-

stood, cannot help coming to the realization that unverifiable statements are neither true nor false.

It should be added that, on this view, only the general Principle of Verifiability is analytic, but each application of it to a particular statement is synthetic and verifiable. It is an empirically verifiable assertion that the statement "This book is the manifestation of the absolute" is unverifiable and indeterminate. As defined in this study, the truth of a statement, its verifiability, the denotation of its terms, and so on, depend, all of them, on the role played by the statement in the patterns of correct linguistic behaviour adopted by its users. Knowledge of the relevant rules of correct linguistic behaviour is certainly empirical. It does not matter that no special observations or experiments are required to ascertain the verifiability or the determinacy of a particular statement, for all the relevant facts are common knowledge. They are just common empirical knowledge.

To put it in other words, the theory of scientific knowledge is largely concerned with the logical analysis of the language used to express such knowledge. If extant scientific knowledge studied by the philosopher may be said to be couched in the object-language, then the language used by the philosopher in his inquiry is a meta-language with regard to the former. In this meta-language, some propositions are analytic, others are synthetic. In particular, the Principle of Verifiability is analytic in the meta-language, but any particular application of the principle to any particular statement of the object-language is itself a synthetic verifiable statement of the meta-language.

(2) The epistemological import of the Principle of Verifiability is substantially maintained in our re-formulation of it. If the fact that there are no synthetic, unverifiable, true propositions is itself an analytic truth, then the existence of limits to human knowledge claimed by the empiricists is an analytic truth as well.

(3) The presence of unverifiable assumptions in science is compatible with the new wording of the principle. In other words, the justified misgivings of the scientist in regard to empirically untestable hypotheses do not and need not prevent him from using in an auxiliary capacity statements which are unverifiable on empirical grounds. He would curtail science beyond repair if he were to renounce the construction of theories with external bases or transcendent theories, since any theory of this type owes its unique ability to organize a potentially infinite supply of relevant observational truth to its unverifiable assumptions. Philosophers may adopt a similar tolerant attitude in dealing with the legitimacy of

metaphysical and related problems. Without renouncing the Principle of Verifiability or the empiricist outlook it implies, they should make use of any empirically unverifiable assumption which proves capable of organizing significant aspects of human experience.

(4) Scientifically solvable problems have turned out to be identical with those whose solutions consist of verifiable statements or verifiable theories; questions answerable by unverifiable statements have proved to be indeterminate. This double result applies in particular to practical problems and those concerned with values. Consequently, the competence of science for solving human problems of this sort should be considered universal; if any of them is solvable at all, then it can be solved in principle by applying scientific methods.

In connection with our previous discussions (cf. §§ 3, 4, 13) the following remarks are in order:

(*a*) Practical problems are concerned with whether some particular action should be taken in order to obtain a particular objective, or whether a whole group of actions should be taken in order to achieve a group of auxiliary objectives which, in turn, can secure a single major objective. If either the objectives involved in a practical problem, or the corresponding actions, or both are describable by unverifiable propositions, then scientific method is obviously incapable of providing a solution. Science has nothing to say to a person who pursues a goal describable by unverifiable statements or who is anxious to attain his goals by means describable by such statements. However, according to our re-formulated Principle of Verifiability, problems involving unverifiable statements can have no solutions. The inapplicability of scientific method to practical problems concerning ends and means describable by unverifiable propositions would therefore not detract from its inherent universality.

(*b*) Value-questions are answerable by statements ascribing some particular value-predicate (cf. § 3) to some particular object or class of objects. It has been reasonably well established in recent investigations that many sentences involving such predicates are not intended to convey any information and aim rather at influencing the listener or the reader in some specifiable way. However, this hardly applies to all value-statements. The question whether value-statements intended to convey reliable information about their subject-matter are actually able to discharge this function depends upon their empirical verifiability. If unverifiable, they will frustrate the speaker's intention to convey reliable (that is, reliably true) information, since no unverifiable information has any truth-value. If verifiable, they are susceptible in principle to validation by scientific methods and the corresponding value-problem is within the range of science.

Thus the scientific solvability of value-problems hinges upon the verifiability of the corresponding answers. We must not expect that all value-statements would be found to meet uniformly the requirement of verifiability. Only a detailed examination of significant groups of value-statements (for example, those pertaining to ethical or aesthetic considerations) can provide the particular relevant answers. One point, however, should be made with regard to all such problems. We have granted that the verifiability of a statement consists in its being a consequence of observational premisses and, moreover, that value-predicates never occur in such premisses. These two circumstances are by no means inconsistent with the assumption that value-statements which always involve value-predicates may nevertheless prove to be verifiable. For a statement to be a consequence of observational premisses, it suffices that it be a logical consequence of such premisses supplemented by a suitable set of additional analytic premisses which may involve the relevant value-predicates (cf. § 31). In particular, we may expect that these additional analytic premisses will contain observational criteria for value-predicates (say, to the effect that an object has positive aesthetic value if its sight is enjoyed by a disinterested onlooker). The claim that some value-statements are empirically verifiable would therefore not presuppose a "naturalistic" view of the relevant values. The existence of observational criteria for a value-predicate does not mean that it has full-fledged definability in observational terms. Only such full-fledged definability would deserve the naturalistic label.

39   IMPLICATIONS OF THE UNIVERSALITY OF SCIENCE

A few final remarks may point out the main philosophical implications of this investigation into the potentialities of science and its conclusion concerning the universality of scientific method. The cognitive monopoly of the scientist was exalted during the nineteenth century by positivistically minded thinkers and those who came under their influence in one way or another. I should like to stress the following points in order to make clear the difference between the logical, inherent universality of science, which forms the subject of our discussions, and the positivistic ideas of the cognitive monopoly of science, which originated mainly with A. Comte and are still extremely influential at present, although they have gone through quite a number of metamorphoses.

(1) Let us notice, first, that in the past the cognitive universality of science has often been purchased at a price which now appears exorbitant and unacceptable. Thus, in many cases, the universality of scientific knowledge has been secured by degrading or curtailing man's cognitive potentialities; science was thought of as dealing with "appearances," "phenomena," or "sense-data" only. Since nothing else was left for man to know, whatever was knowable came, by the same token, within the province of science. Phenomenalism was the price paid, in these cases, for the universality of science.

(2) On the other hand, human knowledge was often held to be but a convenient tool for condensing past information, making it available for predicting future experiences, and thus helping man in his struggle for survival. Since, apparently, science has performed this biological function more efficiently than any other news-agency in man's history, the scientific superiority and monopoly seemed a safe conclusion from historical facts. This utilitarian bias in justifying scientific universality has often resulted in distorted views of the aims and potentialities of science. Thus, according to A. Comte, only astronomy, mathematics, physics, chemistry, biology, and sociology deserve the status of fundamental science. Yet, in spite of this privileged position, they were subjected by him to all sorts of unwarranted restrictions. Since only recording and predicting useful information was left to science, astronomy, for example, had to confine itself to a study of our planetary system and to discard all problems concerned with stellar phenomena. Biology was advised to refrain from investigating too minute details of its subject-matter, and, more particularly, a ban was placed on histology. Similar crippling prohibitions were inflicted on other sciences.

(3) A more insidious way of substantiating the universalistic monopoly of science is to declare that items of information such as statements, theories, or whole disciplines, are literally meaningless unless they fit into a certain narrow concept of significance. This was achieved by laying down a very rigid criterion of meaningful statements, namely by identifying the significance of a statement with its observational verifiability, that is, the possibility of ascertaining whether it is true or false by carrying out some appropriate observations and submitting the latter to an appropriate logical process. The implication was that only the observational core of statements included in scientific theories do meet this requirement. Statements at variance with this criterion of significance, even when contained in a genuine scientific discipline, should be dismissed as meaningless or "metaphysical." Apart from an anti-metaphysical purge within science, this "verificationist" view of mean-

ing implied also a wholesale rejection of ontology and of most of epistemology and ethics.

Unfortunately, such verificationalist tendencies towards removing substantial parts of science and the bulk of philosophy have often led the foremost representatives of the positivistic philosophy of science to block real scientific progress. Thus, E. Mach, the leader of this philosophy of science at the turn of the last century, used his considerable influence to check the spread of Boltzmann's statistical thermodynamics and gas-theory. The threat to the advance of science by a narrow-minded positivistic methodology was thus made once more apparent in a way strangely reminiscent of A. Comte's restrictions. These were already obsolete while Mach was devising his own, which, in their turn, have now become completely obsolete.

Needless to say, the effect of positivistic ideas on the advance of science is by no means confined to obstructive phenomena of the kind we have just mentioned. The removal of futile speculation and the clarification of basic scientific ideas traceable to this philosophy seem to carry, in the long run, more weight than the unwarranted obstruction sometimes associated with positivism. To illustrate this, it suffices to refer to the two most fundamental theories of contemporary science, the Theory of Relativity and the Quantum Theory of atomic phenomena; both embody to a considerable degree the positivistic methodology. Yet it is about time to draw a reasonable line separating the undoubted contributions of positivism to genuine scientific progress from narrow-minded phenomenalism, utilitarianism, and verificationalism, which are actually adventitious to the core of the positivistic philosophy of science. The reformulated Principle of Verifiability (§§ 5, 38) can provide this demarcation.

We have commented in some detail on phenomenalist, verificationalist, and similar distortions of science associated with positivism. To characterize our general frame of reference, let us point out that in suggesting universality as the outcome of our investigation into the method and potentialities of science, we are neither committed to nor prepared to countenance any of the restrictions on scientific aims and powers just mentioned. Our investigation does not entail the reduction of the scientifically knowable to the purely phenomenal, although the requirement of observability has played an important part in discussions concerning the universe of discourse of science. Nor does the investigation imply a crippling and narrow utilitarianism in regard to the aims of scientific inquiry; whatever constitutes a reliable solution to a problem which interests the scientific community is part and parcel of science, even if the only use of such a solution consists in getting rid of an obsessing puzzle or in satisfying the curiosity of a human group. Theoretical problems of explanation and of unification of the scientific world-picture are as legitimate as those of description and prediction. Moreover, "meta-scientific" problems of various levels and their philosophical (ontological, epistemological) generalizations are of as much interest as scientific "object-problems" or "first-level" problems.

Empirically unverifiable assumptions have turned out to lack intrinsic cognitive import but to play, nevertheless, an indispensable rôle in organizing observational data within the framework of scientific theories. That is why no sweeping ban on unverifiable assumptions within or without science follows from the new formulation of the Principle of Verifiability.

One final qualification concerning our conclusions as to the reach of science may be added. The universality of science was often thought of as implying the elimination of all other fields of theoretical endeavour, which were either less spectacularly successful than science, or applied methods different from those of science, or neglected the only aspect of reality which seemed to deserve the investigator's interest, namely the quantitative and measurable aspects of man and of his environment. Man's pre-scientific and extra-scientific activities, for example in philosophy, were implicitly or explicitly sacrificed, once science was granted cognitive monopoly. Even within science proper, a "scientific" bias has unjustifiably attributed exclusive rights to the natural sciences, at the expense of social and humanistic studies. Yet the inherent universality of science does not imply a cognitive monopoly of scientific method, or of science (whether natural or human), let alone of the scientist. The ability of the scientific method to cope, in principle, with every meaningful and solvable problem does not mean that types of investigation usually classified as "ontological" or "metaphysical" are to be dismissed as meaningless or fruitless or as prematurely attacking relevant problems. The usual implication of such anti-metaphysical views is that, should some residue of all metaphysical problems resist the onslaught of critical analysis and come close to a reliable solution, it would have to be incorporated in some special science and lose its metaphysical status. Metaphysics would thus be, by definition, a collection of problems for which no satisfactory solution had been discovered. I do not think that this automatic check on progress in philosophy is more than a terminological twist. Philosophical problems will remain philosophical, even if successfully dealt with by scientific methods. This applies not only to metaphysical questions, but also to those traditionally assigned to the theory of knowledge, ethics, aesthetics, social philosophy, etc. It is to be granted, of course, that more considerable advances in dealing with problems of philosophy have often resulted in their being shifted from philosophy proper to science. But this is by no means a necessary consequence of the universality of the scientific method and has hardly more than a classificatory meaning. When traditionally philosophical problems concerning the nature of space and time, of causality, of the infinite, of life, of thought, of truth, came to be included, respectively, in scientific cosmology, in atomic physics, in abstract set-theory, in theoretical biology, psychology, and logic, everybody felt that these special sciences had been endowed thereby with "philosophical implications." This is but another way of acknowledging the genuinely and enduringly philosophical nature of problems so shifted from philosophy proper to various special sciences. Philosophical problems do not cease to be philosophical when they are transferred from one official department of organized knowledge to another or located on a higher level of inquiry, in order to make them amenable to scientific treatment.

Thus, no attempt at sacrificing philosophy is implicit in claiming the universality of scientific method. The claim rather is that, since this method is universal, it is reasonable to resort to it whenever no headway has been made using other methods in spite of serious and consistent efforts. The theoretically universal applicability of one method does not preclude the legitimate applicability of alternative methods. Universality of science only asserts that if a problem is solvable at all by any method, then its solution can also be reached, at least in principle, by the method of science.

In particular, reliable pre-scientific and para-scientific methods for acquiring knowledge are as legitimate as the method of science, irrespective of its universality. This would apply to telepathy, extra-sensory perception, clairvoyance, and similar unconventional procedures, should their reliability be established; it certainly applies to common-sense procedures. Pre-scientific methods cannot lead to results which are beyond the reach of science, but it may be more economical to use them. In most problems with which every human being is faced at any moment of his life, resorting to the common-sense method for acquiring knowledge is a sheer necessity.

In spite of all these qualifications, the universality of scientific method makes science man's supreme hope and his supreme refuge. The impact of science and scientific technology may tend to increase the role of mechanical contrivances and reduce the importance of human personality in human life. Yet the very example of science shows the limits of these tendencies. In furthering science, the scientist's personal effort is bound to remain paramount. For human science today is a body of reliable and relevant information presently available to man. To obtain such information the scientist has to apply either methods of discovery or merely heuristic procedures. It is true that once a method of discovery is available, the scientist's role is in principle dispensable since he could be replaced by a mechanical contrivance. Still, the discovery of methods of discovery cannot be secured in turn by a supermethod of discovery. Moreover the application of merely heuristic procedures in acquiring scientific information is unthinkable without the scientist's personal effort; accordingly he will remain indispensable and crucial in all cases where only such procedures are available; this area demonstrably includes the overwhelming majority of problems. Furthermore, the scientific relevance or significance of any piece of information is a value-problem to be solved in accordance with the scientist's system of values. This is another reason why the scientist is permanently indispensable in the advance of science.

Finally, the universality of scientific methods establishes that whatever is knowable is scientifically knowable. Man is therefore bound to resort to scientific methods in all of those all too numerous cases where knowledge vitally important to him is not otherwise obtainable. Yet no more than hope is offered by science to man in his present predicament. The problems implicit in this predicament may have no solution and the universality of science, that is, its ability to cope with every solvable problem, would then be of no avail. Moreover, even if the problems can be solved, they may have to be solved now; in spite of its inherent universality, science today may be incapable of providing the required solutions in time.

Index

INDEX

convertibility, 160; Dalton's multiple proportions, 181; Haüy's rational indices, 181; Boyle's, 193; Coulomb's, 202; of conic refraction, 216; of free falling bodies, 220

of Excluded Middle, and the Principle of Verifiability, 44, 256ff; logical and metalogical version of, 258f; and semantic method, 263; and definition of truth, 270

Law, concept of: whether definable in terms of conceptual, spatio-temporal and quantificational universality, 181ff; as essentially quantified statement, 190; as special theory, 191; and true law-like statements, 158

Laws, classifications of: of universal, regional, or individual scope, 60, 161f; of nature vs. natural laws, 157; ultimate vs. derivative, 157, 183, 185, 189; statistical vs. causal, 161, 163, 200; vacuously or genuinely universal, 162, 182ff; finitist vs. infinitist, 163; separately vs. contextually justifiable, 164f; empirical, 165; semi-definitional, 171ff; existential, 182; meta-linguistic equivalents of theories, 227f

functions of: informational (cognitive), 165f; systematizing, 166; predictive, 166; explanatory, 168; definitional, 168ff

validation: and probable inference, 193ff; and Problems of Induction and of Causality, 195f; and conceptual apparatus of science, 197, 229f; of infinitist laws, 197ff; of existential laws, 199f; of statistical laws, 200; of laws of universal scope, 200f

Least count, 140, 219, 296
Leibnitz, G. W., 124, 176, 211
Lepley, R., 99
Le Roy, E., 310
Lesniewski, St., 175
Lewis, C. I., 282
Limiting concepts, 161
Limits of science, 3, 19, 64, 79
Lindenbaum, A., 84, 205, 325
Linguistics, 161
Locke, J., 155
Logic: mathematical vs. formal, 52; classical vs. intuitionist, 60, 263; ap-

plied, 63; inductive, 186, 189, 198; non-Aristotelian, 233
Logical connectives, 57, 283
Logical consequences, 272f
Logical constants, 137, 138
Logical deducibility, 272
Logical range, of science, 245
Logical types, 287
Logicism, 59f
London, F., 179
Löwenheim, L., 84, 325
Lutoslawski, W., 121

MACH, E., 82, 166, 167, 340
Margenau, H., 164, 205, 220
Marx, K., 125
Mathematics: and criterion of scientific status, 51; and logic, 56, 60, 211; formalization vs. reduction to logic of, 56; and observational sciences, 58; consistency of, 90; as science of infinity, 128; and abstract entities, 267

philosophy of: formalist, 52ff; empiricist, 53f; as collection of tautologies, 54; intuitionist, 59, 61, 263; logicist, 60; and Herbrand-Tarski Deduction Theorem, 60

Maxwell, J. L. 13, 45, 47, 48, 49, 159, 188, 197, 202, 205, 211, 218, 219, 300

Meaning: dimensions of, 41; cognitive, 41; verifiability theory of, 41, 43, 256, 296f, 307; holistic view of, 43; primitive empirical, 135, 138, 147, 152, 301; Carnap's recent criterion of, 296; empirical criterion of, 329

Measurement, 74, 128ff; and observation, 112; and advance of science, 119ff; and elimination of human standards in validation, 122, 125; discovery and verification in, 129; direct, indirect or computational, 131ff, 150, 294; and conditional definitions, 134ff; inaccuracy of, 137, 139f; errors of, 141f; and Universe of Discourse of science, 143f; and indirect observation, 144, 150; and finite verifiability, 293f

Mechanics: conventionalism in, 42; and astronomy, 47; condensed in single principle, 124; classical, relativistic, and wave-, 212f; relativity and relativistic, 217

323ff; and indeterminacy of statements, 280; of quantitative concepts, 323f; of ostensively indefinable concepts, 324f; of ostensively definable concepts, 327ff; and connection between unverifiability and indeterminacy, 331f

Validation: and scientific status of information, 48f; and evidence, 48, 70f; methods of, 49, 52, 56, 59ff

Value(s); and needs, 15f, 22; over-all, 15, 16, 23; intrinsic *vs.* instrumental, 16; science of, 20; -predicates, 20, 21, 22, 24; -judgments, 21; -policies, 22; statements about, 24; ideologies and systems of, 30, 34; cognitive, 40; as dispositional term, 177; and verifiability, 337f; naturalistic view of, 338; -problems, 343. *See also* Science, value of

Variables: individual and numerical, 138, 160

Veblen, O., 182

Verification: and scientific method, 37, 317f; direct *vs.* indirect, 320; conclusive *vs.* confirmatory, 320, 322

Verifiability: and cognitive content of theories, 29; of atomic and molecular hypotheses, 39; verifiability theory of meaning, 41, 43, 253, 256, 296f; of mathematical results, 58; definition of, 299, 331

concepts of: private *vs.* public, 248ff, 317; direct *vs.* indirect, 250f, 281ff, 317, 322; positive *vs.* negative, 254f, 299, 312, 334; unilateral *vs.* bilateral, 254, 299, 312, 334; finite *vs.* inductive, 293, 297ff, 317, 322, 333; strong (conclusive) *vs.* weak

(confirmatory), 317, 333; of theories, 299ff

Principle of: conventional formulation, 37; and universality of science, 38; and A. Comte, 38; historical sketch of, 39; re-interpreted as criterion of definiteness of truth-value, 40; versions of, 252ff, 332ff; and Law of Excluded Middle, 257ff; and language of science, 297; deduced from definitions of truth and verifiability, 330ff; epistemological status of, 334ff; and practical problems, 337; and value-problems, 337f

Verificationalism, 310

Vitalism, 211

WAVE MECHANICS, 213

Wave view of light, 13

Weak verifiability, 317, 333

Weber, M., 22

Westermarck, E., 31

Whewell, W., 217

Whitehead, A. N., 57, 121, 124, 163, 233, 308, 309, 310, 312

Wilson, C. T. R., 47

Wisdom, J. O., 231

Wittgenstein, L., 39

Woodger, J. H., 89

World-picture(s): provided by science, 18, 19, 27, 285, 341; pre-scientific, 213, 232; of classical science, 232f; scientific method and four successive world-pictures, 233ff

YONG, Th., 13

ZOOLOGY, 14

www.ingramcontent.com/pod-product-compliance
Lightning Source LLC
Chambersburg PA
CBHW030450210326
41597CB00013B/615